# FREQUENCY AND RISK ANALYSES IN HYDROLOGY

by

## G. W. Kite

*Since 1971*
**WRP**

WATER RESOURCES PUBLICATIONS, LLC.
P. O. BOX 630026
HIGHLANDS RANCH, COLORADO, 80163-0026, USA

*For Information and Correspondence:*

**Water Resources Publications, LLC**
P. O. Box 630026 • Highlands Ranch • Colorado 80163-0026 • USA

Fifth printing 2004
Previous ISBN-13 Number: 978-0-918334-64-0

Sixth Printing 2019

# FREQUENCY AND RISK
# ANALYSES IN HYDROLOGY

by
## G. W. Kite

**ISBN-13 Number: 978-1-887201-94-0**

**Library of Congress Control Number: 2019932778**

# ACKNOWLEDGEMENT

The sample data used are from Water Survey of Canada Records and the author wishes to express his thanks for the permission granted by the late Earl Peterson, ex-director of that organization, to quote from internal reports.

The cover photograph was kindly provided by Gordon Mills, Director of Public Affairs, Souris Basin Development Authority.

# *PLEASE NOTE:*

## FOREWORD TO SIXTH PRINTING

To download the software go to:

https://www.wrpllc.com/software/FRAH_Software.zip

Program listings are in BASIC. All data sets and computer printouts have been changed to SI units.

# TABLE OF CONTENTS

# PREFACE

Every year, floods and droughts cause loss of life and millions of dollars worth of damage in many countries of the world. In many cases, these consequences could be reduced either by nonstructural means such as restricting building in flood plains and by limiting water abstractions, or by better design of regulatory structures to reduce flood peaks and increase low flows. In all these cases, the key is knowledge of the distribution of flows in the river.

Hydrology is the science through which man tries to understand the properties and the distribution of water. Frequency analysis is a set of mathematical and statistical techniques used to describe the probability of occurrence of events.

This book describes some of the methods currently used to apply frequency analysis techniques to hydrological data in order to provide planners and engineers with figures that they can use in practice to reduce the losses caused by flood and drought. Risk analysis is an extension of the technique used to assess the probability that the estimated design event will differ from the actual event.

❦❦❦❦❦❦❦

# Editor's Note to the Sixth Printing

Over the years *FREQUENCY AND RISK ANALYSES IN HYDROLOGY* has proven to be a popular book providing planners and engineers with figures that can be used in reducing losses caused by flood and drought. It is still an important addition to any library that is concerned with these issues. WRP is pleased to include this book as part of our *Classic Resource Edition*.

# CHAPTER 1

## INTRODUCTION

Every year, floods and droughts take lives and damage properties in many parts of the world. It has been estimated (3)
that, over the period 1970 - 1986 floods in Canada caused an aver-
damage of $22 million per year in 1970 dollars. On top of these
direct costs are the loss of life, injury, inconvenience and
other indirect costs.

The most obvious means of reducing the damages caused by
floods and droughts is with a time-specific warning. Warning of
a flood enables people to evacuate a danger area. If sufficient
lead-time is provided, vulnerable possessions can be removed from
the danger zone and preparations, such as sandbagging, can be
made to minimize property damage. Similarly, warning of drought
enables preparations such as provision of alternate food supplies
and, in the extreme, relocation of people.

A second means of reducing damages is by advanced planning
based on the probabilities of events. For example, flood plain
zoning based on the probability of the river or lake reaching
certain levels could ensure that housing or industrial developments are not located in high-risk areas. High-risk areas along
the banks of rivers susceptible to flooding should be used for
activities compatible with this risk such as parks, recreation
areas or some types of crop production or grazing. Similarly,
a knowledge of the probabilities of occurence of the design events should be used in the planning of dams, bridges, culverts,
water-supply works and flood-control structures. The American
Water Works Association (1) has reported that out of 293 dam
failures in the U.S.A. and other countries since 1799, about 20%
were due to inadequate spillway capacity.

Frequency analysis is one of the statistical techniques applied by hydrologists to try and estimate the probabilities associated with design events. Many criticisms can be made of the
method and the assumptions made in it's use but the fact remains
that it is one of the few methods available and it is arguably
better than other non-probabilistic methods.

Frequency analysis is not only of use as an aid in averting
disaster but is also a means of introducing efficient designs.
When a hydraulic structure, through inadequate or inaccurate
data or methods, is underdesigned, the results are regretably
obvious; the dam may fail, the highway may flood or the bridge
collapse. This does not happen very often and so the hydrol-

1

ogist, equating nonfailure with success, is satisfied with his design techniques. Nonfailure, however, does not necessarily mean an efficient design. Frequently, structures are over-designed and hence very safe, but also very expensive. A truly efficient design will be achieved only as the result of studies relating cost to risk and frequency analysis.

The Water Resources Council of the U. S. Government (2) has recently noted that because of the range of uncertainty in design flood analysis there is a need for continued research and development to solve the many unresolved problems. Current methods of providing design floods for hydraulic structures include the deterministic use of meteorological data in tech-niques such as dynamic flow equations and the so-called Probable Maximum Flood method (PMF), and the stochastic use of frequency analysis techniques. The PMF and similar methods suffer from the major disadvantages of being entirely subjective and of having no associated probability level. This latter is partic-ularly important since to non-technical people it implies that no risk is involved, that the maximum flood cannot exceed this certain limit. This, of course, is untrue and can sometimes have disasterous consequences. Yevjevich (4) has characterised the difference between the PMF method and the frequency analysis method as being between "expediency" and "truth". While many hy-drologists would not go that far, the difference between the two techniques is certainly between having one estimated event with unknown probability and having a series of estimated events with probabilities calculated from reasonable assumptions.

Neglecting the PMF approach, this report discusses only the techniques to be used in flood frequency analysis and, to a lesser degree, drought analysis, together with an analysis of the associated risk. All frequency techniques are totally data-dependent. Chapter 2 discusses the data requirements and the basic assumptions involved in the subsequent analyses.

An assumption must be made of a theoretical frequency dis-tribution for the population of events and the statistical parameters of the distribution must be computed from the sample data. Chapters 3-11 describe, for some of the distributions commonly used in hydrology, the form of the distribution, estim-ation of parameters, estimation of events at given return periods and estimation of confidence limits. The objective of these descriptions is to provide sufficient background information to enable an hydrologist to intelligently select a distribution to use in frequency analysis.

A flood frequency relation for a complete region may be preferable to one developed for a specific site for two reasons: (a) Because of the sample variation possible at a single station, any single station analysis is subject to large error. This

2

error can be reduced by combining data from many sites. (b) There are many more sites where hydrologic data are needed, than there are sites at which data are collected. This means that some form of analysis is required which can transfer data from gauged sites to ungauged sites. Chapter 13 discusses some of the methods of regional analysis presently available.

Both single station and regional frequency analyses involve risks. In determining the design flood for a project, the length of data available, the project life and the permissible probability of failure are all factors to be considered. Chapter 14 reviews some of the statistical methods available to analyse these risks.

## References

1. American Water Works Association, 1966, Spillway Design Practice, AWWA Manual M13, New York.

2. United States Water Resources Council, 1967, A Uniform Technique for Determining Flood Flow Frequencies, Hydrol. Comm. Bull. No. 15.

3. Naik, H. & D. Bjonback, 1988, Role of Disaster Assistance in Floodplain Management. Proceedings of Sixth IWRA World Congress on Water Resources, Ottawa.

4. Yevjevich, V., 1972, New Vistas for Flood Investigations, Academia Nationale Dei Lincei, Roma, Quaderno N. 169, pp. 515-546.

# CHAPTER 2
## DATA

Annual Series and Partial-Duration Series

Starting with the original recorded hydrograph, or with the tabulated data abstracted from the hydrograph, there are two ways in which these data may be used in a frequency analysis. The first method, direct frequency analysis, is to select from the total data only that information which is required in the design process, e.g. maximum instantaneous flow in each year (annual series), all instantaneous flows above a certain base flow (partial-duration), minimum 7-day flow in each year (annual series), etc.. This represents a considerable loss of information, however, since such a few data points are utilised. Dyhr-Nielson (12) has discussed the effects of this loss of information on the estimation of probabilities of extremes.

The second method of using the basic hydrologic data is to design a mathematical model which will describe the observed hydrograph. This model can then be used to generate (simulate) many sets of data from which the required events can be abstracted. As an example, if the objective of the frequency analysis is to define the flow having a magnitude such that it will occur on the average once every 50 years, then 50 years of hydrologic data can be generated and the maximum flow during that period can be found. By generating many different sets of 50-year records, a distribution of 50-year flows can be found and confidence limits can be set on the mean (19). The disadvantage of the data generation method is that it is very unwieldy. It is practicable when dealing with monthly data, less so for daily means and becomes difficult when the need is to generate flows that can be considered instantaneous. In addition, most models cannot successfully generate the extreme peaks and lows of a variable; they only operate well for average conditions.

This report describes only the first method of analysis, direct frequency analysis. Within this method there are two ways in which the required data may be abstracted from the original recorded or tabulated data. As indicated earlier, these are known as annual series and partial-duration series.

An annual series takes one event, and only one event, from each year of the record. A disadvantage of this abstraction technique is that the second or third, etc., highest events in a particular year may be higher than the maximum event in another year and yet they are totally disregarded. This disadvantage is remedied in the partial-duration series method in which all events above a certain base magnitude are included in the

4

analysis. The base is generally selected low enough that at least one event in each year is included. Each event to be included in the partial-duration series must be separate and distinct,i.e. including two consecutive daily flows caused by the same meteorologic event is not valid. If the total number of events which occurred during the entire period of record are ranked without regard to the year in which they occurred and then the n top ranking events are selected, where n is the number of years of record, the events are termed annual exceedences (34).

The recurrence interval of an event of given magnitude is defined as the average length of time between occurrences of that event. This is a probabilistic term and contains no inference of periodicity. The recurrency intervals of annual series and partial duration series have different meanings. In the first case the recurrence interval means the average number of years between the occurrence of an event of a given magnitude as an annual maximum. In the second case the recurrence interval carries no implication of annual maximum.

Chow (8) investigated the theoretical relationship between these two recurrence intervals and their corresponding probabilities. If $P_E$ is the probability of an event in a partial-duration series being equal to or greater than x and if the number of events in the partial-duration series is Nm, where N is the number of years and m is the average number of events per year, then $P_E/m$ is the annual probability of an event being equal to or greater than x. The probability of an event x being the largest of the m events in a year must then be

$$P_M = (1 - P_E/m)^m \qquad (2-1)$$

But $P_M$ is then the probability of an annual event of magnitude x and corresponds to the annual series. Substituting the approximation $(1 - P_E/m)^m$ equal to $e^{-P_E}$ and letting $T_E = 1/P_E$ and $T_M = 1/P_M$ where $T_E$ and $T_M$ are the recurrence intervals of the partial-duration and annual series respectively, then

$$T_E = \frac{1}{\ln T_M - \ln (T_M - 1)} \qquad (2-2)$$

The following table from Dalrymple (10) compares the recurrence intervals of the two types of series. The difference amounts to about 10% when $P_E$ is 5 years and about 5% when $P_E$ is 10 years. The distinction is only of importance at low recurrence intervals.

To some extent the decision to use an annual series or a partial-duration series depends on the use to which the frequency analysis will be put. Some types of structure, for example erosion protection works, are susceptible not particularly to

Table 2-1

Comparison of Recurrence Intervals for Annual and Partial-Duration Series

| Recurrence intervals (years) | |
|---|---|
| Partial-Duration | Annual Series |
| 0.5 | 1.16 |
| 1 | 1.58 |
| 1.44 | 2 |
| 2 | 2.54 |
| 5 | 5.52 |
| 10 | 10.5 |
| 20 | 20.5 |
| 50 | 50.5 |
| 100 | 100.5 |

one peak flow but more to closely repeated high flows and so a partial-duration series analysis might be more suitable. If the design flood is likely to have a low recurrence interval, then again the partial duration series may be more suitable since the smallest recurrence interval for the annual maximum series is one year.

If flood flows are being investigated, they should preferably be maximum instantaneous flows derived from a continuous hydrograph. Usually, however, instantaneous flows are only available for a comparatively few years, but older data may be available as maximum mean daily flows. In this case, by correlation between the two series, it may be possible to extend the instantaneous maxima back in time (20).

Most data series are incomplete. For various reasons such as mechanical failure, inaccessibility, flood damage to the recorder, etc., some flood peaks are usually missing. This may have a large effect on the recurrence intervals assigned to the floods on record. Dalrymple (10) has described a method of overcoming this problem. By regression with a gauged stream in the same hydrologic region it is possible to estimate the magnitudes of the missing floods at the stream being analysed. As an example of a technique, Langbein (21) has correlated the logarithms of flows standardized on a monthly basis. Dalrymple (10) used the computed flows as an aid in sorting the observed flows by magnitude and assigning recurrence intervals. The computed flows were then discarded and not used in the further analysis.

Sample Data Set

Table 2-2 contains a sample set of annual maximum and minimum daily discharges for the St Mary's River at Stillwater, Nova Scotia. Data are correct except for the maximum daily discharge for 1915 which has been estimated from an incomplete data-year and is included only for completeness. Computer outputs for the two-parameter lognormal, the three-parameter lognormal, the type I extremal, the Pearson type III and the log-Pearson type III distributions reproduced in the book are derived from the annual maximum daily flows of the St Mary's River.

Output for the type III extremal distribution is derived from the annual minimum daily flows.

6

## Samples From Different Populations

For some stations it may be necessary to carry out more than one frequency analysis on the data. Stoddart and Watt (31) have described how watersheds in southern Ontario have two distinct types of flood; those due to rainfall only, generally occurring in the summer, and those due to snowmelt (sometimes combined with rainfall) occurring in the winter or spring. The hydrographs of these two types of flood are quite different and cannot be considered to be from the same population.

Suppose a flood of a given magnitude x would have a recurrence interval of $T_p$ if due to rainfall and $T_S$ if due to snowmelt. The probabilities of not equalling or exceeding x are given as

and

$$q_p = 1 - 1/T_p \qquad (2-3)$$

$$q_S = 1 - 1/T_S \qquad . \qquad (2-4)$$

So that the probability of not equalling or exceeding x in any year, $q_A$, is given by

$$q_A = q_p q_S \qquad (2-5)$$

and the recurrence interval for the annual flood equalling or exceeding x is

$$T_A = 1/(1 - q_A) \qquad (2-6)$$

$$T_A = \frac{T_p T_S}{(T_p + T_S - 1)} \qquad . \qquad (2-7)$$

The frequency curve of the annual series will be asymptotic to both the rainfall and snowmelt frequency curves. Stoddart and Watt list four possible combinations of these frequency curves.

Various statistical tests are available to test for the presence of data from different distributions. In general, the procedure used in applying these tests is:

(a)  State the null hypothesis ($H_o$) that there is no difference between the two data sets.
(b)  Choose a statistical test.
(c)  Select a significance level ($\alpha$).
(d)  Assume the sampling distribution of the statistical test under $H_o$.
(e)  On the basis of (b), (c) and (d) define the region of rejection.
(f)  Compute the value of the statistical test. If the

7

value is in the region of rejection, reject $H_o$.

There are two types of statistical test, parametric and non-parametric. Parametric tests are associated with certain assumptions about the sample being tested. These assumptions are (Seigel, 27):

(a) The observations must be independent.
(b) The observations must be drawn from normally distributed populations.
(c) These populations must have the same variance.

Generally, non-parametric tests need only assumption (a). In addition, although it is not important here, non-parametric tests do not require as high a level of data measurement as do parametric tests.

The most common parametric test used to test whether or not two samples are from the same population is the t test,

$$t = \frac{|\bar{x}_1 - \bar{x}_2|}{\sigma \sqrt{1/n_1 + 1/n_2}} \qquad (2\text{-}8)$$

where $\bar{x}_1$ and $\bar{x}_2$ are the arithmetic means of the two sub-samples size $n_1$ and $n_2$ and $\sigma$ is the unknown population standard deviation estimated from the sample variances $S_1^2$ and $S_2^2$

$$\sigma = \sqrt{\frac{n_1 S_1^2 + n_2 S_2^2}{n_1 + n_2 - 2}} \qquad (2\text{-}9)$$

If t is greater than the tabulated value of "Students" distribution at significance level $\alpha$ and $n_1 + n_2 - 2$ degrees of freedom and if assumptions (a)-(c) are correct, then it cannot be confirmed that the sub-samples are from the same population. However, the applicability of assumptions (b) and (c) must be in doubt; extreme events are not likely to be normally distributed and there is no reason why the variance of events from, say, snowmelt should be the same as the variance from thunderstorm events. For these reasons the t test is not commonly applied in frequency analysis.

Turning to non-parametric tests, amongst those used are the Terry test (33), the Mann and Whitney (24) U test, the Kruskal-Wallis (27) test and the Kolmogorov-Smirnov test (27). To give an example of the use of each of these tests, consider the annual maximum daily discharges of the St. Mary's River at Stillwater, Nova Scotia, for the period 1915 to 1986, listed in Table 2-2. These data will be used in later chapters to test

8

# NOVA SCOTIA
## ST. MARYS RIVER AT STILLWATER - STATION NO. 01E0001

| YEAR | MAXIMUM DAILY DISCHARGE (m³/s) | MINIMUM DAILY DISCHARGE (m³/s) | YEAR | MAXIMUM DAILY DISCHARGE (m³/s) | MINIMUM DAILY DISCHARGE (m³/s) |
|---|---|---|---|---|---|
| 1915 | 565 ON DEC ? | 4.05 ON SEPT 21 | 1960 | 360 ON APR 7 | 0.221 ON SEP 12 |
| 1916 | 294 ON APR 7 | 1.05 ON AUG 26 | 1961 | 371 ON DEC 30 | 0.541 ON AUG 26 |
| 1917 | 303 ON APR 8 | 1.05 ON SEP 17 | 1962 | 544 ON APR 3 | 5.44 ON JUL 1 |
| 1918 | 569 ON JAN 5 | 2.52 ON AUG 25 | 1963 | 552 ON MAY 3 | 3.37 ON AUG 13 |
| 1919 | 232 ON APR 19 | 3.09 ON JUL 10 | 1964 | 651 ON APR 17 | 7.65 ON AUG 23 |
| 1920 | 405 ON MAR 15 | 3.09 ON JUL 31 | 1965 | 190 ON APR 18 | 1.96 ON OCT 1 |
| 1921 | 228 ON MAR 12 | 0.651 ON SEP 22 | 1966 | 202 ON APR 2 | 0.963 ON SEP 4 |
| 1922 | 232 ON JUL 6 | 2.78 ON JUN 19 | 1967 | 405 ON NOV 25 | 3.74 ON JUL 30 |
| 1923 | 394 ON OCT 3 | 2.01 ON MAR 16 | 1968 | 583 ON AUG 31 | 0.507 ON AUG 20 |
| 1924 | 238 ON JUN 17 | 1.59 ON AUG 5 | 1969 | 725 ON NOV 10 | 0.838 ON SEP 5 |
| 1925 | 524 ON APR 1 | 1.22 ON SEP 8 | 1970 | 232 ON JUL 13 | 4.96 ON JAN 28 |
| 1926 | 368 ON DEC 2 | 1.05 ON SEP 20 | 1971 | 974 ON AUG 16 * | 2.34 ON JUL 14 |
| 1927 | 464 ON OCT 20 | 6.71 ON SEP 18 | 1972 | 456 ON MAY 17 | 6.43 ON FEB 13 |
| 1928 | 411 ON DEC 11 | 1.78 ON SEP 12 | 1973 | 289 ON APR 30 | 1.92 ON OCT 4 |
| 1929 | 368 ON SEP 20 | 2.24 ON JUL 14 | 1974 | 348 ON DEC 4 | 1.21 ON AUG 18 |
| 1930 | 487 ON MAR 13 | 1.15 ON SEP 14 | 1975 | 564A ON DEC 12 | 0.238 ON AUG 22 |
| 1931 | 394 ON APR 3 | 2.75 ON AUG 13 | 1976 | 479 ON NOV 7 | 0.719 ON AUG 29 |
| 1932 | 337 ON DEC 2 | 1.70 ON JUL 21 | 1977 | 303 ON DEC 27 | 6.88 ON AUG 16 |
| 1933 | 385 ON OCT 9 | 1.59 ON AUG 11 | 1978 | 603 ON JAN 16 | 0.705 ON SEP 7 |
| 1934 | 351 ON APR 15 | 0.453 ON SEP 15 | 1979 | 514 ON DEC 18 | 3.01 ON JUL 27 |
| 1935 | 518 ON JAN 12 | 1.04 ON AUG 22 | 1980 | 377 ON MAY 10 | 0.839 ON SEP 14 |
| 1936 | 365 ON MAR 14 | 3.45 ON AUG 6 | 1981 | 318 ON DEC 4 | 2.84 ON SEP 15 |
| 1937 | 515 ON NOV 21 | 0.564 ON SEP 8 | 1982 | 342 ON JAN 6 | 2.30 ON JUL 20 |
| 1938 | 280 ON NOV 26 | 3.96 ON JUN 27 | 1983 | 593 ON MAR 23 | 5.21 ON SEP 17 |
| 1939 | 289 ON APR 21 | 0.674 ON SEP 6 | 1984 | 378 ON MAR 17 | 1.90 ON AUG 6 |
| 1940 | 255 ON DEC 10 | 1.04 ON AUG 19 | 1985 | 255 ON MAR 14 | 1.90 ON OCT 5 |
| 1941 | 334 ON APR 29 | 2.24 ON JUL 8 | 1986 | 292 ON JAN 29 | 2.72 ON JUL 28 |
| 1942 | 456 ON OCT 27 | 0.150 ON SEP 9 * | | | |
| 1943 | 479 ON MAY 5 | 3.03 ON FEB 3 | | | |
| 1944 | 334 ON NOV 7 | 0.589 ON AUG 25 | | | |
| 1945 | 394 ON JAN 18 | 1.28 ON AUG 21 | | | |
| 1946 | 348 ON DEC 23 | 0.589 ON JUL 21 | | | |
| 1947 | 428 ON MAY 3 | 0.453 ON AUG 24 | | | |
| 1948 | 337 ON APR 3 | 3.34 ON AUG 6 | | | |
| 1949 | 311 ON NOV 27 | 1.78 ON AUG 3 | | | |
| 1950 | 453 ON NOV 29 | 0.405 ON OCT 12 | | | |
| 1951 | 328 ON MAY 30 | 2.10 ON AUG 9 | | | |
| 1952 | 564 ON JAN 24 | 0.728 ON SEP 17 | | | |
| 1953 | 527 ON FEB 9 | 0.728 ON AUG 4 | | | |
| 1954 | 510 ON FEB 25 | 1.15 ON SEP 30 | | | |
| 1955 | 371 ON FEB 13 | 1.50 ON AUG 18 | | | |
| 1956 | 824E ON JAN 7 | 1.17 ON OCT 6 | | | |
| 1957 | 292 ON NOV 4 | 1.01 ON SEP 2 | | | |
| 1958 | 345 ON NOV 30 | 2.35 ON JUL 29 | | | |
| 1959 | 442 ON APR 5 | 7.36 ON JAN 16 | | | |

\* - EXTREME RECORDED FOR THE PERIOD OF RECORD

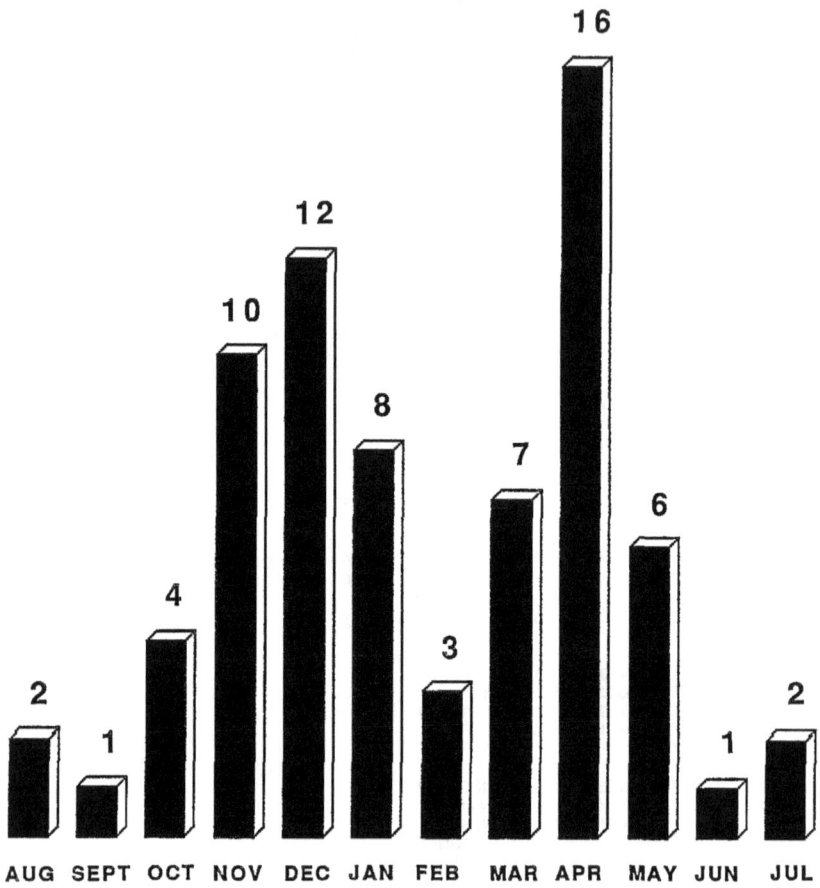

**Figure 2-1.** *St. Marys River at Stillwater Monthly Distribution of Annual Maximum Daily Discharge 1915 - 1986.*

various frequency distributions so it is as well to check their
consistency and to see if all the data are from the same population.

Looking at the column of maximum daily discharges in Table
2-2 it is apparent that the floods occur in all months of the
year but that November and April are the most popular months;
perhaps the November floods are rainfall and the April floods
are snowmelt. Figure 2-1 shows the histogram of annual maximum
daily discharges by month. The abscissa starts with August and
ends with July to emphasize the double peaks of the distribution.
The null hypothesis, $H_o$, is that the flood peaks in the months
March, April and May are from the same population as the flood
peaks in the months June to February and the level of significance, $\alpha$, is 5%.

*Terry Test* - The 72 annual flood events are ranked in increasing order with the 1965 event of 190 $m^3/s$ ranked 1 and the
1971 event of 974 $m^3/s$ ranked 72. The events are then divided
into two samples according to date of occurence, 29 events in
March-May and 43 events in the second sample, June-February. For
each event in the first sample the expected value of that order
statistic in a sample of 72 from a standardized normal population,
$E(Z_{Ni})$, is found from Harter's tables (17).

For example, the event of April 17, 1964, is 651 $m^3/s$, ranks 4th
out of the 72 events, and has an expected value of 1.64742. Similarly, the event of April 7, 1916, is 294 $m^3/s$, ranks 58 out of
72 and has an expected normal value of -.83366 (corresponding to
the rank working from the lowest event - the other end of the normal distribution). The sum of the expected normal values for the
29 events in the first sample is denoted by c and the test statistic is

$$t = \frac{c}{S} \qquad (2\text{-}10)$$

where S is the standard deviation of c defined as

$$S = \sqrt{\frac{mn}{N(N-1)} \sum_{i=1}^{N} [E(Z_{Ni})]^2} \qquad (2\text{-}11)$$

where m and n are the sub-sample sizes, N is m+n, the total
sample size, and $E(Z_{Ni})$ is the expected ith order statistic from
a sample size N from a standardized normal population. For the
St. Mary's River data, N = 72, m = 29, n = 43, c = 6.33 and
S = 4.11 so that t = 1.54. The sampling distribution of t
under $H_o$ is normal (0,1) so that the critical value of t in a
two-tail test at 5% and N-2 degrees of freedom is 1.96. Since
t calculated is less than t critical then $H_o$ cannot be rejected.

11

*Mann and Whitney Test* - The Mann and Whitney U test (24) is one of the most powerful non-parametric tests and is frequently used (27) as an alternative to the parametric t test. The sample size N is ranked in increasing order as before and the two statistics

$$U_1 = mn + \frac{m}{2} (m + 1) - R_m \qquad (2\text{-}12)$$

and

$$U_2 = mn - U_1 \qquad (2\text{-}13)$$

are calculated where m and n are the subsample sizes and $R_m$ is the sum of the ranks assigned to the sample of size m. The smaller of $U_1$ and $U_2$, say U, is then used to calculate the test statistic, Z, as

$$Z = \frac{U - \frac{mn}{2}}{\sqrt{\frac{1}{12} [mn (m + n + 1)]}} \qquad (2\text{-}14)$$

The Mann and Whitney test is based on a continuous distribution and if tied observations are present a correction must be made. If t is the number of observations tied for a given rank and

$$T = \frac{1}{12} (t^3 - t) \qquad (2\text{-}15)$$

then the test statistic is

$$Z = \frac{U - \frac{mn}{2}}{\sqrt{\frac{mn}{N(N + 1)} (\frac{N^3 - N}{12} - \Sigma T)}} \qquad (2\text{-}16)$$

where $\Sigma T$ is the sum of T's over all groups of tied observations. The T's are computed only for those tied observations which contain observations in both samples. For the sample data, $U_1 = 736$, $U_2 = 511$, and $Z = 1.29$. For subsamples both greater than 20 items the sampling distribution of Z under $H_0$ is normal (0,1). The one-tail probability of Z may be obtained from tables of the area under the normal curve as 0.10. Since this is greater than $\alpha$ the null hypothesis cannot be rejected.

*Kruskal-Wallis Test* - The statistic used in the Kruskal-Wallis test (27) is H, where

$$H = \frac{12}{N(N + 1)} \sum_{j=1}^{k} \frac{R_j^2}{n_j} - 3 (N + 1) \qquad (2\text{-}17)$$

k is the number of samples (2 in this case), $R_j$ is the sum of

the ranks in the jth sample, $n_j$ is the number of observations
in the jth sample and $N = \sum_{j=1}^{k} n_j$, the total number of observations. For all $n_j > 5$, H is distributed as chi-square with k-1
degrees of freedom. Ties are corrected for by dividing H by

$$1 - \frac{\Sigma T}{N^3 - N} \qquad (2\text{-}18)$$

where $T = t^3 - t$ and t is the number of observations tied for
a given rank. For the sample data H is 1.64. From tables of chi-
square the critical value at (0.05, 1) is 3.84.

Since the calculated value of H is less than the critical value,
the null hypothesis cannot be rejected.

*Kolmogorov-Smirnov* - If two subsamples are from the same
population, then the cumulative probability distributions of both
subsamples should be fairly close to each other. If the two
cumulative distributions are "too far apart" at any point, then
the subsamples may be from different populations. Thus a large
enough deviation between the two cumulative distributions is
evidence for rejecting $H_o$. Using the same class intervals, a
cumulative probability distribution is prepared for each sub-
sample and the largest difference in cumulative functions is
found

$$D = \max |S_1 (x) - S_2 (x)| \qquad (2\text{-}19)$$

for a two-tail test. Siegel (27) gives tables of critical values
of D for both small (< 40) and large sample sizes. Alternately,
for large samples, the fact that D follows a chi-square distribution can be used. For the sample data, using 17 class intervals, D = 0.153. The value of chi-square is then computed as

$$\chi^2 = \frac{4 D^2 m n}{N} \qquad (2\text{-}20)$$

which for the sample data is 1.62. From tables of chi-square it
is seen that this is less than the critical value of 5.99 for
2 degrees of freedom and 0.05 significance and therefore $H_o$ may
not be rejected.

In summary, all four non-parametric tests have shown that
the differences between the two subsamples, March-May and June-
February, are not large enough to reject the hypothesis of a
single population.

## Use of Historic Data

Properly authenticated historic events, antedating periods

of consecutive records, can be used in frequency analysis to increase the accuracy of the analysis. Discharge estimated on the basis of authenticated stages of historic floods that occurred prior to the modern continuous stage records may be used in conjunction with the modern records to obtain a more accurate probability curve (32).

Benson (3) studying the Susquehanna River at Harrisburg, Pa., found 7 historic floods with stages greater than 18.0 feet in the period 1786 to 1873 and a record of continuous annual maxima ranging from 14.3 to 30.3 feet for the period 1874 to 1947. Since the period of historical floods may have contained an unknown number of events of less than 18.0 feet stage, the problem was to combine the two types of record in the proper proportions to obtain an array of events properly representative of the total period. Benson arrayed all events, historic and recent, in order of descending magnitude using the plotting position m/(n+1). The order number of those flood peaks which were lower than the lowest historical event (the base event, 18.0 feet on the Susquehanna) were then adjusted as

$$m_c = t_b + \frac{n - t_b}{t - t_b} (m - t_b) \qquad (2\text{-}21)$$

where $m_c$ is the corrected order number, m is the original order number, n is the period in years from the first historical event to the most recent recorded event, $t_b$ is the number of events equalling or exceeding the base event, and t is the total number of events.

If a single historic event is known and there is not much difference in magnitude between that event and the highest event in the recent record, then the Equation 2.21 can be used with confidence. If there is a large gap in magnitude then there should be some reasonable certainty that there were few or no events during the ungauged interval which exceeded the highest in the recent period. With a large gap in magnitude and no intimation of what may have happened in between, then Equation 2.21 is no longer applicable. Dalrymple (9) has included further examples of the application of Benson's (3) method.

Glos and Krause (14) have described the augmentation of recorded data with historical data for the Rivers Dnepr in the USSR, Elbe in Czechoslovakia, Main in West Germany and the Spree and Werra in East Germany. As an example, the Elbe River at Decin, Czechoslovakia, has a record of annual floodpeaks for 111 years (1851 to 1962). For purposes of flood frequency analysis historical data can be used to extend this record to 530 years (back to 1432). In this case only the relative magnitudes of the historical floods were used since obviously the absolute magnitudes were not measured.

Glos and Krause used the historical data to increase the accuracy of estimation of the sample mean, $\bar{x}$, and variance; $S^2$, as;

$$\bar{x} = \Sigma x_i p_i \qquad (2-22)$$

$$S^2 = \Sigma (x_i - \bar{x})^2 p_i \qquad (2-23)$$

where $p_i$ is a weight given to each event, $\Sigma p_i = 1$. For historical floods the weights were assigned as:

$$p = 1/N \qquad (2-24)$$

where N is the length of the historical period. The events of the annual record were weighted as:

$$p = \frac{N - k}{N \, n} \qquad (2-25)$$

where k is the number of historical floods and n is the number of years in the annual record. The computed mean and standard deviation were then used in the standard frequency equation:

$$x_T = \bar{x} + K \, S \qquad (2-26)$$

where $x_T$ is the event magnitude at return period T and K, the frequency factor, depends upon the return period required and the distribution characteristics.

Leese (22) has shown how historical flood marks may be used to achieve greater precision in the estimates of flood events and hence greater efficiency in design. The Type I extremal distribution (see Chapter 8 for description) was used on the 29-year record of annual maximum discharges of the River Avon at Bath to estimate floods with return periods of 10, 25, 50, 100 and 1000 years. The procedure was then repeated adding to the data 13 historic floods determined from old flood marks. It was found that the sampling error of the flood estimates was reduced in the second case by between 8% for the 10-year flood and 18% for the 1000-year flood. By applying expressions for the benefits and costs of an imaginary structure Leese found that if the structure were designed on the basis of a 50-year flood then the incorporation of the historic data would reduce the cost of the structure by approximately 1%. More importantly it was found that to obtain the same cost reduction from a continuous record would require a further 20 years of streamflow data.

The use of historical records is apparently quite common in the People's Republic of China. In a recent paper (7) Chen et al. describe some of the advantages of using old records in the estimation of spillway design floods. In historical times the local people would put up monuments marking extreme water

levels; for example, along the Chang Jiang or Yangtse River
there are a large number of stone sculptures and monuments with
inscriptions of the great floods occurring in 1153, 1227, 1560,
1788, 1796, 1860 and 1870.  China has an extremely long written
history with some flood descriptions being up to 2000 years old.
During the Ming and Qing dynasties (a period of nearly 600 years)
flood records are complete and in detail.  As a further example,
the manuscript Shui Jing Zhu (completed in 527 AD) records the
level of the flood of 223 AD on the Yi He River.  Since the
channel is in rock, Chen (7) was able to compute the corresponding discharge from the modern rating curve.  The authors point
out the necessity of accounting for changes in the local climate
and in the physical conditions of the watersheds.

Errors

Three assumptions are implicit in any frequency analysis
(29):

(a) That the data to be analysed describe random events.
(b) That the natural processes involved are stationary
with respect to time.
(c) That the population parameters can be estimated from
the sample.

This chapter has already discussed the implications of assumption (a) on data selection, emphasizing the fact that all events
used must be independent.  Assumption (b) is more difficult to
guarantee.  The earth is in a constant state of flux with innu-
merable processes affecting the hydrologic cycle and its various
components.  Statistical tests are available (36) to check for
stationarity of time series.  Non-stationarity in hydrologic
time series is generally due to one of two basic causes:

(a) A slow change in hydrologic parameters such as might
be caused by the gradual urbanization of watersheds
or (on a different time scale) long-term changes in
temperature or precipitation distribution.
(b) Rapid change in parameters caused by, for example,
earthquakes, landslides, building of dams.

In a recent paper, Yevjevich (37) gave a plot of annual
maximum flows of the Danube River at Orshova in Romania which
shows a pronounced upward trend.  This trend in discharge is
mainly due to the construction of flood protection levees along
the Danube and its major tributaries.  Non-homogeneities such as
this shift the mean value of the distribution and increase the
variance.

Another source of error is basic inconsistency in the data
due to systematic measurement and computational errors.  This
subject has been discussed in detail by Dickinson (11), Robertson

16

(24), and Herschy (18), and the particular errors involved in determination of winter flows in Canada have been discussed by Rosenberg and Pentland (26).  The major emphasis in this book is on flood flows and these are the very events subject to the maximum measurement error.  In fact maximum flows are seldom, if ever, measured because of the difficulties involved firstly in predicting the time at which maximum flow will occur, secondly the difficulty in getting to the gauging site at that time and finally the difficulties of actually carrying out the gauging at the high stage and corresponding high velocities.  As a result high flows are normally estimated by extrapolation of the rating curve, estimation of mean velocity from an isolated surface velocity measurement, use of the slope-area method (28) or other similar procedures.  The resulting estimates of peak discharge contain a high error component which has been estimated by Blench (5) to be at least ±25%.

    Yen and Ang (35) have termed these errors "subjective uncertainty" and have shown that if the individual uncertainties are denoted by $\sigma_1$, $\sigma_2$... then the overall subjective uncertainty $\sigma_n$, can be written as:

$$\sigma_n = [\sigma_1^2 + \sigma_2^2 + \dots]^{1/2} \qquad (2-27)$$

    A third assumption made in any frequency analysis, and a very important one, is that the sample data available can provide good estimates of the population parameters.  This assumption is necessary so that estimates of population statistics such as mean, variance, skew, etc. may be derived from the sample. Benson (4) used a theoretical frequency curve to obtain 1000 random events.  These base data were then divided into shorter records, e.g. 40 records of 25 events each, 20 records of 50 events each, etc. and Benson investigated the variability of the frequency curves of these samples compared to the original theoretical curve.  Benson's conclusions provide estimates of the number of events needed in a sample before estimates of magnitudes at various return periods based on sample statistics are comparable to the values computed from population statistics.

Plotting Position

    Consider a series of 10 annual maximum discharges.  Suppose that it is desired to plot these annual maxima on a graph in order to better interpret the data, perhaps detect errors, or to get an idea of which probability distribution to use to describe the data.  The ordinate of such a graph conventionally contains the event magnitudes either on a linear or logarithmic scale while the abscissa will be some measure of the probability of occurrence of each event or the average time interval between occurrences (since this is simply the inverse of the probability of occurrence).  Frequently the scale of the abscissa will be arranged so that events distributed according to a given distribution will plot as a straight line.

The question then arises as to how to derive the probability of occurrence or average return period of each of the set of annual maximum floods. Sorting the annual events in order of magnitude it is apparent that the largest flood occurs once in the 10 years. But is 10 years the true average return period (i.e. the average interval between occurrences) of this flood? We do not know. In the particular sample a flood of this magnitude occurred only once but in other samples of annual maxima of equal length the same magnitude of flood might occur several times or not at all. That is, the maximum flood in the 10-year sample may have a true return period of 5 years, 50 years, 500 years or any other number of years. And yet, for practical reasons, some probability of occurrence must be assigned to this flood.

For the maximum of a series of 10 annual maximum values it can be shown (34) that the true return period, T, has a 10% probability of being as low as 5 years and a 10% probability of being as high as 100 years. In theory, if the basic data are truly representative, a flood having a return period of 10 years will be equalled or exceeded in a great length of time, on an average of once in 10 years. In 1000 years of record there would be 1000/10, or 100 such floods. However, if these 1000 years were divided into 100 periods of 10 years each, about 37% of such periods would not experience a flood of that magnitude; about 37% of the periods would experience 1 such flood; about 18% would experience 2; about 6% would experience 3; about 1.5% would experience 4; and about 0.5% would experience 5 or more such floods.

As an illustration of the problem consider a large rain storm occurring simultaneously over several adjacent streamflow basins. Imagine that one stream has been gauged for 5 years, another for 10 years and another for 15 years. If the resulting flood runoff is the maximum in the 15 years, it will have an apparent return period of 5 years at one gauge, 10 years at another and 15 years at the third. Which is correct? Obviously the estimate based on 15 years is more correct than those based on 10 or 5 years. In the absence of infinite records some method of correcting for this variation in apparent return period with available record must be derived. The problem of plotting position is to locate the "correct" return period for each event so that the ranges of variation can be replaced by point positions.

The large variation possible in return period, T, can be shown more formally by considering T as an independent random variable which can take the values 1, 2, ... t-1, t, t+1...∞ where t is the number of years from the occurrence of one event until the occurrence of the next event. The distribution of T is then of the form:

$$P(T = t) = p(1 - p)^{t-1} \qquad (2-28)$$

where p is the probability of the event occuring in any given year. The average value of T or return period is then given by:

$$E(T) = \sum_{t=1}^{\infty} t\, P(T = t) = 1/p \qquad (2-29)$$

The variance of T can similarly be expressed as:

$$\text{var } T = E(T - E(T))^2 \qquad (2-30)$$

Following Lloyd (23) this expression reduces to:

$$\text{var } T = (1 - p)/p^2 \qquad (2-31)$$

For small p the variance of the return period can be approximated as 1/p i.e. the same as the expected value. This explains the large observed variations in values of T. Lloyd has also shown that for non-independent sequences of events the variance of the return period is even larger.

The problem of apparent return period varying with sample length is approached in practice by defining a plotting position for the frequency of occurrence, p. The plotting position is generally based on some assumption of the position of the sample estimate of the frequency within a population distribution of frequencies. A few general requirements of any plotting position have been given by Gumbel (15) as:

(a) The plotting position should be such that all observations can be plotted.
(b) The plotting positions should lie between the observed
frequencies (m - 1)/n and m/n and should be distribution free (m is the order of the particular event in
the series of n maximum events. For the largest of
the n events, m = 1).
(c) The return period of a value equal to or larger than
the largest observation should approach n, the number
of observations.
(d) The observations should be equally spaced on the frequency scale, i.e. the difference between the plotting
positions of the (m + 1)th and the mth observations
should be a function of n only and be independent of m.
(e) The plotting position should have an intuitive meaning
and should be analytically simple.

*Plotting Position as an Extreme* - The simplest assumption regarding the sample frequency in its population is that the events correspond directly to their observed frequency i.e. the

maximum event in a series of 10 independent maxima would have a
return period of 10 and a frequency of occurrence of 0.1. This
is known as the California method (23) and is given by the general
equation

$$p = \frac{m}{n} \qquad (2\text{-}32)$$

That is, considering the frequency interval 0.1 - 0, the California methods uses a plotting position at the upper extreme of
this interval. By first considering the observed maxima arranged
in a decreasing order of magnitude the observed frequency is
equally legitimately given by

$$p = \frac{m-1}{n} \qquad (2\text{-}33)$$

which corresponds to the lower extreme of the frequency interval
noted above. Since the frequencies zero and unity do not exist
for an unlimited variate the largest observation of the series
cannot be plotted using the function $(m-1)/n$ and the smallest
observation cannot be plotted using the function $m/n$. These
two functions therefore fail Gumbel's conditions and are not
acceptable as plotting positions unless an upper or lower limit
to the population can be envisaged.

   *Plotting Position as the Mean* - Foster (13) has pointed out
that there is no reasonable basis for assuming that the maximum
event in a sample represents exactly the simple (California)
frequency. Instead he contended that this maximum event is
representative of the whole class of possible events occurring
with frequencies less than that of the maximum in the sample,
i.e. for a sample of 10 the maximum event represents the interval
0.10 to 0, and therefore should be plotted at the mean of this
class interval (i.e. at 0.05 in our example). The Foster (or
sometimes called Hazen) plotting position is given by the general
equation

$$p = \frac{2m - 1}{2n} \qquad (2\text{-}35)$$

Equation 2-35 is a compromise between Equations 2-32 and 2-33.
When m = 1, the largest event of the sample, this plotting position claims that an event which has already happened once in n
years will occur, in the mean, once in 2n years.

   Similarly if the observed frequency of an event is assumed
to be the mean of the population of frequencies for that event,
$\bar{p}$, then

$$p = \frac{\int_0^1 (1-z^{1/n}) \, dZ}{\int_0^1 dZ} \qquad (2\text{-}36)$$

20

where Z is the frequency of occurrence of the event in the n year period. For the maximum value, $\bar{p} = 1/(n+1)$ so that for a 10-year sample the maximum value would have an assigned average return period of 11 years. Use of the mean frequency leads to the general equation

$$p = \frac{m}{n+1} \qquad (2-37)$$

where m is the order of the flood, m being 1 for the largest and n for the smallest event in the n years of record. The probability p is thus the average of the probabilities of all events with rank m in a series of periods each of n years.

*Plotting Position as the Mode* - Another assumption possible is that the observed frequency is the mode (by definition the event which occurs most frequently) of the population of frequencies. Equation 2-28 has no mode and so the type of probability distribution must be incorporated. As an example Equation 2-28 can be used with the Gumbel or type I extremal distribution (15) to yield:

$$Z = (e^{-e^{-y}})^n \qquad (2-38)$$

where y is a linear function of discharge. For the mode of any distribution the probability density is a maximum i.e. $dZ/dy = 0$ and $d^2Z/dy^2 < 0$, which for equation 2-38 yields

$$p = 1 - (\frac{1}{e})^{1/n} \qquad (2-39)$$

For the maximum annual event in a 10-year record this equation indicates an average return period of 10-1/2 years. The plotting positions for the maximum events in samples of different size are given in the following table:

It is clear that as the sample size gets larger, p is not significantly different from the simple plotting position $p = m/n$.

Table 2-3

Plotting Positions for Maximum Event Under Modal Assumption

| Sample Size (n) | Plotting Position (p) |
|---|---|
| 2 | 0.393 |
| 5 | 0.181 |
| 10 | 0.0951 |
| 20 | 0.0488 |
| 50 | 0.0198 |
| 100 | 0.00995 |
| 200 | 0.00499 |
| 500 | 0.00199 |
| 1000 | 0.000999 |

Blythe (6) computed the mode of a distribution of extreme values as

$$mode = \mu - \sigma \frac{\beta_1^{\frac{1}{2}}(\beta_2 + 3)}{2(5\beta_2 - 6\beta_1 - 9)} \qquad (2-40)$$

21

where $\mu$, $\sigma$, $\beta_1$ and $\beta_2$ are respectively the mean, standard deviation, coefficient of skew and coefficient of kurtosis. From the normal distribution the probability of occurrence of this mode can be determined. The maximum values of samples of different sizes are given in Table 2-4. No plotting positions are available by this method for events other than the maximum of the sample, thus Blythe recommended that for practical applications the Foster plotting positions (Equation 2-35) should be used.

*Plotting Position as the Median* - A further possible assumption is that at any given order of magnitude within a sample set of events, the return period used should be the median of all possible return periods obtained from a population of equally sized samples. The return period to be used at any given order of magitude, j, within a sample of size n events is then given by

$$T = 1/p_j \qquad (2-41)$$

where $p_j$ is the solution of the binomial equation

$$\sum_{i=0}^{j-1} \binom{n}{i} p_j^i (1 - p_j)^{n-1} = 0.5$$

$$(2-42)$$

the is, $p_j$ is a solution of a polynomial of order equal to the sample size n and having a number of terms equal to the numerical value of j.

Table 2-4

Plotting Positions for Maximum Event of Extremal Distribution

| Sample Size (n) | Plotting Position (p) |
|---|---|
| 2 | 0.306 |
| 5 | 0.143 |
| 10 | 0.077 |
| 20 | 0.041 |
| 60 | 0.014 |
| 100 | 0.0087 |
| 200 | 0.0046 |
| 500 | 0.0018 |
| 1000 | 0.00092 |

In effect, this method results in a return period of approximately 1.44 n for the largest flood in a period of n years. This occurs because, for example, the event with a return period of 144 years has an equal chance of being exceeded or not exceeded in any period of 100 years.

As an example of the use of the median assumption in assigning plotting positions, the maximum event in a sample of 10 maxima would have a computed return period, t, of 14.92 years. Beard (2) gave the plotting positions for a sequence of 10 maxima as in Table 2-5.

22

This occurs because, for example, the event with a return period of 144 years has an equal chance of being exceeded or not exceeded in any period of 100 years.

As an example of the use of the median assumption in assigning plotting positions, the maximum event in a sample of 10 maxima would have a computed return period, t, of 14.92 years. Beard (2) gave the plotting positions for a sequence of 10 maxima as in Table 2-5.

Table 2-5

Plotting Positions Under Median Assumption

| Magnitude Order (m) | Plotting Position (p) |
|---|---|
| 1 | 0.067 |
| 2 | 0.164 |
| 3 | 0.258 |
| 4 | 0.355 |
| 5 | 0.452 |
| 6 | 0.548 |
| 7 | 0.645 |
| 8 | 0.742 |
| 9 | 0.836 |
| 10 | 0.933 |

The theoretical values of the plotting positions are tedious to compute requiring the solution of a polynomial of degree equal to the sample size and Beard has recommended that the Foster equation (Equation 2-35) which can be transformed to

$$p_j = p_1 + (j-1) \cdot (1-2p_1)/n-1$$

$$(2-43)$$

where $p_1 = 1/2n$ should be used as an approximate solution.

Banerji and Gupta (1) proposed an alternate solution to Equation 2-43 as

$$p_1 = 1 - 0.5^{1/n} \qquad (2-44)$$

In practice the procedure advocated by Banerji and Gupta is to compute $p_1$ from Equation 2-44, compute the increment $(1-2p_1)/n-1$ and then compute $p_j$ for j = 2, 3... n/2 from Equation 2-43. The remaining plotting positions $p_j$, j = n/2, (n/2+1)...n can be calculated as

$$p_{(n+1)/2} = 0.5 \qquad (2-45)$$

$$p_{n-j+1} = 1.0 - p_j \qquad (2-46)$$

Beard (2) and Hardison and Jennings (16) have demonstrated that for a normal distribution the average exceedence probability $\bar{p}$, for an event with a 10-year return period estimated from a sample of 10 events is 0.1261. As the length of the sample increases so the average exceedence probability decreases until as $n \to \infty$, $\bar{p} \to p$ where p is the simple probability exceedence estimated as p = m/n.

## References

1.  Banerji, S., and D. K. Gupta, 1967, On a General Theory of Duration Curve and its Application to Evaluate the Plotting Position of Maximum Probable Precipitation or Discharge, Proc. Symposium on Floods and their Computations, Leningrad, pp. 183-193.

2.  Beard, L. R., 1943, Statistical Analysis in Hydrology, Trans., ASCE, Vol. 69, No. 8, Pt. 2, pp. 1110-1160.

3.  Benson, M. A., 1950, Use of Historical Data in Flood Frequency Analysis, Trans. Am. Geophys. Union, V. 31, pp. 419-424.

4.  Benson, M. A., 1960, Characteristics of Frequency Curves Based on a Theoretical 1000 Year Record, USGS Water Supply Paper No. 1543-A, pp. 51-73.

5.  Blench, T., 1959, Empirical Methods, Proceedings of Symposium No. 1, Spillway Design Floods, NRC, Ottawa, pp. 36-48.

6.  Blythe, R. H., 1943, Discussion of Paper No. 2201, Trans. ASCE, Vol. 69, No. 8, Pt. 2, pp. 1137-1138.

7.  Chen, J. Q., Ye, Y-Y., and W-Y. Tan, 1975, The Important Role of Historical Flood Data in the Estimation of Spillway Design Floods, Scientia Sinica, Vol. XXVIII, No. 5, pp. 669-680.

8.  Chow, V. T., 1964, Handbook of Hydrology, McGraw-Hill.

9.  Dalrymple, T., 1956, Measuring Floods, Proceedings of IASH Symposia, Darcy, Dijon, Vol. 3, pp. 380-404.

10. Dalrymple, T., 1960, Flood Frequency Analyses, USGS Water Supply Paper No. 1543-A, pp. 1-47.

11. Dickinson, W. T., 1967, Accuracy of Discharge Determinations, Hydrology Paper No. 20, Colorado State University, Fort Collins, Colorado.

12. Dyhr-Nielson, M., 1972, Loss of Information by Discretizing Hydrology Series, Hydrology Paper No. 54, Colorado State University, Fort Collins, Colorado.

13. Foster, H. A., 1936, Methods for Estimating Floods, USGS Water Supply Paper 771, Washington, D. C.

14. Glos, E., and R. Krause, 1967, Estimating the Accuracy of Statistical Flood Values by Means of Long-Term Dis-

charge Records and Historical Data, Proc. Leningrad Symposium on Floods and their Computations, pp. 144-151.

15. Gumbel, E. J., 1958, Statistics of Extremes, Columbia University Press.

16. Hardison, C. H., and M. E. Jennings, 1972, Bias in Computed Flood Risk, Proc. ASCE, Vol. 98, No. HY3, pp. 415-427.

17. Harter, H. L., 1961, Expected Values of Normal Order Statistics, Biometrika, Vol. 48, Nos. 1 and 2, pp. 151-165.

18. Herschy, R. W., 1969, The Evaluation of Errors at Flow Measurement Stations, Technical Note No. 11, Water Resources Board, Reading, England.

19. Kite, G. W., and R. L. Pentland, 1971, Data Generating Methods in Hydrology, Technical Bulletin No. 36, Inland Waters Directorate, Department of the Environment, Ottawa.

20. Kite, G. W., 1974, Case Study of Regional Analysis Techniques for Design Flood Estimation, Can. Jour. Earth Sciences, Vol. 11, No. 6, pp. 801-808.

21. Langbein, W. B., 1960, Plotting Positions in Frequency Analysis, USGS Water Supply Paper No. 1543-A, pp. 48-51.

22. Leese, M. N., 1973, The Use of Censored Data in Estimating T-Year Floods, Proceedings of the UNESCO/WMO/IAHS Symposium on the Design of Water Resources Projects with Inadequate Data, Madrid, Vol. 1, pp. 235-247.

23. Lloyd, E. H., 1970, Return Periods in the Presence of Persistence, Journal of Hydrology, Vol. 10, No. 3, pp. 291-298.

24. Mann, H. B., and D. R. Whitney, 1947, On a Test of Whether One of Two Random Variables is Stochastically Larger than the Other, Ann. Math. Stat., Vol. 18, pp. 50-61.

25. Robertson, A.I.G.S., 1966, The Magnitude of Probable Errors in Water Level Determination at a Gauging Station, Technical Note, No. 7, Water Resources Board, Reading, England.

26. Rosenberg, H. B., and R. L. Pentland, 1966, Accuracy of Winter Streamflow Records, Proc. Eastern Snow Conference, Hartford, Connecticut.

27. Siegel, S., 1956, Nonparametric Statistics for the Behavioral Sciences, McGraw-Hill Series in Psychology, New York, 311 p.

28. Smith, A. G., 1974, Peak Flows by the Slope-Area Method, Technical Bulletin No. 79, Inland Waters Directorate, Canada Dept. of the Environment, Ottawa, 31 pp.

29. Spence, E. S., 1973, Theoretical Frequency Distributions for the Analysis of Plains Streamflow, Can. J. Earth Sci., V. 10, pp. 130-139.

30. Stall, J. B., and J. C. Neill, 1961, A Partial Duration Series for Low-Flow Analyses, Journal of Geophysical Research, Vol. 66, No. 12, pp. 4219-4225.

31. Stoddart, R. B. L., and W. E. Watt, 1970, Flood Frequency Prediction for Intermediate Drainage Basins in Southern Ontario, C. E. Research Report No. 66, Queen's University at Kingston.

32. Subcommittee of the Joint Division Committee of Floods, 1953, Review of Flood Frequency Methods, Trans. ASCE, Vol. 118, pp. 1220-1231.

33. Terry, M. E., 1952, Some Rank Order Tests Which are Most Powerful Against Specified Parameti.. Alternatives, Ann. Math. Stat., Vol. 23, pp. 346-366.

34. USGS, 1936, Floods in the United States, Water Supply Paper No. 771, Washington, D. C.

35. Yen, B. C., and A. H.-S. Ang, 1971, Risk Analysis in Design of Hydraulic Projects, Proceedings of Symposium on Stochastic Hydraulics, Univ. of Pittsburg, pp. 694-709.

36. Yevjevich, V., 1972, Probability and Statistics in Hydrology, Water Resources Publications, Fort Collins, Colorado.

37. Yevjevich, V., 1972, New Vistas for Flood Investigations, Academia Nationale Dei Lincei, Roma, Quaderno N. 169, pp. 515-546.

# CHAPTER 3
## FREQUENCY DISTRIBUTIONS (GENERAL)

Introduction

One of the most common problems faced in hydrology is the estimation of a design flood or drought from a fairly short record of streamflows. Plotting the magnitude of the measured events (annual maxima for example), some kind of pattern is generally apparent. The question is how to use this pattern to extend the available data and estimate the design event.

If a large number of observed or measured events are available from a period of record at least as long as the return period of the required design event then the problem is simplified. In the extreme if a large enough sample were available (say one million events) then the design event and its confidence interval could be derived directly from the sample data. This amount of data will not be available, however, and so the sample data are generally used to fit a frequency distribution which in turn is used to extrapolate from the recorded events to the design events either graphically or by estimating the parameters of some standard frequency distribution.

Graphical methods have the advantages of simplicity and visual presentation and the fact that no assumption of distribution type is made. These advantages are outweighed, however, by the disadvantage that, given twenty engineers to fit a curve through a set of points, it is highly probable that at least twenty different curves would result. In other words the method is highly subjective and is not compatible with the other phases of engineering design. It must be noted, however, that subjective information need not be totally eliminated and that techniques such as Bayesian analysis may permit its use in the statistical procedure (13).

Numerous different probability or frequency distributions have been used in hydrology (1). Discrete distributions such as the binomial and Poisson have been used to define the average intervals between events (5) and to evaluate risks (7). Continuous distributions such as the normal and lognormal have been used for both annual series (6) and partial duration series (1) to define the magnitude of an event corresponding to a given probability of occurrence. The two types of distribution have also been combined (4), (11), (12) to give models of frequency of occurrence and frequency of magnitude of extreme events.

There are two sources of error in using a frequency distribution to estimate event magnitudes. The first source of error is that it is not known which of the many distributions available is the "true" distribution, i.e., which distribution, if any, the events naturally follow. This is important because the sample events available are usually for relatively low return periods (i.e. around the centre of the probability distribution) while the events it is required to estimate are generally of large return period (i.e. in the tail of the distribution). Many distributions have similar shape in their centers but differ widely in the tails. It is thus possible to fit several distributions to the sample data and end up with several different estimates of the T-year event. Chi-square and similar tests of goodness of fit can be used to choose the distribution which best describes the sample data but this does not overcome the basic problem.

Once a distribution has been chosen then the second source of error becomes apparent. The statistical parameters of the probability distribution must be estimated from the sample data. Since the sample data is subject to error the method of fitting must minimize these errors and must therefore be as efficient as possible.

Parameter Estimation

There are four parameter estimation techniques in current use:

1.  method of moments,
2.  method of maximum likelihood,
3.  least squares, and
4.  graphical.

*Method of Moments* - The method of moments utilizes either the general equation for calculation of the rth moment about the origin of a distribution, $p(x)$:

$$\mu_r' = \int_{-\infty}^{\infty} x^r \; p(x) \; dx \qquad (3-1)$$

or the corresponding equation for central moments of the distribution

$$\mu_r = \int_{-\infty}^{\infty} (x - \mu_1')^r \; p(x) \; dx \qquad (3-2)$$

where $\mu_1'$ is the first moment about the origin. The method of moments then relates the derived moments to the parameters of the distribution.

As an example consider the application of the method of moments to the normal distribution with two parameters, say $\alpha$

and $\beta$, for which:

$$p(x) = \frac{1}{\beta\sqrt{2\pi}} \, e^{-\frac{1}{2}(\frac{x-\alpha}{\beta})^2} \qquad (3\text{-}3)$$

so that

$$\mu_r' = \int_{-\infty}^{\infty} x^r \cdot \frac{1}{\beta\sqrt{2\pi}} \, e^{-\frac{1}{2}(\frac{x-\alpha}{\beta})^2} \, dz \qquad (3\text{-}4)$$

For the first moment about the origin

$$\mu_1' = \int_{-\infty}^{\infty} \frac{x}{\beta\sqrt{2\pi}} \, e^{-\frac{1}{2}(\frac{x-\alpha}{\beta})^2} \, dz \qquad (3\text{-}5)$$

Substituting z for $(x-\alpha)/\beta$

$$\mu_1' = \frac{1}{\sqrt{2\pi}} \int_{-\infty}^{\infty} (z\beta + \alpha) \, e^{-z^2/2} \, dz \qquad (3\text{-}6)$$

but $ze^{-z^2/2}$ is an odd function and the integral of an odd function between symmetric limits is zero so

$$\mu_1' = \frac{\alpha}{\sqrt{2\pi}} \int_{-\infty}^{\infty} e^{-z^2/2} \, dx = \alpha \qquad (3\text{-}7)$$

i.e. for this distribution the parameter $\alpha$ is the arithmetic mean (the first moment about the origin), say $\bar{x}$.

Utilizing Equation 3-2 for the second central moment

$$\mu_2 = \int_{-\infty}^{\infty} (x - \bar{x})^2 \, \frac{1}{\beta\sqrt{2\pi}} \, e^{-\frac{1}{2}(\frac{x-\alpha}{\beta})^2} \, dx \qquad (3\text{-}8)$$

but from 3-7 $\alpha = \bar{x}$. Substituting z for $(x - \bar{x})/\beta$

$$\mu_2 = \frac{\beta^2}{\sqrt{2\pi}} \int_{-\infty}^{\infty} z^2 \, e^{-z^2/2} \, dx \qquad (3\text{-}9)$$

Since the function $z^2 e^{-z^2/2}$ is even, and substituting y for $z^2/2$

$$\mu_2 = \frac{2\beta^2}{\sqrt{2\pi}} \int_{0}^{\infty} y^{\frac{1}{2}} \, e^{-y} \, dy \qquad (3\text{-}10)$$

$$\mu_2 = \frac{2\beta^2}{\sqrt{2\pi}} \, \frac{\sqrt{\pi}}{2} = \beta^2 \qquad (3\text{-}11)$$

i.e. parameter $\beta$ is the square root of the second central moment.

To compute higher order central moments it is frequently easier to compute the moments about the origin, $\mu_r'$, and then convert to central moments, $\mu_r$, using the expression

$$\mu_r = \sum_{j=0}^{r} \binom{r}{j} \mu_{r-j}' \left(-\mu_1'\right)^j \tag{3-12}$$

so that, for example,

$$\mu_2 = \mu_2' - \mu_1'^2 \tag{3-13}$$

*Method of Maximum Likelihood* - The principle of maximum likelihood states that for a distribution with a probability density function $p(x;\alpha,\beta,...)$ where $\alpha,\beta...$ are the distribution parameters to be estimated, then the probability of obtaining a given value of x, $x_i$, is proportional to $p(x_i;\alpha,\beta,...)$ and the joint probability, L, of obtaining a sample of n values $x_1$, $x_2$,... $x_n$ is proportional to the product

$$L = \prod_{i=1}^{n} p(x_i;\alpha,\beta,...) \tag{3-14}$$

This is called the likelihood. The method of maximum likelihood is to setimate, $\alpha$, $\beta$,..., such that L is maximized. This is obtained by partially differentiating L with respect to each of the parameters and equating to zero. Frequently ln L is used instead of L to simplify computations.

As an example of the application of the maximum likelihood technique consider again the normal distribution with parameters $\alpha$ and $\beta$:

$$p(x) = \frac{1}{\beta\sqrt{2\pi}} e^{-\frac{1}{2}\left(\frac{x-\alpha}{\beta}\right)^2} \tag{3-15}$$

so that

$$L = \left\{\frac{1}{\beta\sqrt{2\pi}}\right\}^n e^{\frac{\sum_{i=1}^{n}(x_i-\alpha)^2}{2\beta^2}} \tag{3-16}$$

Taking logarithms

$$\ln L = -\frac{n}{2}\ln 2\pi - \frac{n}{2}\ln \beta^2 - \frac{\sum_{i=1}^{n}(x_i-\alpha)^2}{2\beta^2} \tag{3-17}$$

Differentiating with respect to the parameters $\alpha$ and $\beta^2$ and equating to zero

$$\frac{\partial \ln L}{\partial \alpha} = \frac{\sum\limits_{i=1}^{n} (x_i - \alpha)}{\beta^2} = 0 \qquad (3\text{-}18)$$

so that

$$\sum_{i=1}^{n} x_i - \sum_{i=1}^{n} \alpha = 0 \qquad (3\text{-}19)$$

but

$$\sum_{i=1}^{n} \alpha = n\alpha \qquad (3\text{-}20)$$

so that

$$\alpha = \frac{\sum\limits_{i=1}^{n} x_i}{n} \qquad (3\text{-}21)$$

Also

$$\frac{\partial \ln L}{\partial \beta^2} = -\frac{n}{2\beta^2} + \frac{\sum\limits_{i=1}^{n} (x_i - \alpha)^2}{2\beta^4} = 0 \qquad (3\text{-}22)$$

so that

$$\beta^2 = \frac{\sum\limits_{i=1}^{n} (x_i - \alpha)^2}{n} \qquad (3\text{-}23)$$

i.e. parameters $\alpha$ and $\beta$ are the mean and standard deviation of the distribution.

*Least Squares* - The least squares estimation method (14) consists of fitting a theoretical function to an empirical distribution. The sum of squares of all deviations of observed points from the fitted function is then minimized. Thus to fit a function

$$\hat{y} = p(x; \alpha, \beta, \ldots) \qquad (3\text{-}24)$$

the sum to be minimized is

$$S = \sum_{i=1}^{n} (y_i - \hat{y}_i)^2 \qquad (3\text{-}25)$$

or

$$S = \sum_{i=1}^{n} (y_i - f(x_i; \alpha, \beta, \ldots))^2 \qquad (3\text{-}26)$$

where $x_i$ and $y_i$ are coordinates of observed points, $\alpha$, $\beta$,... are parameters and n is the sample size. To obtain the minimum sum of squares Equation 3-25 is partially differentiated with respect to the parameter estimates a, b,...

31

$$\frac{\partial \sum\limits_{i=1}^{n} (y_i - \hat{y}_i)^2}{\partial a} = 0 \qquad (3\text{-}27)$$

$$\frac{\partial \sum\limits_{i=1}^{n} (y_i - \hat{y}_i)^2}{\partial b} = 0 \qquad (3\text{-}28)$$

These partial derivatives give a number of equations equal to the number of parameters to be estimated. In order that the least squares method be an efficient estimator three conditions must be satisfied:

1. The deviations $y_i - \hat{y}_i$ be normally or at least symmetrically distributed.
2. The population variance of the deviations be independent of the magnitude of $y_i$.
3. The population variance of the deviations along the least squares curve be constant.

   *Graphical Method* - The graphical method consists of fitting a function

$$\hat{y} = p(x; \alpha, \beta, \ldots) \qquad (3\text{-}29)$$

visually through the set of coordinate pairs. To estimate m parameters, m points on the curve are selected giving m equations to solve. The process may be simplified by trying various types of graph paper using transformed coordinates until a straight line fit is possible.

In ascending order of efficiency the four methods of estimation may be listed as graphical, least squares, method of moments, maximum likelihood. To offset its greater efficiency, however, the method of maximum likelihood is somewhat more difficult to apply.

These four methods will, more or less efficiently, compute the parameters of a distribution from a particular data sample. As discussed earlier this sample may or may not be typical of the underlying population and use of the sample estimates of parameters may bias the results. A method of Bayesian analysis has been recommended (13) which would use distributions of parameter values rather than point values. These distributions might be obtained from regional analysis.

There are also general criteria with which a distribution should comply before being used in hydrology. As an example, since negative flows are inacceptable a distribution should be bounded on the lower tail. This criterion would eliminate both

the normal and the double exponential or Gumbel distributions.
In fact, both of these distributions are used in hydrology by
ignoring negative flows, replacing them with zeros, or treating
the probability of zero flow as a probability mass (14).

## Frequency Factor

Having selected a distribution and estimated its parameters
the question is how to use this distribution in the frequency
analysis. Chow (1) has proposed a general equation:

$$x_T = \mu + K \sigma \tag{3-30}$$

or

$$x_T = m_1' + K\sqrt{m_2} \tag{3-31}$$

where $x_T$ is the event magnitude at a given return period, T, $\mu$
and $\sigma$ are the population mean and standard deviation estimated
by sample moments $m_1'$ and $\sqrt{m_2}$ and K is a frequency factor which
is a function of the return period and the distribution parameters. For any chosen distribution a relationship can be derived
between the return period and the frequency factor.

## Standard Error

A measure of the variability of the resulting event magnitudes is the standard error of estimate. Each method of estimating the parameters of a distribution can also be used to derive
the standard error. The standard error of estimate is defined
as:

$$S = \left\{ \left[ \sum_{i=1}^{n} (x_i - \hat{x}_i)^2 \right] / n \right\}^{\frac{1}{2}} \tag{3-32}$$

where $\hat{x}_i$ is the computed estimate of recorded event $x_i$.

The differences between the recorded and the computed events
may have two origins:

1. The choice of a theoretical population distribution
   from the sample may be wrong, and
2. The errors in the parameters of the chosen population
   distribution may be inaccurate due to the shortness
   of the sample data.

The standard error of estimate accounts for only the second of
these causes.

*Standard Error by Moments* - In general the variance of the
T-year event, $x_T$, can be derived from moment estimates as follows.
If $x_T$ depends upon the first three moments of a distribution and

the return period, T, as

$$x_T = f(m_1', m_2, m_3, T) \tag{3-33}$$

then, since T is not a variable, the variance of $x_T$ is given by

$$S_T^2 = \left(\frac{\partial x_T}{\partial m_1'}\right)^2 \text{var } m_1' + \left(\frac{\partial x_T}{\partial m_2}\right)^2 \text{var } m_2 + \left(\frac{\partial x_T}{\partial m_3}\right)^2 \text{var } m_3$$

$$+ 2\frac{\partial x_T}{\partial m_1'} \cdot \frac{\partial x_T}{\partial m_2} \text{ cov } (m_1', m_2) + 2\frac{\partial x_T}{\partial m_1'} \cdot \frac{\partial x_T}{\partial m_3} \text{ cov } (m_1', m_3)$$

$$+ 2\frac{\partial x_T}{\partial m_2} \cdot \frac{\partial x_T}{\partial m_3} \text{ cov } (m_2, m_3) \tag{3-34}$$

The actual relationship between $x_T$ and the moments has been given in 3-30 as

$$x_T = m_1' + K\sqrt{m_2} \tag{3-30}$$

where K is a function of the coefficient of skew, g, and $g = m_3/m_2^{3/2}$ so that:

$$\frac{\partial x_T}{\partial m_1'} = 1 \tag{3-35}$$

$$\frac{\partial x_T}{\partial m_2} = \frac{K}{2\sqrt{m_2}} + \sqrt{m_2}\frac{\partial K}{\partial m_2} = \frac{K}{2\sqrt{m_2}} - \frac{3g}{2\sqrt{m_2}} \cdot \frac{\partial K}{\partial g} \tag{3-36}$$

and

$$\frac{\partial x_T}{\partial m_3} = \frac{\partial x_T}{\partial K} \cdot \frac{\partial K}{\partial g} \cdot \frac{\partial g}{\partial m_3} = \frac{1}{m_2} \cdot \frac{\partial K}{\partial g} \tag{3-37}$$

so that

$$S_T^2 = \text{var } m_1' + \left[\frac{1}{2\sqrt{m_2}}\left(K - 3g\frac{\partial K}{\partial g}\right)\right]^2 \text{var } m_2 + \frac{1}{m_2^2}\left(\frac{\partial K}{\partial g}\right)^2 \text{var } m_3$$

$$+ \frac{1}{\sqrt{m_2}}\left(K - 3g\frac{\partial K}{\partial g}\right) \text{cov } \left(m_1', m_2\right) + \frac{2}{m_2} \cdot \frac{\partial K}{\partial g} \text{cov } \left(m_1', m_3\right)$$

$$+ \frac{1}{m_2^{3/2}} \cdot \frac{\partial K}{\partial g} \left( K - 3g \frac{\partial K}{\partial g} \right) \text{cov} \left( m_2, m_3 \right) \qquad (3\text{-}38)$$

but from Kendall and Stuart (8) pp. 230-232

$$\text{var } m_1' = \mu_2/n \qquad (3\text{-}39)$$

$$\text{var } m_2 = \frac{1}{n} \left\{ \mu_4 - \mu_2^2 \right\} \qquad (3\text{-}40)$$

$$\text{var } m_3 = \frac{1}{n} \left\{ \mu_6 - \mu_3^2 - 6\mu_4\mu_2 + 9\mu_2^3 \right\} \qquad (3\text{-}41)$$

$$\text{cov} \left( m_1', m_2 \right) = \mu_3/n \qquad (3\text{-}42)$$

$$\text{cov} \left( m_1', m_3 \right) = \frac{1}{n} \left\{ \mu_4 - 3\mu_2^2 \right\} \qquad (3\text{-}43)$$

and

$$\text{cov} \left( m_2, m_3 \right) = \frac{1}{n} \left\{ \mu_5 - 4\mu_3\mu_2 \right\} \qquad (3\text{-}44)$$

Substituting these results into Equation 3-38, using population parameters throughout, and simplifying by substituting

$$\gamma_1 = \mu_3/\mu_2^{3/2} \qquad (3\text{-}45)$$

$$\gamma_2 = \mu_4/\mu_2^2 \qquad (3\text{-}46)$$

$$\gamma_3 = \mu_5/\mu_2^{5/2} \qquad (3\text{-}47)$$

and

$$\gamma_4 = \mu_6/\mu_2^3 \qquad (3\text{-}48)$$

the following expression results:

$$S_T^2 = \frac{\mu_2}{n} \left\{ 1 + K\gamma_1 + \frac{K^2}{4} \left[ \gamma_2 - 1 \right] + \frac{\partial K}{\partial \gamma_1} \left[ 2\gamma_2 - 3\gamma_1^2 \right. \right.$$

$$- 6 + K \left( \gamma_3 - 6\gamma_1\gamma_2/4 - 10\gamma_1/4 \right) \bigg] + \left( \frac{\partial K}{\partial \gamma_1} \right)^2 \bigg[ \gamma_4$$

$$- 3\gamma_3\gamma_1 - 6\gamma_2 - 9\gamma_1^2\gamma_2/4 + 35\gamma_1^2/4 + 9 \bigg] \bigg\} \qquad (3\text{-}49)$$

The standard error of estimate, $S_T$, at any given return period, T, can now be found by substituting n, $\mu_2$, $\gamma_1 - \gamma_4$, K and $\partial K/\partial \gamma_1$ for a particular distribution and sample and taking the square root of 3-49.

In later chapters the expression

$$S_T = \delta \sqrt{\frac{\mu_2}{n}}$$

(3-50)

is used so that tables of $\delta$ can be prepared independently of $\mu_2$ and n. Note that for a distribution in which the frequency factor, K, is not dependent on the coefficient of skew the expression for $\delta$ simplifies to:

$$\delta = \left\{ 1 + K\gamma_1 + \frac{K^2}{4} \left[ \gamma_2 - 1 \right] \right\}^{\frac{1}{2}}$$

(3-51)

*Standard Error by Maximum Likelihood* - Assume that for a particular distribution the three parameters $\alpha$, $\beta$ and $\gamma$ have been estimated by the method of maximum likelihood. The T-year event, $x_T$, is then some function of these parameters and the return period T:

$$x_T = f(\alpha, \beta, \gamma, T)$$

(3-52)

Since T is not a variable the standard error of $x_T$, $S_T$, may be written as

$$S_T^2 = \left( \frac{\partial x}{\partial \alpha} \right)^2 \text{var } \alpha + \left( \frac{\partial x}{\partial \beta} \right)^2 \text{var } \beta + \left( \frac{\partial x}{\partial \gamma} \right)^2 \text{var } \gamma$$

$$+ 2\frac{\partial x}{\partial \alpha}\frac{\partial x}{\partial \beta} \text{ cov } (\alpha,\beta) + 2\frac{\partial x}{\partial \alpha} \cdot \frac{\partial x}{\partial \gamma} \text{ cov } (\alpha,\gamma)$$

$$+ 2\frac{\partial x}{\partial \beta}\frac{\partial x}{\partial \gamma} \text{ cov } (\beta,\gamma)$$

(3-53)

The partial derivatives may be obtained directly from the likelihood equation and Fisher (3), quoted in Kendall and Stuart (9), has shown that the variance/covariance matrix

$$\begin{bmatrix} \text{var } \alpha & \text{cov } (\alpha,\beta) & \text{cov } (\alpha,\gamma) \\ & \text{var } \beta & \text{cov } (\beta,\gamma) \\ & & \text{var } \gamma \end{bmatrix}$$

(3-54)

is the inverse of the symmetric matrix

$$
\begin{bmatrix}
-\dfrac{\partial^2 L}{\partial\alpha^2} & -\dfrac{\partial^2 L}{\partial\alpha\partial\beta} & -\dfrac{\partial^2 L}{\partial\alpha\partial\gamma} \\[4mm]
 & -\dfrac{\partial^2 L}{\partial\beta^2} & -\dfrac{\partial^2 L}{\partial\beta\partial\gamma} \\[4mm]
 & & -\dfrac{\partial^2 L}{\partial\gamma^2}
\end{bmatrix}
\qquad (3\text{-}55)
$$

where L is the likelihood equation for the particular distribution. Analytical expressions for each of the variances and covariances may therefore be derived in terms of the minors and determinant of the symmetric matrix above, as for example:

$$
\mathrm{var}\ \alpha = \left[\frac{\partial^2 L}{\partial\beta^2}\cdot\frac{\partial^2 L}{\partial\gamma^2} - \left\{\frac{\partial^2 L}{\partial\beta\partial\gamma}\right\}^2\right]\Big/ D
\qquad (3\text{-}56)
$$

where D, the determinant, is given by

$$
D = -\frac{\partial^2 L}{\partial\alpha^2}\left[\frac{\partial^2 L}{\partial\beta^2}\cdot\frac{\partial^2 L}{\partial\gamma^2} - \left\{\frac{\partial^2 L}{\partial\beta\partial\gamma}\right\}^2\right] + \frac{\partial^2 L}{\partial\alpha\partial\beta}\left[\frac{\partial^2 L}{\partial\alpha\partial\beta}\cdot\frac{\partial^2 L}{\partial\gamma^2}\right.
$$

$$
\left. -\frac{\partial^2 L}{\partial\beta\partial\gamma}\cdot\frac{\partial^2 L}{\partial\alpha\partial\gamma}\right] - \frac{\partial^2 L}{\partial\alpha\partial\gamma}\left[\frac{\partial^2 L}{\partial\alpha\partial\beta}\cdot\frac{\partial^2 L}{\partial\beta\partial\gamma} - \frac{\partial^2 L}{\partial\beta^2}\frac{\partial^2 L}{\partial\alpha\partial\gamma}\right]
\qquad (3\text{-}57)
$$

Evaluation of the standard error of estimate is then simple.

For distribution such as the lognormal, extremal type III and log-Pearson type III, which are logarithmic transformations of simple distributions there are two possible methods of computing event magnitudes and standard errors of event magnitudes. The first method is to develop analytical relationships for the frequency factor K and the parameter $\delta$ of the transformed distributions and use these with the mean and standard deviation of the original data. The second method is to logarithmically transform the original data and use the mean and standard deviation of the logarithms together with the K and $\delta$ values for the simple untransformed distribution.

If the distribution of the T-year event were known, then confidence limits could be derived for the event. Two methods can be used to find this distribution: analytical and empirical.

37

The analytical approach utilizes the probability distribution fitted to the observed data. Cramer (2) has shown that for a sample of n values fitted with a distribution having a probability density function p(x) and a cumulative probability function P(x) then the density function, g(x), of a random variable $x_T$ is given by

$$g(x) = \binom{n}{m} (n-m)(P(x))^m (1-P(x))^{n-m-1} p(x) \qquad (3-58)$$

where m is n P and P is the cumulative probability associated with event magnitude $x_T$. Now if G(x) represents the cumulative probability function of the random variable $x_T$

$$G(x) = \int_0^{x_0} g(x) \, dx \qquad (3-59)$$

then the upper or lower confidence limits, $x_0$, for the T-year event, $x_T$, may be found by solving Equation 3-59 for different levels of significance. For example, for the 95% upper confidence limits $x_0$ must be found for G(x) = 0.95.

Unfortunately the evaluation of Equation 3-59 for $x_0$ involves a lot of approximations and still must be carried out by a numerical iteration process. Because of these computational difficulties the second method of deriving confidence limits, the empirical method, is frequently used. In the empirical method Equation 3-30 is used to compute the mean T-year event, $x_T$, and the standard error of the T-year event, $S_T$ is computed by either moments or maximum likelihood. The assumption is then made (10) that the distribution of T-year events is normal so that the confidence interval is given by

$$x_T \pm t \, S_T \qquad (3-60)$$

where t is the standard normal deviate corresponding to the required confidence level.

## Computer Programs

Each chapter describing the application of a frequency distribution in hydrology contains sample computer program listings. The first listing of each chapter was written in FORTRAN IV for use on a CDC* Cyber 6400 mainframe computer. The second listing of each chapter is written in BASIC and is suitable for use on IBM** PC, XT and AT and compatible microcomputers.

While each of the programs has been tested on several data

38

sets, it is possible that errors exist or that there are limitations on the applicability of some or any of the programs. Use of non-original programs always involves some risk. The programs have, in each case, been written as simply as possible and may not give solutions for data sets having multiple roots.

It has been found that for use of some of the programs in this book on IBM** mainframes it is necessary to use double precision because of the short word length. This means specifying REAL*8 for all real variables and changing functions such as ALOG, ABS and SQRT to DLOG, DABS, and DSQRT. Similarly, the BASIC versions of the programs should be run using the double precision option (/D) on all microcomputers.

To save effort in coding the programs, 5 1/4" diskettes containing all the BASIC source code and executable versions of each program, together with sample data sets, are available from Water Resources Publications.


* CDC is a registered trademark of Control Data Corporation
** IBM is a registered trademark of International Business Machines.


## References

1.  Chow, V. T., 1964, Editor-in-Chief, Handbook of Applied Hydrology, McGraw-Hill.

2.  Cramer, H., 1946, Mathematical Methods of Statistics, Princeton University Press.

3.  Fisher, R. A., 1921, On the Mathematical Foundations of Theoretical Statistics, Phil. Trans. Roy. Soc., Series A, Vol. 222, pp. 309-368.

4.  Frost, J., and R. T. Clarke, 1972, Estimating the T-year Flood by the Extension of Records of Partial Duration Series, Bull. IAHS, Vol. XVIII, No. 1, pp. 209-217.

5.  Hall, W. A., and D. T. Howell, 1963, Estimating Flood Probabilities within Specific Time Intervals, J. Hydrology, Vol. 1, No. 1, pp. 265-271.

6.  Hardison, C. H., 1969, Accuracy of Streamflow Characteristics, USGS Professional Paper No. 650-D, pp. D210-D214.

7.  Kalinin, G. P., 1960, Calculation and Forecasts of Stream-

flow from Scanty Hydrometric Readings, Trans. Inter-regional Seminar on Hydrologic Networks and Methods, Bangkok, 1959, WMO Flood Control Series No. 15, pp. 42-52.

8.  Kendall, M. G., and A. Stuart, 1963, The Advanced Theory of Statistics, Vol. I, Distribution Theory, Griffin, London.

9.  Kendall, M. G., and A. Stuart, 1967, The Advanced Theory of Statistics, Vol. II, Inference and Relationship, Griffin, London.

10. Kite, G. W., 1975, Confidence Limits for Design Events, Wat. Res. Res., Vol. 11, No. 1, pp. 48-53.

11. Shane, R. M., 1966, A Statistical Analysis of Base-Flow Flood Discharge Data, Cornell University, PhD Thesis.

12. Todorovic, P., and D. A. Woolhiser, 1972, On the Time When the Extreme Flood Occurs, Wat. Res. Res., Vol. 8, No. 6, pp. 1433-1438.

13. Wood, E. F., and I. Rodriquez-Iturbe, 1975, Bayesian Inference and Decision Making for Extreme Hydrologic Events, Wat. Res. Res., Vol. 11, No. 4, pp. 533-542.

14. Yevjevich, V., 1972, Probability and Statistics in Hydrology, Water Resources Publications, Fort Collins, Colorado.

# CHAPTER 4
## DISCRETE DISTRIBUTIONS

Binomial

Tossing a coin or drawing a card from a pack are examples of a Bernoulli trial. Bernoulli trials operate under three conditions:

1. Any trial can have only one or two possible outcomes; success or failure, true or false, rain or no rain, etc.
2. Successive trials are independent.
3. Probabilities are stable.

Under these conditions the probability of x successes in n trials is given by the binomial distribution as

$$p(x) = \binom{n}{x} p^x q^{n-x} \qquad (4-1)$$

where $\binom{n}{x}$, sometimes written as nCx or $C_x^n$, is the number of combinations of n events taken x at a time,

$$\binom{n}{x} = \frac{n!}{x! \, (n-x)!} \qquad (4-2)$$

p is the probability of occurrence of an event, for example the probability of success in tossing a coin, q is the probability of failure,

$$q = 1-p \qquad (4-3)$$

and x is the variate or the number of successful trials.

As an example of the use of the binomial distribution suppose that a dam has a projected life of 50 years and we wish to evaluate the probability that a flood with a return period of 100 years will occur once during the life of the dam. Then $p = 1/T = 0.01$, $q = 1-p = 0.99$, $x = 1$ and $n = 50$, so that

$$p(1) = \binom{50}{1} (0.1)^1 (.99)^{49} = 0.306 \qquad (4-4)$$

i.e., there is about a 31% chance that an event of that magnitude will occur once in the life of the dam.

Poisson

The terms of a binomial expansion are a little inconvenient

to compute in any large number.  Provided that p is small (say < 0.1) and n is large (say n > 30) and the mean np is constant well defined, it can be shown (3) that as $p \to 0$, $q \to 1$ and $n \to \infty$,

$$(p+q)^n \to e^{-\lambda} \cdot e^{\lambda} = e^{-\lambda} + \lambda e^{-\lambda} + \frac{\lambda^2 e^{-\lambda}}{2!} + \ldots \ldots \quad (4-5)$$

This is known as the Poisson expansion and is generally written

$$p(x) = \frac{\lambda^x e^{-\lambda}}{x!} \quad (4-6)$$

where $\lambda = np$ is the mean.  The finite binomial distribution can thus be approximated by the infinite Poisson distribution provided that the following four conditions apply:

1. The number of events is discrete.
2. Two events cannot coincide.
3. The mean number of events in unit time is constant.
4. Events are independent.

Repeating the previous example, the probability that a 100 year return period flood will occur once in a 50 year period is seen to be

$$p(1) = \frac{0.5^1 e^{-0.5}}{1.} = 0.303 \quad (4-7)$$

which agrees well with the result obtained from the binomial expansion.

Equation 4-6 will give not only the probability of one event occurring in a given time but also the probability that two events may occur in that time, that three may occur, etc., etc.  The probability that one or more events occur will therefore be given as a summation of Equation 4-6.

$$P(1,2\ldots\infty) = \sum_{x=1}^{\infty} p(x) \quad (4-8)$$

but

$$\sum_{x=1}^{\infty} p(x) = \sum_{x=0}^{\infty} p(x) - p(0) \quad (4-9)$$

which, from Equation 4-5, gives

$$P(1,2\ldots\infty) = e^{-\lambda} e^{\lambda} - e^{-\lambda} \quad (4-10)$$

or

$$P(1,2\ldots\infty) = 1 - e^{-\lambda} \quad (4-11)$$

Table 4-1 shows, for $\lambda = 0.5$, the values of $P(x)$ at $x = 1, 2, 3, 4$ and at $x = \infty$.

Table 4-1

Some Values of Probability of
One or More Events for a Poisson
Distribution with λ = 0.5

| Variate Value, x | Probability P(1...x) |
|---|---|
| 1 | 0.30326 |
| 2 | 0.37908 |
| 3 | 0.39171 |
| 4 | 0.39329 |
| ⋮ | ⋮ |
| ∞ | 0.39347 |

Abbreviating the probability of one or more occurrences $P(1,2,...\infty)$ to P and replacing λ, the average number of events per time period, by $\Delta t/T$ where $\Delta t$ is the time interval being considered (e.g. project life) and T is the event return period, then Equation 4-11 becomes:

$$P = 1 - e^{-\Delta t/T} \qquad (4-12)$$

Table 4-2 shows values of P for different time intervals and return periods.

Table 4-2

Probabilities of One or More Occurrences of Events With Different Return Periods in Different Time Intervals

| Time Interval Δt | Return Period T | | | | | | | | | | |
|---|---|---|---|---|---|---|---|---|---|---|---|
| | .1 | .2 | .4 | 1 | 2 | 4 | 10 | 20 | 40 | 100 | 200 |
| .1 | .632 | .393 | .221 | .095 | .049 | .025 | .010 | .005 | .003 | .001 | .0005 |
| .2 | .865 | .632 | .393 | .181 | .095 | .049 | .020 | .010 | .005 | .002 | .001 |
| .4 | .982 | .865 | .632 | .380 | .181 | .095 | .039 | .020 | .010 | .004 | .002 |
| 1 | 1.000 | .993 | .918 | .632 | .393 | .221 | .095 | .049 | .025 | .010 | .005 |
| 2 | 1.000 | 1.000 | .993 | .865 | .632 | .393 | .181 | .095 | .049 | .020 | .010 |
| 4 | 1.000 | 1.000 | 1.000 | .982 | .865 | .632 | .330 | .181 | .095 | .039 | .020 |
| 10 | 1.000 | 1.000 | 1.000 | 1.000 | .993 | .918 | .632 | .393 | .221 | .095 | .049 |
| 20 | 1.000 | 1.000 | 1.000 | 1.000 | 1.000 | .993 | .865 | .632 | .393 | .181 | .095 |
| 40 | 1.000 | 1.000 | 1.000 | 1.000 | 1.000 | 1.000 | .982 | .865 | .632 | .330 | .181 |
| 100 | 1.000 | 1.000 | 1.000 | 1.000 | 1.000 | 1.000 | 1.000 | .993 | .918 | .632 | .393 |
| 200 | 1.000 | 1.000 | 1.000 | 1.000 | 1.000 | 1.000 | 1.000 | 1.000 | .993 | .865 | .632 |

Note that, in this table, both Δt and T must be measured in the same time units. Hall and Howell (1) have extended this type of table to time intervals of 5 to 35 days. Since the probabilities in such tables are cumulative, by taking differences it is possible to compute probabilities of one or more occurrences of events with return periods between different values. For example the probability of one or more occurrences within 1 year of flood events with return periods between 10 and 100 years is 0.095-0.010 is 0.085.

Under the assumptions that flood exceedences are independent identically distributed random variables and that the counting process for exceedence is a nonhomogeneous Poisson process, Todorovic and Woolhiser (2) have derived a one-dimensional distribution function for the time of occurrence of the largest event in some time interval.

If $z(t)$ represents the number of flood peak exceedences in the time interval $(0,t)$ and $\Lambda(t)$ is the expected value of $z(t)$

$$\Lambda(t) = E \left\{ z(t) \right\} \qquad (4-13)$$

then the probability that the time of occurrence $T(t)$ of the largest momentary flood exceedence in the time interval $(0,t)$ will be less than or equal to U, $P(T(t) \leq U)$, is given by

$$P(T(t) \leq U) = \exp \left\{ -\Lambda(t) \right\} + \frac{\Lambda(U)}{\Lambda(t)} (1 - \exp \left\{ -\Lambda(t) \right\} ) \quad (4-14)$$

The expression $\Lambda(t)$ was derived by Todorovic and Woolhiser as a finite Fourier series.

Risk is discussed in more detail in a later chapter but it may be pointed out in passing that since from Equation 4-6, using $\lambda = \Delta t/T$,

$$p(o) = e^{-\Delta t/T} \qquad (4-15)$$

then the return period, T, of a design flood with a risk of failure $p(0)$ in a project life $\Delta t$ is:

$$T = \Delta t/\ln p(o) \qquad (4-16)$$

e.g. for a 5% risk of failure (i.e. risk of an event of given magnitude occurring) in a 50 year life the project must be designed for a flood with return period of 975 years which is considerably more than might be thought of at first glance.

## References

1. Hall, W. A., and D. T. Howell, 1963, Estimating Flood Probabilities Within Specific Time Intervals, J. Hydrol., Vol. 1, No. 1, pp. 265-271.

2. Todorovic, P., and D. A. Woolhiser, 1972, On the Time when the Extreme Flood Occurs, Wat. Res. Res., Vol. 8, No. 6, pp. 1433-1438.

3. Weatherburn, C. E., 1962, A First Course in Mathematical Statistics, Cambridge University Press.

# CHAPTER 5
## NORMAL DISTRIBUTION

### Introduction

A distribution is said to be normal if the variable can take any value from $-\infty$ to $+\infty$ and the probability density function is defined as:

$$p(x) = \frac{1}{\sigma\sqrt{2\pi}} e^{\frac{-(x-\mu)^2}{2\sigma^2}} \qquad (5-1)$$

where $\mu$ and $\sigma$ are the distribution parameters shown later to be the population mean and standard deviation of the variable. The normal distribution is applicable if:

1. The variable is continuous.
2. Consecutive values are independent.
3. Probabilities are stable.

The normal distribution can be shown (5) to be a limiting case of the binomial when $p \to q \to \frac{1}{2}$ and $n \to \infty$. One of the features of the normal distribution is that the mean, mode and median are all the same.

If the variable, x, is standardized, i.e. forced to a mean of zero and unit variance by subtracting the mean and dividing by the standard deviation, and is denoted by t then Equation 5-1 becomes

$$p(t) = \frac{1}{\sqrt{2\pi}} e^{-t^2/2} \qquad (5-2)$$

which is known as the standard normal distribution. Equation 5-2 has been approxomated (accuracy $> 2.27 \times 10^{-3}$) by a series of polynomials (1) such as:

$$p(t) = (a_0 + a_1 t^2 + a_2 t^4 + a_3 t^6)^{-1} \qquad (5-3)$$

where

$$a_0 = 2.490895, \quad a_2 = -0.024393,$$
$$a_1 = 1.466003, \quad a_3 = 0.178257$$

Tables of the ordinates of the normal curve are also available such as Table 5-1.

The probability corresponding to any interval in the range of the variate is represented by the area under the probability

Table 5-1

## Ordinates of the Normal Curve

| t | 0.00 | 0.01 | 0.02 | 0.03 | 0.04 | 0.05 | 0.06 | 0.07 | 0.08 | 0.09 |
|---|------|------|------|------|------|------|------|------|------|------|
| 0.0 | .3989 | .3989 | .3989 | .3988 | .3986 | .3984 | .3982 | .3980 | .3977 | .3973 |
| 0.1 | .3970 | .3965 | .3961 | .3956 | .3951 | .3945 | .3939 | .3932 | .3925 | .3918 |
| 0.2 | .3910 | .3902 | .3894 | .3885 | .3876 | .3867 | .3857 | .3847 | .3836 | .3825 |
| 0.3 | .3814 | .3802 | .3790 | .3778 | .3765 | .3752 | .3739 | .3725 | .3712 | .3697 |
| 0.4 | .3683 | .3668 | .3653 | .3637 | .3621 | .3605 | .3589 | .3572 | .3555 | .3538 |
| 0.5 | .3521 | .3503 | .3485 | .3467 | .3448 | .3429 | .3410 | .3391 | .3372 | .3352 |
| 0.6 | .3332 | .3312 | .3292 | .3271 | .3251 | .3230 | .3209 | .3187 | .3166 | .3144 |
| 0.7 | .3123 | .3101 | .3079 | .3056 | .3034 | .3011 | .2989 | .2966 | .2943 | .2920 |
| 0.8 | .2897 | .2874 | .2850 | .2827 | .2803 | .2780 | .2756 | .2732 | .2709 | .2685 |
| 0.9 | .2661 | .2637 | .2613 | .2589 | .2565 | .2541 | .2516 | .2492 | .2468 | .2444 |
| 1.0 | .2420 | .2396 | .2371 | .2347 | .2323 | .2299 | .2275 | .2251 | .2227 | .2203 |
| 1.1 | .2179 | .2155 | .2131 | .2107 | .2083 | .2059 | .2036 | .2012 | .1989 | .1965 |
| 1.2 | .1942 | .1919 | .1895 | .1872 | .1849 | .1826 | .1804 | .1781 | .1758 | .1736 |
| 1.3 | .1714 | .1691 | .1669 | .1647 | .1626 | .1604 | .1582 | .1561 | .1539 | .1518 |
| 1.4 | .1497 | .1476 | .1456 | .1435 | .1415 | .1394 | .1374 | .1354 | .1334 | .1315 |
| 1.5 | .1295 | .1276 | .1257 | .1238 | .1219 | .1200 | .1182 | .1163 | .1145 | .1127 |
| 1.6 | .1109 | .1092 | .1074 | .1057 | .1040 | .1023 | .1006 | .0989 | .0973 | .0957 |
| 1.7 | .0940 | .0925 | .0909 | .0893 | .0878 | .0863 | .0848 | .0833 | .0818 | .0804 |
| 1.8 | .0790 | .0775 | .0761 | .0748 | .0734 | .0721 | .0707 | .0694 | .0681 | .0669 |
| 1.9 | .0656 | .0644 | .0632 | .0620 | .0608 | .0596 | .0584 | .0573 | .0562 | .0551 |
| 2.0 | .0540 | .0529 | .0519 | .0508 | .0498 | .0488 | .0478 | .0468 | .0459 | .0449 |
| 2.1 | .0440 | .0431 | .0422 | .0413 | .0404 | .0395 | .0387 | .0379 | .0371 | .0363 |
| 2.2 | .0355 | .0347 | .0339 | .0332 | .0325 | .0317 | .0310 | .0303 | .0297 | .0290 |
| 2.3 | .0283 | .0277 | .0270 | .0264 | .0258 | .0252 | .0246 | .0241 | .0235 | .0229 |
| 2.4 | .0224 | .0219 | .0213 | .0208 | .0203 | .0198 | .0194 | .0189 | .0184 | .0180 |
| 2.5 | .0175 | .0171 | .0167 | .0163 | .0158 | .0154 | .0151 | .0147 | .0143 | .0139 |
| 2.6 | .0136 | .0132 | .0129 | .0126 | .0122 | .0119 | .0116 | .0113 | .0110 | .0107 |
| 2.7 | .0104 | .0101 | .0099 | .0096 | .0093 | .0091 | .0088 | .0086 | .0084 | .0081 |
| 2.8 | .0079 | .0077 | .0075 | .0073 | .0071 | .0069 | .0067 | .0065 | .0063 | .0061 |
| 2.9 | .0060 | .0058 | .0056 | .0055 | .0053 | .0051 | .0050 | .0048 | .0047 | .0046 |
| 3.0 | .0044 | .0043 | .0042 | .0040 | .0039 | .0038 | .0037 | .0036 | .0035 | .0034 |
| 3.1 | .0033 | .0032 | .0031 | .0030 | .0029 | .0028 | .0027 | .0026 | .0025 | .0025 |
| 3.2 | .0024 | .0023 | .0022 | .0022 | .0021 | .0020 | .0020 | .0019 | .0018 | .0018 |
| 3.3 | .0017 | .0017 | .0016 | .0016 | .0015 | .0015 | .0014 | .0014 | .0013 | .0013 |
| 3.4 | .0012 | .0012 | .0012 | .0011 | .0011 | .0010 | .0010 | .0010 | .0009 | .0009 |
| 3.5 | .0009 | .0008 | .0008 | .0008 | .0008 | .0007 | .0007 | .0007 | .0007 | .0006 |
| 3.6 | .0006 | .0006 | .0006 | .0005 | .0005 | .0005 | .0005 | .0005 | .0005 | .0004 |
| 3.7 | .0004 | .0004 | .0004 | .0004 | .0004 | .0004 | .0003 | .0003 | .0003 | .0003 |
| 3.8 | .0003 | .0003 | .0003 | .0003 | .0003 | .0002 | .0002 | .0002 | .0002 | .0002 |
| 3.9 | .0002 | .0002 | .0002 | .0002 | .0002 | .0002 | .0002 | .0002 | .0001 | .0001 |

Note: $t = (x-\mu)/\sigma$

density curve,

$$P(x) = \int_{-\infty}^{x} \frac{1}{\sigma\sqrt{2\pi}}\, e^{\dfrac{-(x-\mu)^2}{2\sigma^2}}\, dx \qquad (5-4)$$

Standardizing, the cumulative probability corresponding to
Equation 5-2 is:

$$P(t) = \int_{-\infty}^{t} \frac{1}{\sqrt{2\pi}} e^{-t^2/2} \, dt \qquad (5-5)$$

Similarly, Abramowitz and Stegun (1, p. 932) list several
approximations for Equation 5-5.  A convenient polynomial
approximation with an error term less than $1 \times 10^{-5}$ is

$$P(t) = 1-f(t)(a_1 q + a_2 q^2 + a_3 q^3) \qquad (5-6)$$

where q is $1.0/(1.0 + a\ t)$, t is the positive standard normal
deviate and a, $a_1$, $a_2$ and $a_3$ are constants with values

$$a = 0.33267 \qquad a_2 = -0.12017$$

$$a_1 = 0.43618 \qquad a_3 = 0.93730$$

A similar polynomial approximation has been programmed for computer by IBM (3) as subroutine NDTR.  Tables of the area under
the standard normal curve are also available such as Table 5-2.

## Estimation of Parameters

The estimations of parameters for the normal distribution
by the methods of moments and maximum likelihood have been used
as examples of techniques in Chapter 3.  The results may be
summarized as

$$\mu = \mu_1' \qquad (5-7)$$

$$\sigma = \sqrt{\mu_2} \qquad (5-8)$$

All odd central moments are zero and all even central moments
may be expressed in terms of $\mu_2$ as

$$\mu_{2r} = \frac{(2r)!}{2^r r \, !} \mu_2^r \qquad (5-9)$$

e.g.

$$\mu_4 = 3\mu_2^2 = 3\sigma^4 \qquad (5-10)$$

## Frequency Factor

In the case of the normal distribution the frequency factor,
K, in the standard equation

$$x_T = \mu + K\sigma \qquad (5-11)$$

is the standard normal deviate, t.

Table 5-2

Area Under the Standard Normal Curve

| t | 0.00 | 0.01 | 0.02 | 0.03 | 0.04 | 0.05 | 0.06 | 0.07 | 0.08 | 0.09 |
|---|------|------|------|------|------|------|------|------|------|------|
| 0.0 | .0000 | .0040 | .0080 | .0120 | .0159 | .0199 | .0239 | .0279 | .0319 | .0359 |
| 0.1 | .0398 | .0438 | .0478 | .0517 | .0557 | .0596 | .0636 | .0675 | .0714 | .0753 |
| 0.2 | .0793 | .0832 | .0871 | .0910 | .0948 | .0987 | .1026 | .1064 | .1103 | .1141 |
| 0.3 | .1179 | .1217 | .1255 | .1293 | .1331 | .1368 | .1406 | .1443 | .1480 | .1517 |
| 0.4 | .1554 | .1591 | .1628 | .1664 | .1700 | .1736 | .1772 | .1808 | .1844 | .1879 |
| 0.5 | .1915 | .1950 | .1985 | .2019 | .2054 | .2088 | .2123 | .2157 | .2190 | .2224 |
| 0.6 | .2257 | .2291 | .2324 | .2357 | .2389 | .2422 | .2454 | .2486 | .2518 | .2549 |
| 0.7 | .2580 | .2611 | .2642 | .2673 | .2704 | .2734 | .2764 | .2794 | .2823 | .2852 |
| 0.8 | .2881 | .2910 | .2939 | .2967 | .2995 | .3023 | .3051 | .3078 | .3106 | .3133 |
| 0.9 | .3159 | .3186 | .3212 | .3238 | .3264 | .3289 | .3315 | .3340 | .3365 | .3389 |
| 1.0 | .3413 | .3438 | .3461 | .3485 | .3508 | .3531 | .3554 | .3577 | .3599 | .3621 |
| 1.1 | .3643 | .3665 | .3686 | .3708 | .3729 | .3749 | .3770 | .3790 | .3810 | .3830 |
| 1.2 | .3849 | .3869 | .3888 | .3907 | .3925 | .3944 | .3962 | .3980 | .3997 | .4015 |
| 1.3 | .4032 | .4049 | .4066 | .4082 | .4099 | .4115 | .4131 | .4147 | .4162 | .4177 |
| 1.4 | .4192 | .4207 | .4222 | .4236 | .4251 | .4265 | .4279 | .4292 | .4306 | .4319 |
| 1.5 | .4332 | .4345 | .4357 | .4370 | .4382 | .4394 | .4406 | .4418 | .4430 | .4441 |
| 1.6 | .4452 | .4463 | .4474 | .4485 | .4495 | .4505 | .4515 | .4525 | .4535 | .4545 |
| 1.7 | .4554 | .4564 | .4573 | .4582 | .4591 | .4599 | .4608 | .4616 | .4625 | .4633 |
| 1.8 | .4641 | .4649 | .4656 | .4664 | .4671 | .4678 | .4686 | .4693 | .4699 | .4706 |
| 1.9 | .4713 | .4719 | .4726 | .4732 | .4738 | .4744 | .4750 | .4756 | .4762 | .4767 |
| 2.0 | .4772 | .4778 | .4783 | .4788 | .4793 | .4798 | .4803 | .4808 | .4812 | .4817 |
| 2.1 | .4821 | .4826 | .4830 | .4835 | .4838 | .4842 | .4846 | .4850 | .4854 | .4857 |
| 2.2 | .4861 | .4865 | .4868 | .4871 | .4875 | .4878 | .4881 | .4884 | .4887 | .4890 |
| 2.3 | .4893 | .4896 | .4898 | .4901 | .4904 | .4906 | .4909 | .4911 | .4913 | .4916 |
| 2.4 | .4918 | .4920 | .4922 | .4925 | .4927 | .4929 | .4931 | .4932 | .4934 | .4936 |
| 2.5 | .4938 | .4940 | .4941 | .4943 | .4945 | .4946 | .4948 | .4949 | .4951 | .4952 |
| 2.6 | .4953 | .4955 | .4956 | .4957 | .4959 | .4960 | .4961 | .4962 | .4963 | .4964 |
| 2.7 | .4965 | .4966 | .4967 | .4968 | .4969 | .4970 | .4971 | .4972 | .4973 | .4974 |
| 2.8 | .4974 | .4975 | .4976 | .4977 | .4977 | .4978 | .4979 | .4980 | .4980 | .4981 |
| 2.9 | .4981 | .4982 | .4983 | .4983 | .4984 | .4984 | .4985 | .4985 | .4986 | .4986 |
| 3.0 | .4986 | .4987 | .4987 | .4988 | .4988 | .4989 | .4989 | .4989 | .4990 | .4990 |
| 3.1 | .4990 | .4991 | .4991 | .4991 | .4992 | .4992 | .4992 | .4992 | .4993 | .4993 |

Note: $t = (x-\mu)/\sigma$

The easiest way of obtaining the standard normal deviate is from tables of the area under the normal curve such as Table 5-2. As an example consider the determination of t for an event with a mean return period of 100 years. The cumulative probability of non-exceedence, as a percentage, associated with this T value is 99%, $(1-1/T)$. Of this, 50% is contributed by the integral of the standard normal density curve from $-\infty$ to 0 leaving 49% from the integral from 0 to x. Looking up 0.49 in the body of the table it is found that t is given as 2.33. The corresponding event magnitude is therefore $\mu + 2.33\sigma$.

This is using the standard normal curve in a one-tail manner. Statistical tests often use the standard normal deviate corresponding to a two-tail situation; that is, one in which the event might occur at either tail of the distribution. Figures 5-1 and 5-2 compare these two concepts for an area under the curve of 0.95.

As a convenience, the frequency factors (standard normal deviates) for the normal (and lognormal) distribution are given in Table 5-3 for some commonly used cumulative probabilities. Table 5-3 also shows the standard normal deviates for the two-tail situation.

Table 5-3

Frequency Factor for Use in Normal and Lognormal Distributions

| Cumulative Probability, P, % | | | | | |
|---|---|---|---|---|---|
| 50 | 80 | 90 | 95 | 98 | 99 |
| Corresponding Return, Period, T, Years | | | | | |
| 2 | 5 | 10 | 20 | 50 | 100 |
| Frequency Factor | | | | | |
| 0 | 0.8416 | 1.2816 | 1.6449 | 2.0538 | 2.3264 |
| Two-Tail Standard Normal Deviate | | | | | |
| 0.6745 | 1.2816 | 1.6449 | 1.9600 | 2.3264 | 2.5758 |

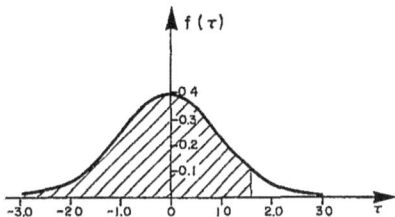

Figure 5-1. Area under the standard normal curve, one-tail.

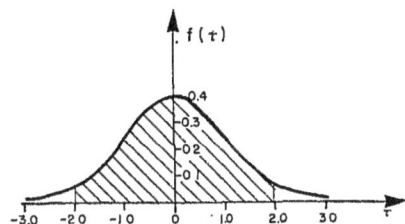

Figure 5-2. Area under the standard normal curve, two-tail.

It is sometimes useful to combine tables of the standard normal deviate with the plotting position $m/(n+1)$ in which m is the rank of the event in decreasing order of magnitude. Table 5-4 is a sample table of this type.

Polynomial approximations are also available (1) to obtain the standard normal deviate corresponding to a given probability level. As an example, if the cumulative probability is $P(t)$, then for $0 < P(t) \leq 0.5$

$$t \approx w - \frac{c_0 + c_1 w + c_2 w^2}{1 + d_1 w + d_2 w^2 + d_3 w^3}$$

(5-12)

where
$c_0 = 2.515517$; $d_1 = 1.432788$
$c_1 = 0.802853$; $d_2 = 0.189269$
$c_2 = 0.010328$; $d_3 = 0.001308$

49

## Table 5-4

### Plotting Positions, P, and Standard Normal Deviate, t, for a Range of Samples of Size n Events

| Event Rank No. (m) | n = 19 P % | t | n = 20 P % | t | n = 21 P % | t | n = 22 P % | t | n = 23 P % | t |
|---|---|---|---|---|---|---|---|---|---|---|
| 1 | 5.0 | 1.64 | 4.8 | 1.67 | 4.5 | 1.69 | 4.3 | 1.71 | 4.3 | 1.73 |
| 2 | 10.0 | 1.28 | 9.5 | 1.31 | 9.1 | 1.34 | 8.7 | 1.36 | 8.3 | 1.38 |
| 3 | 15.0 | 1.04 | 14.3 | 1.07 | 13.6 | 1.10 | 13.0 | 1.12 | 12.5 | 1.15 |
| 4 | 20.0 | .84 | 19.0 | .87 | 18.2 | .91 | 17.4 | .94 | 16.7 | .97 |
| 5 | 25.0 | .67 | 23.8 | .71 | 22.7 | .75 | 21.7 | .78 | 20.8 | .81 |
| 6 | 30.0 | .52 | 28.6 | .57 | 27.3 | .61 | 26.1 | .64 | 25.0 | .67 |
| 7 | 35.0 | .39 | 33.3 | .43 | 31.8 | .47 | 30.4 | .51 | 29.2 | .55 |
| 8 | 40.0 | .25 | 38.1 | .30 | 36.4 | .35 | 34.8 | .39 | 33.3 | .43 |
| 9 | 45.0 | .13 | 42.9 | .18 | 40.9 | .23 | 39.1 | .27 | 37.5 | .32 |
| 10 | 50.0 | .00 | 47.6 | .06 | 45.5 | .12 | 43.5 | .16 | 41.7 | .21 |
| 11 | 55.0 | - .13 | 52.4 | - .06 | 50.0 | .00 | 47.8 | .05 | 45.8 | .10 |
| 12 | 60.0 | - .25 | 57.1 | - .18 | 54.5 | - .12 | 52.2 | - .05 | 50.0 | .00 |
| 13 | 65.0 | - .39 | 61.9 | - .30 | 59.1 | - .23 | 56.5 | - .16 | 54.2 | - .10 |
| 14 | 70.0 | - .52 | 66.7 | - .43 | 63.6 | - .35 | 60.9 | - .27 | 58.3 | - .21 |
| 15 | 75.0 | - .67 | 71.4 | - .57 | 68.2 | - .47 | 65.2 | - .39 | 62.5 | - .32 |
| 16 | 80.0 | - .84 | 76.2 | - .71 | 72.7 | - .61 | 69.6 | - .51 | 66.7 | - .43 |
| 17 | 85.0 | -1.04 | 81.0 | - .87 | 77.3 | - .75 | 73.9 | - .64 | 70.8 | - .55 |
| 18 | 90.0 | -1.28 | 85.7 | -1.07 | 81.8 | - .91 | 78.3 | - .78 | 75.0 | - .67 |
| 19 | 95.0 | -1.64 | 90.5 | -1.31 | 86.4 | -1.10 | 82.6 | - .94 | 79.2 | - .81 |
| 20 | | | 95.2 | -1.67 | 90.9 | -1.34 | 87.0 | -1.12 | 83.3 | - .97 |
| 21 | | | | | 95.5 | -1.69 | 91.3 | -1.36 | 87.5 | -1.15 |
| 22 | | | | | | | 95.7 | -1.71 | 91.7 | -1.38 |
| 23 | | | | | | | | | 95.8 | -1.73 |

Notes: $P = m/(n+1)$, $t = (x-\mu)/\sigma$

and where

$$w = \sqrt{\ln \left(1/P(t)^2\right)} \tag{5-13}$$

This approximation is particularly useful when digital computers are used since it avoids the reverse integration in Equation 5-5. The error term in the approximation is stated [1] to be less than $4.5 \times 10^{-4}$. For values of $P(t) > 0.5$ use $1.0 - P(t)$ in the approximation and change the sign of the calculated value of t. This approximation has been programmed for computer as subroutine NDTRI in the IBM Scientific Subroutine Package [3].

### Standard Error

_Method of Moments_ - The general equation for the parameter δ in the standard error of estimate of a 2 parameter distribution from the method of moments (see Equation 3-51) is:

$$\delta = \left\{ 1 + \gamma_1 K + \left[ \gamma_2 - 1 \right] \frac{K^2}{4} \right\}^{\frac{1}{2}} \qquad (5\text{-}14)$$

For the normal distribution the central moments are given by:

$$\mu_2 = \sigma^2 \qquad (5\text{-}15)$$

$$\mu_3 = 0 \qquad (5\text{-}16)$$

and

$$\mu_4 = 3\sigma^4 \qquad (5\text{-}17)$$

Substituting these moment expressions into 5-14 and using t as frequency factor results in

$$\delta = [1 + t^2/2]^{\frac{1}{2}} \qquad (5\text{-}18)$$

so that the standard error is

$$S_T = \delta \sqrt{\frac{\sigma^2}{n}} \qquad (5\text{-}19)$$

Table 5-5 provides values of the parameter $\delta$ for some common cumulative probabilities.

Table 5-5

Parameter $\delta$ for Use in Standard Error
of Normal and Lognormal Distributions

| Cumulative Probability, P, % | | | | | |
|---|---|---|---|---|---|
| 50 | 80 | 90 | 95 | 98 | 99 |
| Corresponding Return Period, T, Years | | | | | |
| 2 | 5 | 10 | 20 | 50 | 100 |
| 1.0000 | 1.1638 | 1.3497 | 1.5340 | 1.7634 | 1.9249 |

An alternate method of tabulating standard error is as the ratio $S_T/\sigma$. Hardison (2) has provided tables of this type from which Table 5-6 has been derived. Since both $\delta$ and the ratio of deviations $S_T/\sigma$ are dimensionless, Tables 5-5 and 5-6 can equally well be used with the lognormal distribution.

If $v(K)$ is the expected coefficient of variation of the estimate $x_T$ and $V(K)$ is the coefficient of variation of the true event $y_T$ defined respectively as

$$v(K) = S_T/y_T \qquad (5\text{-}20)$$

## Table 5-6

### Dimensionless Ratio of the Standard Error of the T-Year Event to the Standard Deviation of the Annual Events for Normal and Lognormal Distributions

| Return Period T | Sample Length, n | | | | | |
|---|---|---|---|---|---|---|
| | 2 | 5 | 10 | 20 | 50 | 100 |
| 2 | 0.707 | 0.447 | 0.316 | 0.224 | 0.141 | 0.100 |
| 5 | 0.782 | 0.495 | 0.350 | 0.247 | 0.156 | 0.116 |
| 10 | 0.954 | 0.604 | 0.427 | 0.302 | 0.191 | 0.135 |
| 20 | 1.083 | 0.685 | 0.484 | 0.342 | 0.217 | 0.153 |
| 50 | 1.208 | 0.764 | 0.540 | 0.382 | 0.242 | 0.176 |
| 100 | 1.364 | 0.863 | 0.610 | 0.431 | 0.273 | 0.193 |

and

$$V(K) = \sigma/\mu \tag{5-21}$$

then from Equation 5-19

$$v(K) = \frac{V(K)}{\sqrt{n}} \sqrt{\frac{1 + t^2/2}{(1 + t \cdot V(K))}} \tag{5-22}$$

Nash and Amorocho (4) have shown that as $K \to \infty$, $v(K) \to 1/\sqrt{2n}$. Graphs of $v(K)\sqrt{2n}$ versus $t$ for different values of $V(K)$ show that the coefficient of variation has a minimum value between $t = 0$ and $t = 2$ and that for values of $V(K) > 0.2$ the mean annual event ($t = 0$) is less well defined than the event corresponding to $t = 1$.

*Maximum Likelihood* - For the normal distribution with two parameters (using $\alpha$ and $\beta^2$ for generality) the T-year event is a function of these parameters and $t$:

$$x_T = f(\alpha, \beta^2, t) \tag{5-23}$$

So that the variance of $x_T$, $s_T^2$, is

$$s_T^2 = \left(\frac{\partial x}{\partial \alpha}\right)^2 \text{var } \alpha + \left(\frac{\partial x}{\partial \beta^2}\right)^2 \text{var } \beta^2 + 2\frac{\partial x}{\partial \alpha} \frac{\partial x}{\partial \beta} \text{cov } (\alpha, \beta^2) \tag{5-24}$$

The result is obvious but for completeness we follow the procedure through.

From the logarithm of the likelihood equation (Equation 3-17) the required partial derivatives are

$$\frac{\partial^2 \ln L}{\partial \alpha^2} = -\frac{n}{\beta^2} \tag{5-25}$$

52

$$\frac{\partial^2 \ln L}{\partial (\beta^2)^2} = \frac{n}{2\beta^4} - \frac{\sum\limits_{i=1}^{n} (x_i - \alpha)^2}{2\beta^6} \tag{5-26}$$

$$\frac{\partial^2 \ln L}{\partial \alpha \partial \beta} = - \frac{\sum\limits_{n=1}^{n} (\bar{x}_i - \alpha)}{\beta^4} \tag{5-27}$$

but from the maximum likelihood solutions

$$\sum_{i=1}^{n} (x_i - \alpha) = 0 \tag{5-28}$$

and

$$\sum_{i=1}^{n} (x_i - \alpha)^2 = n\beta^2 \tag{5-29}$$

so that

$$\frac{\partial^2 \ln L}{\partial (\beta^2)^2} = - \frac{n}{2\beta^4} \tag{5-30}$$

and

$$\frac{\partial^2 \ln L}{\partial \alpha \partial \beta} = 0 \tag{5-31}$$

Following the procedure of taking minors and determinants of the likelihood matrix the variances and covariances of the parameters are

$$\text{var } \alpha = \beta^2/n \tag{5-32}$$

$$\text{var } \beta^2 = 2\beta^4/n \tag{5-33}$$

and

$$\text{cov } (\alpha, \beta^2) = 0 \tag{5-34}$$

Now the partial derivatives in Equation 5-24 can be obtained from the frequency equation for the normal distribution as:

$$\frac{\partial x}{\partial \alpha} = 1 \tag{5-35}$$

$$\frac{\partial x}{\partial \beta^2} = t/2\beta \tag{5-36}$$

The standard error of the T-year event is then:

$$S_T^2 = \frac{\beta^2}{n} \left[ 1 + t^2/2 \right] \tag{5-37}$$

which is the same result as by method of moments.

As an example of the computation of a confidence interval for a normal distribution consider the estimate of the 100 year event from a sample record of 50 years. The cumulative probability of non-exceedence of the 100 year event (area under the normal curve) is 99% and so from Table 5-2 the standard normal deviate, t, is 2.33. In Equation 5-19, $S_T$ is computed as $0.273\sigma$, where $\sigma$ is the sample estimate of the population standard deviation. Using a 95% confidence level (two-tail) the confidence interval around the 100 year event, $x_{100}$, is given as $x_{100} \pm 1.96 * 0.273\sigma$.

## References

1.  Abramowitz, M., and I. A. Stegun, 1965, Handbook of Mathematical Functions, Dover Publications, New York.

2.  Hardison, C. H., 1969, Accuracy of Streamflow Characteristics, USGS Professional Paper No. 650-D, pp. D210-D214.

3.  IBM, 1968, System/360 Scientific Subroutine Package (360A-CM-03X) Version III, Programmer's Manual, White Plains, New York.

4.  Nash, J. E., and J. Amorocho, 1966, The Accuracy of the Prediction of Floods of High Return Period, Wat. Res. Res., Vol. 2, No. 2, pp. 191-198.

5.  Weatherburn, C. E., 1962, A First Course in Mathematical Statistics, Cambridge University Press, Cambridge, England.

# CHAPTER 6
## TWO-PARAMETER LOGNORMAL DISTRIBUTION

Introduction

If the logarithms, ln x, of a variable x are normally distributed, then the variable x is said to be logarithmic-normally distributed so that

$$p(x) = \frac{1}{x\sigma_y \sqrt{2\pi}}\, e^{-\frac{[\ln x - \mu_y]^2}{2\sigma_y^2}}$$  (6-1)

where $\mu_y$ and $\sigma_y$ are the mean and standard deviation of the natural logarithms of x.

Chow (2) has provided a theoretical justification for the use of the lognormal distribution. The causative factors for many hydrologic variables act multiplicatively rather than additively and so the logarithms of these factors will satisfy the four basic conditions for normal distributions. The hydrologic variable will then be the product of these causative factors.

Singh and Sinclair (9) described a mixed, or compound, probability distribution made up of two lognormal distributions, as:

$$P(x) = a_1 P_1(x) + a_2 P_2(x)$$  (6-2)

where

$$P_1(x) = \frac{1}{\sigma_1 \sqrt{2\pi}} \int_{-\infty}^{x} e^{-(x - \mu_1)^2/2\sigma_1^2}\, dx$$  (6-3)

$$P_2(x) = \frac{1}{\sigma_2 \sqrt{2\pi}} \int_{-\infty}^{x} e^{-(x - \mu_2)^2/2\sigma_2^2}\, dx$$  (6-4)

and

$$a_1 + a_2 = 1$$  (6-5)

The mean, variance and coefficient of skew of the distribution P(x) may be estimated from the sample, or computed from

$$\mu = a_1\mu_1 + a_2\mu_2$$  (6-6)

$$\sigma^2 = a_1\sigma_1^2 + a_2\sigma_2^2 + a_1 a_2(\mu_2 - \mu_1)^2$$  (6-7)

$$\gamma_1 = \frac{3a_1 a_2 (\mu_1 - \mu_2)(\sigma_1^2 - \sigma_2^2)}{\sigma^3} + \frac{a_1 a_2 (a_2 - a_1)(\mu_1 - \mu_2)^3}{\sigma^3} \quad (6\text{-}8)$$

The advantage of this method (9) is that it has the versatility of high parameter models without the errors and uncertainties which result from the use of higher order sample moments.

### Estimation of Parameters

*Method of Moments* - Applying the general equation for moments about the origin to the pdf of the lognormal distribution:

$$\mu_r' = \int_0^\infty x^r \frac{1}{x\sigma_y \sqrt{2\pi}} \exp\left[-(\ln x - \mu_y)^2/2\sigma_y^2\right] dx \quad (6\text{-}9)$$

Complete the square in the exponent by substituting

$$\omega = (\ln x - r\sigma_y^2 - \mu_y)/\sigma_y \quad (6\text{-}10)$$

so that

$$\mu_r' = e^{r\mu_y + r^2\sigma_y^2/2} \cdot \frac{1}{\sqrt{2\pi}} \int_{-\infty}^\infty e^{-\omega^2/2} d\omega \quad (6\text{-}11)$$

and since the latter integral is unity

$$\mu_r' = e^{r\mu_y + r^2\sigma_y^2/2} \quad (6\text{-}12)$$

for which

$$\mu_1' = \mu = e^{\mu_y + \sigma_y^2/2} \quad (6\text{-}13)$$

and

$$\mu_2 = \sigma^2 = (e^{\sigma_y^2} - 1) \cdot \mu_1'^2 \quad (6\text{-}14)$$

Similarly the third central moment, $\mu_3$, can be shown to be

$$\mu_3 = (e^{3\sigma_y^2} - 3e^{\sigma_y^2} + 2) e^{3\mu_y + 3\sigma_y^2/2} \quad (6\text{-}15)$$

The coefficient of variation of the recorded events $\mu_2^{\frac{1}{2}}/\mu_1'$, is then

$$z = (e^{\sigma_y^2} - 1)^{\frac{1}{2}} \quad (6\text{-}16)$$

and the coefficient of skew of the recorded events $\mu_3/\mu_2^{3/2}$, is

$$\gamma_1 = \frac{e^{3\sigma_y^2} - 3e^{\sigma_y^2} + 2}{(e^{\sigma_y^2} - 1)^{3/2}} \tag{6-17}$$

From Equations 6-16 and 6-17 the relationship between the coefficients of variation and skew is

$$\gamma_1 = 3z + z^3 \tag{6-18}$$

*Maximum Likelihood* - Following the maximum likelihood procedure the logarithm of the likelihood expression for Equation 6-1 is

$$\ln L = -\frac{n}{2}\ln 2\pi - \frac{n}{2}\ln \sigma_y^2 + \sum_{i=1}^{n} \ln (1/x_i) - \sum_{i=1}^{n} (\ln x_i - \mu_y)^2/2\sigma_y^2 \tag{6-19}$$

Differentiating Equation 6-19 with respect to $\mu_y$ and $\sigma_y^2$ and equating to zero yields the maximum likelihood estimates

$$\mu_y = \sum_{i=1}^{n} \ln x_i/n \tag{6-20}$$

$$\sigma_y^2 = \sum_{i=1}^{n} (\ln x_i - \mu_y)^2/n \tag{6-21}$$

In practice a problem frequently arises in applying either the method of maximum likelihood or the method of moments. Many streamflow records have the occasional zero flow and, when taking logarithms, this becomes $-\infty$ and cannot be processed. Several simple solutions to this problem have been proposed (7) such as

1.  Add 1.0 to all data.
2.  Add small positive value (such as 0.1, 0.01, 0.001, etc.) to all data.
3.  Substitute 1.0 in place of all zero readings.
4.  Substitute small positive value in place of all zero readings.
5.  Ignore all zero observations.

All of these solutions affect the parameters of the distribution (1 and 2 affect the mean, 3, 4 and 5 affect both the mean and variance). Substitution of small positive values is to be avoided because of the large effect this has in a logarithmic scale.

An alternate solution is to consider the probability distribution as the sum of a probability mass at zero and a probability density distribution over the remainder of the range (10), so that:

$$P[0 \leq x \leq x_o] = P[x=0] + P[0 < x \leq x_o] \qquad (6\text{-}22)$$

In this way, if $p_o$ is the probability of occurrence of $x = 0$ in any one year and $p_x$ is the conditional probability of occurence of x in any one year given that x is not zero, then:

$$P[0 < x < x_o] = p_x(1 - p_o) \qquad (6\text{-}23)$$

Jennings and Benson (6) have described an operational procedure of applying this conditional probability. Given a data sample containing zero events (or events below any base flow $Q_b$) to which it is required to fit a log-probability distribution, first of all fit the probability distribution to those events above $Q_b$. Then multiply the resulting probabilities by the ratio of the number of events greater than zero (or $> Q_b$) to the total number of events in the sample. The result will be the required probabilities of exceedence.

Frequency Factor

If $y = \ln x$ is normally distributed, the general frequency equation (e.g. Equation 3-30) can be written in terms of logarithms as:

$$\ln x_T = y_T = \mu_y + t\sigma_y \qquad (6\text{-}24)$$

where, t is the standard normal deviate.

Alternately to avoid computation of the mean and standard deviation of logarithms, the general frequency equation, Equation 3-30, may be modified by substituting for $\mu$ and $\sigma$ from Equations 6-13 and 6-14 and using $e^{y_T}$ in place of $x_T$:

$$e^{y_T} = e^{\mu_y + \sigma_y^2/2}\,[1 + K(e^{\sigma_y^2} - 1)^{\frac{1}{2}}] \qquad (6\text{-}25)$$

From Equation 6-25 the frequency factor for the lognormal distribution, K, is given by

$$K = \frac{e^{y_T - \mu_y - \sigma_y^2/2} - 1}{(e^{\sigma_y^2} - 1)^{\frac{1}{2}}} \qquad (6\text{-}26)$$

But, from Equation 6-24

$$y_T - \mu_y = t\sigma_y \qquad (6\text{-}27)$$

which, when substituted in Equation 6-26, yields:

$$K = \frac{e^{\sigma_y t - \sigma_y^2/2} - 1}{(e^{\sigma_y^2} - 1)^{\frac{1}{2}}} \qquad (6\text{-}28)$$

where, as before, t is the standard normal deviate. Now, by using Equation 6-16 relating the coefficient of variation, z, of the observed events to the standard deviation of the logarithms, $\sigma_y$, all logarithmic parameters can be avoided:

$$K = \frac{e^{[\ln(1+z^2)]^{\frac{1}{2}}t - [\ln(1+z^2)]/2} - 1}{z} \qquad (6\text{-}29)$$

Table 6-1 shows values of K computed from Equation 6-29 for some commonly used return periods and various values of z, the coefficient of variation. Chow (1,2) gives more comprehensive tables

Table 6-1

Frequency Factor for Lognormal Distribution

| Coefficient of Variation z | Cumulative Probability, P, % | | | | | |
| --- | --- | --- | --- | --- | --- | --- |
| | 50 | 80 | 90 | 95 | 98 | 99 |
| | Corresponding Return Period, T, Years | | | | | |
| | 2 | 5 | 10 | 20 | 50 | 100 |
| 0.0500 | -0.0250 | 0.8334 | 1.2965 | 1.6863 | 2.1341 | 2.4370 |
| 0.1000 | -0.0496 | 0.8222 | 1.3078 | 1.7247 | 2.2130 | 2.5489 |
| 0.1500 | -0.0738 | 0.8085 | 1.3156 | 1.7598 | 2.2899 | 2.6607 |
| 0.2000 | -0.0971 | 0.7926 | 1.3200 | 1.7911 | 2.3640 | 2.7716 |
| 0.2500 | -0.1194 | 0.7746 | 1.3209 | 1.8183 | 2.4348 | 2.8805 |
| 0.3000 | -0.1406 | 0.7547 | 1.3183 | 1.8414 | 2.5016 | 2.9866 |
| 0.3500 | -0.1604 | 0.7333 | 1.3126 | 1.8602 | 2.5638 | 3.0890 |
| 0.4000 | -0.1788 | 0.7106 | 1.3037 | 1.8746 | 2.6212 | 3.1870 |
| 0.4500 | -0.1957 | 0.6870 | 1.2920 | 1.8848 | 2.6734 | 3.2799 |
| 0.5000 | -0.2111 | 0.6626 | 1.2778 | 1.8909 | 2.7202 | 3.3673 |
| 0.5500 | -0.2251 | 0.6379 | 1.2613 | 1.8931 | 2.7615 | 3.4488 |
| 0.6000 | -0.2375 | 0.6129 | 1.2428 | 1.8915 | 2.7974 | 3.5241 |
| 0.6500 | -0.2485 | 0.5879 | 1.2226 | 1.8866 | 2.8279 | 3.5930 |
| 0.7000 | -0.2582 | 0.5631 | 1.2011 | 1.8786 | 2.8532 | 3.6556 |
| 0.7500 | -0.2667 | 0.5387 | 1.1784 | 1.8677 | 2.8735 | 3.7118 |
| 0.8000 | -0.2739 | 0.5148 | 1.1548 | 1.8543 | 2.8891 | 3.7617 |
| 0.8500 | -0.2801 | 0.4914 | 1.1306 | 1.8388 | 2.9002 | 3.8056 |
| 0.9000 | -0.2852 | 0.4686 | 1.1060 | 1.8212 | 2.9071 | 3.8437 |
| 0.9500 | -0.2895 | 0.4466 | 1.0810 | 1.8021 | 2.9103 | 3.8762 |
| 1.0000 | -0.2929 | 0.4254 | 1.0560 | 1.7815 | 2.9098 | 3.9035 |

which, however, require somewhat awkward interpolation to use for values of z. It is worth noting that in Chow's tables the first line, for a zero coefficient of skew, is equivalent to the normal distribution. Since the relationship between the coefficients of variation and skew (Equation 6-18) does not hold for the normal distribution the frequency factors are independent of the coefficient of variation (8) at that particular coefficient of skew.

There are then two possible procedures for using a lognormal distribution to estimate T-year event magnitudes. The standard normal deviate, t, may be used with the mean and standard deviation of the logarithms of the recorded events, or the frequency factor, K, may be used with the mean and standard deviation of the recorded events.

As an example of these two procedures consider the following data from Collier (3). Annual maximum discharges of the Saint John River at Fort Kent, New Brunswick for the 37 years 1927 to 1963 yield a mean discharge of 81,000 cfs (2,300 $m^3$/s) a standard deviation of 22,800 cfs (646 $m^3$/s) and a coefficient of variation of 0.28. The mean and standard deviation of the logarithms of these 37 events are 11.263 in cfs units (7.6811 in SI units) and 0.284.

Using the first method, the magnitude of the 100-year event (t = 2.326) is given by

$$x_{100} = e^{11.263 + 2.326 \times 0.284} = 151,000 \text{ cfs} \qquad (6\text{-}30)$$

Using the second method the frequency factor, K, from Table 6-1 is 2.944 so that

$$x_{100} = 81,000 + 2.944 \times 22,800 = 148,000 \text{ cfs} \qquad (6\text{-}31)$$

The two methods thus produce comparable results and so, if the lognormal distribution is to be used, there is nothing to gain from the extra work involved in taking logarithms and computing $\mu_y$ and $\sigma_y$.

## Standard Error

*Method of Moments* - The general equation for the parameter δ in the standard error of estimate of a 2-parameter distribution by the method of moments is:

$$\delta = \left\{ 1 + \gamma_1 K + \left[ \gamma_2 - 1 \right] \frac{K^2}{4} \right\}^{\frac{1}{2}} \qquad (6\text{-}32)$$

If the moment ratios, $\gamma_1$ and $\gamma_2$, are now taken for logarithms so that the frequency factor is t, the standard normal deviate, then:

$$\delta = [1 + t^2/2]^{\frac{1}{2}} \qquad (6\text{-}33)$$

and the standard error of estimate, $S_T$, in logarithmic units is:

$$S_T = \delta\sigma_y/\sqrt{n} \qquad (6\text{-}34)$$

Values of the parameter δ are given in Table 6-2 for various commonly used return periods. From the standard error in logarithmic

Table 6-2

Parameter $\delta$ for Use in Standard Error
of Normal and Lognormal Distributions

| Cumulative Probability, P, % | | | | | |
|---|---|---|---|---|---|
| 50 | 80 | 90 | 95 | 98 | 99 |
| Corresponding Return Period, T, Years | | | | | |
| 2 | 5 | 10 | 20 | 50 | 100 |
| 1.0000 | 1.1638 | 1.3495 | 1.5339 | 1.7632 | 1.9251 |

units, $S_T$, the positive and negative standard errors can be
derived as

$$PSE = x_T(e^{S_T} - 1) \qquad (6\text{-}35)$$

and

$$NSE = -x_T(e^{-S_T} - 1) \qquad (6\text{-}36)$$

Hardison (5) has published graphs showing the variation of
$S_T$ with the standard deviation of the logarithms, $\sigma_y$, and has
provided tables to convert $S_T$ from logarithmic units back to the
units of the basic data.

Alternatively, to avoid the computation of the standard
deviation of the logarithms, use can be made of the moment ratios
of the original events

$$\gamma_1 = \frac{\mu_3}{\mu_2^{3/2}} = \frac{e^{3\sigma_y^2} - 3e^{\sigma_y^2} + 2}{(e^{\sigma_y^2} - 1)^{3/2}} \qquad (6\text{-}37)$$

and

$$\gamma_2 = \frac{\mu_4}{\mu_2^2} = \left(e^{\sigma_y^2/2}\right)^8 + 2\left(e^{\sigma_y^2/2}\right)^6 + 3\left(e^{\sigma_y^2/2}\right)^4 - 3 \qquad (6\text{-}38)$$

Substituting Equations 6-37 and 6-38 into Equation 6-32 and
making use of the relationship between the coefficient of variation, z, and the standard deviation of the logarithms, $\sigma_y$, (Equation 6-16) results in:

$$\delta_y = [1 + (z^3 + 3z)K + (z^8 + 6z^6 + 15z^4 + 16z^2 + 2)K^2/4]^{\frac{1}{2}} \qquad (6\text{-}39)$$

where K, the frequency factor, is given by 6-29 so that

$$S_T = \delta_y \sigma/\sqrt{n} \qquad (6\text{-}40)$$

where $S_T$ is now in linear units. Table 6-3 provides values of $\delta_y$ for some commonly used values of return period and coefficient of variation. Confidence limits on the event magnitude may be computed by this method as:

$$x_T \pm t\delta_y \sigma/\sqrt{n} \qquad (6\text{-}41)$$

assuming the normality of the distribution of $x_T$.

### Table 6-3

### Parameter $\delta_y$ for Use in Standard Error of Lognormal Distribution

| Coefficient of Variation Z | Cumulative Probability, P, % | | | | | |
|---|---|---|---|---|---|---|
| | 50 | 80 | 90 | 95 | 98 | 99 |
| | Corresponding Return Period, T, Years | | | | | |
| | 2 | 5 | 10 | 20 | 50 | 100 |
| .05 | .9983 | 1.2161 | 1.4323 | 1.6442 | 1.9086 | 2.0967 |
| .10 | .9932 | 1.2698 | 1.5222 | 1.7682 | 2.0765 | 2.2977 |
| .15 | .9848 | 1.3240 | 1.6187 | 1.9054 | 2.2674 | 2.5296 |
| .20 | .9733 | 1.3782 | 1.7210 | 2.0556 | 2.4817 | 2.7937 |
| .25 | .9589 | 1.4322 | 1.8288 | 2.2184 | 2.7199 | 3.0914 |
| .30 | .9420 | 1.4853 | 1.9416 | 2.3936 | 2.9826 | 3.4242 |
| .35 | .9229 | 1.5376 | 2.0590 | 2.5811 | 3.2704 | 3.7937 |
| .40 | .9021 | 1.5887 | 2.1810 | 2.7811 | 3.5841 | 4.2017 |
| .45 | .8801 | 1.6387 | 2.3073 | 2.9935 | 3.9246 | 4.6501 |
| .50 | .8575 | 1.6873 | 2.4381 | 3.2187 | 4.2929 | 5.1409 |
| .55 | .8351 | 1.7348 | 2.5733 | 3.4570 | 4.6903 | 5.6764 |
| .60 | .8138 | 1.7811 | 2.7133 | 3.7090 | 5.1182 | 6.2592 |
| .65 | .7945 | 1.8263 | 2.8582 | 3.9753 | 5.5781 | 6.8920 |
| .70 | .7784 | 1.8705 | 3.0083 | 4.2566 | 6.0719 | 7.5778 |
| .75 | .7669 | 1.9139 | 3.1642 | 4.5538 | 6.6013 | 8.3199 |
| .80 | .7615 | 1.9566 | 3.3261 | 4.8678 | 7.1685 | 9.1217 |
| .85 | .7635 | 1.9986 | 3.4946 | 5.1996 | 7.7758 | 9.9870 |
| .90 | .7746 | 2.0402 | 3.6702 | 5.5503 | 8.4255 | 10.9198 |
| .95 | .7959 | 2.0815 | 3.8533 | 5.9210 | 9.1202 | 11.9242 |
| 1.00 | .8284 | 2.1225 | 4.0445 | 6.3129 | 9.8625 | 13.0046 |

*Maximum Likelihood* - The general expression for the maximum likelihood estimate of the standard error of the T-year event for the 2 parameter lognormal distribution is

$$S_T^2 = \left(\frac{\partial x_T}{\partial \mu_y}\right)^2 \operatorname{var} \mu_y + \left(\frac{\partial x_T}{\partial \sigma_y^2}\right)^2 \operatorname{var} \sigma_y^2 + 2\frac{\partial x_T}{\partial \mu_y}\frac{\partial x_T}{\partial \sigma_y^2} \operatorname{cov}(\mu_y, \sigma_y^2)$$

$$(6\text{-}42)$$

and from Equation 6-24

$$x_T = e^{\mu_y + t\sigma_y} \qquad (6\text{-}43)$$

so that

$$\frac{\partial x_T}{\partial \mu_y} = e^{\mu_y + t\sigma_y} \tag{6-44}$$

$$\frac{\partial x_T}{\partial \sigma_y^2} = \frac{t \, e^{\mu_y + t\sigma_y}}{2\sigma_y} \tag{6-45}$$

so that

$$S_T^2 = \omega^2 \text{ var } \mu_y + \frac{t^2 \omega^2}{4\sigma_y^2} \text{ var } \sigma_y^2 + \frac{t\omega^2}{\sigma_y} \text{ cov } (\mu_y, \sigma_y^2) \tag{6-46}$$

where $\omega$ represents $e^{\mu_y + t\sigma_y}$.  The logarithm of the likelihood equation (Equation 6-19) is

$$\ln L = -\frac{n}{2}\ln 2\pi - \frac{n}{2}\ln \sigma_y^2 - \sum_{i=1}^{n} \ln x_i - \sum_{i=1}^{n} (\ln x_i - \mu_y)^2 / 2\sigma_y^2 \tag{6-47}$$

so that

$$\frac{\partial \ln L}{\partial \mu_y} = \frac{\sum_{i=1}^{n} (\ln x_i - \mu_y)}{\sigma_y^2} \tag{6-48}$$

$$\frac{\partial \ln L}{\partial \sigma_y^2} = -\frac{n}{2\sigma_y^2} + \frac{\sum_{i=1}^{n} (\ln x_i - \mu_y)^2}{2\sigma_y^4} \tag{6-49}$$

$$\frac{\partial^2 \ln L}{\partial \mu_y^2} = -\frac{n}{\sigma_y^2} \tag{6-50}$$

$$\frac{\partial^2 \ln L}{\partial (\sigma_y^2)^2} = \frac{n}{2\sigma_y^4} - \frac{\sum_{i=1}^{n} (\ln x_i - \mu_y)^2}{\sigma_y^6} = -\frac{n}{2\sigma_y^4} \tag{6-51}$$

$$\frac{\partial^2 \ln L}{\partial \mu_y \partial \sigma_y^2} = -\frac{\sum_{i=1}^{n} (\ln x_i - \mu_y)}{\sigma_y^4} = 0 \tag{6-52}$$

The information matrix is

$$I = \frac{n}{\sigma_y^2} \begin{bmatrix} 1 & 0 \\ \\ 0 & \frac{1}{2\sigma_y^2} \end{bmatrix} \qquad (6\text{-}53)$$

with a determinant, D, of

$$D = \frac{1}{2\sigma_y^2} \qquad (6\text{-}54)$$

Inverting this matrix gives the variances and covariances as

$$\text{var } \mu_y = \sigma_y^2/n \qquad (6\text{-}55)$$

$$\text{var } \sigma_y^2 = 2\sigma_y^4/n \qquad (6\text{-}56)$$

$$\text{cov } (\mu_y, \sigma_y^2) = 0 \qquad (6\text{-}57)$$

Substituting these expressions in 6-46 yields

$$S_T^2 = \frac{\sigma_y^2 \, e^{2(\mu_y + t\sigma_y)}}{n} \left[ 1 + t^2/2 \right] \qquad (6\text{-}58)$$

where $S_T$ is now in linear units. However,

$$e^{\mu_y + t\sigma_y} = x_T \qquad (6\text{-}59)$$

and

$$x_T = \mu (1 + Kz) \qquad (6\text{-}60)$$

where K, the frequency factor is given by Equation 6-29 and z is the coefficient of variation. Substituting these relationships in 6-58 gives

$$S_T^2 = \frac{\sigma^2}{n} \left\{ \frac{[\ln (z^2 + 1)](1 + Kz)^2 (1 + t^2/2)}{z^2} \right\} \qquad (6\text{-}61)$$

Simplifying to

$$S_T = \delta_z \sigma/\sqrt{n} \qquad (6\text{-}62)$$

Table 6-4 gives values of $\delta_z$ for some values of z and t.

## Table 6-4

### Parameter $\delta_z$ for Use in Standard Error of Lognormal Distribution

| Coefficient of Variation Z | Cumulative Probability, P, % | | | | | |
|---|---|---|---|---|---|---|
| | 50 | 80 | 90 | 95 | 98 | 99 |
| | Corresponding Return Period, T, Years | | | | | |
| | 2 | 5 | 10 | 20 | 50 | 100 |
| .05 | .9981 | 1.2114 | 1.4361 | 1.6622 | 1.9502 | 2.1584 |
| .10 | .9926 | 1.2562 | 1.5222 | 1.7940 | 2.1481 | 2.4099 |
| .15 | .9834 | 1.2975 | 1.6068 | 1.9280 | 2.3557 | 2.6786 |
| .20 | .9710 | 1.3348 | 1.6890 | 2.0629 | 2.5714 | 2.9632 |
| .25 | .9555 | 1.3679 | 1.7679 | 2.1974 | 2.7935 | 3.2617 |
| .30 | .9373 | 1.3963 | 1.8426 | 2.3301 | 3.0201 | 3.5720 |
| .35 | .9167 | 1.4201 | 1.9126 | 2.4597 | 3.2491 | 3.8918 |
| .40 | .8942 | 1.4391 | 1.9773 | 2.5850 | 3.4786 | 4.2184 |
| .45 | .8702 | 1.4535 | 2.0362 | 2.7052 | 3.7065 | 4.5493 |
| .50 | .8450 | 1.4634 | 2.0892 | 2.8192 | 3.9312 | 4.8819 |
| .55 | .8190 | 1.4690 | 2.1360 | 2.9264 | 4.1509 | 5.2137 |
| .60 | .7925 | 1.4706 | 2.1768 | 3.0263 | 4.3642 | 5.5424 |
| .65 | .7658 | 1.4686 | 2.2115 | 3.1187 | 4.5699 | 5.8659 |
| .70 | .7390 | 1.4632 | 2.2405 | 3.2032 | 4.7670 | 6.1823 |
| .75 | .7126 | 1.4549 | 2.2639 | 3.2799 | 4.9547 | 6.4901 |
| .80 | .6865 | 1.4440 | 2.2820 | 3.3489 | 5.1324 | 6.7879 |
| .85 | .6610 | 1.4308 | 2.2953 | 3.4103 | 5.2998 | 7.0747 |
| .90 | .6362 | 1.4156 | 2.3040 | 3.4644 | 5.4567 | 7.3496 |
| .95 | .6120 | 1.3987 | 2.3085 | 3.5114 | 5.6030 | 7.6121 |
| 1.00 | .5887 | 1.3805 | 2.3093 | 3.5518 | 5.7387 | 7.8618 |

## References

1. Chow, V. T., 1954, The Log-Probability Law and its Engineering Applications, Proc. ASCE, Vol. 80, pp. 1-25.

2. Chow, V. T., 1964, Editor-in-Chief, Handbook of Applied Hydrology, McGraw-Hill.

3. Collier, E. P., 1965, Flood Frequency Curves - Single Station Analysis, Water Resources Branch, Dept. Environment, Ottawa.

4. Condie, R. and G. Nix, 1976, Flood Damage Reduction Frequency Analysis Program, Inland Waters Directorate, Dept. Environment, Ottawa.

5. Hardison, C. H., 1969, Accuracy of Streamflow Characteristics, USGS Professional Paper No. 650-D, pp. D210-D214.

6. Jennings, M. E. and M. A. Benson, 1969, Frequency Curves for Annual Flood Series with Some Zero Events or Incomplete Data, Wat. Res. Res., Vol. 5, No. 1, pp. 276-280.

7. Kilmartin, R. F. and J. R. Peterson, 1972, Rainfall-Runoff Regression with Logarithmic Transforms and Zeros in the Data, Wat. Res. Res., Vol. 8, No. 4, pp. 1096-1099.

8.  Sangal, B. P. and A. K. Biswas, 1970, The 3-Parameter Log-normal Distribution and its Applications in Hydrology, Wat. Res. Res., Vol. 6, No. 2, pp. 505-515.

9.  Singh, K. P. and R. A. Sinclair, 1972, Two-Distribution Method for Flood-Frequency Analysis, Proc. ASCE, Vol. 98, No. HY1, pp. 29-45.

10. WMO, 1969, Estimation of Maximum Floods, Technical Note No. 98, WMO No. 233.TP.126, Geneva.

```
      PROGRAM LN2(INPUT,OUTPUT,TAPE5=INPUT,TAPE6=OUTPUT)         A    1
C                                                                A    2
C                                                                A    3
C     COMPUTES METHOD OF MOMENTS AND MAXIMUM LIKELIHOOD ESTIMATES FOR  A    4
C     T YEAR EVENTS AND STANDARD ERRORS FOR 2 PARAMETER LOGNORMAL      A    5
C     DISTRIBUTION                                               A    6
C     INPUT                                                      A    7
C     TITLE                                                      A    8
C     N NUMBER OF ANNUAL MAXIMUM EVENTS                          A    9
C     X SERIES OF EVENTS                                         A   10
C                                                                A   11
C                                                                A   12
      DIMENSION SND(6), X(100)                                   A   13
      DIMENSION XT(6), SX(6), TITLE(80)                          A   14
      REAL K,M1,M2,M3                                            A   15
      SND(1)=0.0                                                 A   16
      SND(2)=0.8416                                              A   17
      SND(3)=1.2816                                              A   18
      SND(4)=1.6449                                              A   19
      SND(5)=2.0538                                              A   20
      SND(6)=2.3264                                              A   21
      READ (5,8) TITLE                                           A   22
      READ (5,9) N                                               A   23
      XN=N                                                       A   24
      READ (5,10) (X(I),I=1,N)                                   A   25
      WRITE (6,11) TITLE                                         A   26
      WRITE (6,12)                                               A   27
      A=0.0                                                      A   28
      B=0.0                                                      A   29
      C=0.0                                                      A   30
      DO 1 I=1,N                                                 A   31
      A=A+X(I)                                                   A   32
      B=B+X(I)**2                                                A   33
      C=C+X(I)**3                                                A   34
1     CONTINUE                                                   A   35
      M1=A/XN                                                    A   36
      M2=(B/XN)-(A/XN)**2                                        A   37
      M3=(C/XN)+2.0*M1**3-3.0*M1*(B/XN)                          A   38
      M2=M2*XN/(XN-1.0)                                          A   39
      G=M3/(M2**1.5)                                             A   40
      WRITE (6,5) M1                                             A   41
      WRITE (6,6) M2                                             A   42
      WRITE (6,7) G                                              A   43
      Z=(SQRT(M2))/M1                                            A   44
      A=ALOG(1.0+Z**2)                                           A   45
      DO 2 J=1,6                                                 A   46
      T=SND(J)                                                   A   47
      K=(EXP(SQRT(A)*T-A/2.0)-1.0)/Z                             A   48
      XT(J)=M1+K*SQRT(M2)                                        A   49
      DELTA=SQRT(1.0+((Z**3+3.0*Z)*K)+((Z**8+6.0*Z**6+15.0*Z**4+16.0*Z**  A   50
     12+2.0)*K**2)/4.0)                                          A   51
2     SX(J)=DELTA*SQRT(M2/XN)                                    A   52
      WRITE (6,13)                                               A   53
      WRITE (6,14) (XT(J),J=1,6)                                 A   54
      WRITE (6,15) (SX(J),J=1,6)                                 A   55
      DO 3 J=1,6                                                 A   56
      T=SND(J)                                                   A   57
      K=(EXP(SQRT(A)*T-A/2.0)-1.0)/Z                             A   58
      DELTA=SQRT(((A*((1.0+K*Z)**2)*(1.0+(T**2)/2.0))/Z**2)      A   59
3     SX(J)=DELTA*SQRT(M2/XN)                                    A   60
      A=0.0                                                      A   61
      B=0.0                                                      A   62
      C=0.0                                                      A   63
      DO 4 I=1,N                                                 A   64
      X(I)=ALOG(X(I))                                            A   65
      A=A+X(I)                                                   A   66
      B=B+X(I)**2                                                A   67
      C=C+X(I)**3                                                A   68
4     CONTINUE                                                   A   69
      M1=A/XN                                                    A   70
      M2=(B/XN)-(A/XN)**2                                        A   71
      M3=(C/XN)+2.0*M1**3-3.0*M1*(B/XN)                          A   72
      M2=M2*XN/(XN-1.0)                                          A   73
      G=M3/(M2**1.5)                                             A   74
      WRITE (6,17) M1                                            A   75
      WRITE (6,18) M2                                            A   76
      WRITE (6,19) G                                             A   77
      WRITE (6,16)                                               A   78
      WRITE (6,13)                                               A   79
      WRITE (6,14) (XT(J),J=1,6)                                 A   80
      WRITE (6,15) (SX(J),J=1,6)                                 A   81
      STOP                                                       A   82
```

```
C
5      FORMAT (20X,9HMEAN OF X,16X,E12.5)                                    A  83
6      FORMAT (20X,13HVARIANCE OF X,12X,E12.5)                              A  84
7      FORMAT (20X,9HSKEW OF X,16X,E12.5,/)                                 A  85
8      FORMAT (80A1)                                                        A  86
9      FORMAT (I5)                                                          A  87
10     FORMAT (8F10.0)                                                      A  88
11     FORMAT (1H1,/,80A1,//,21X,38H _ TWO PARAMETER LOGNORMAL DISTRIBUTIO  A  89
       1N,/)                                                                A  91
12     FORMAT (31X,17HMETHOD OF MOMENTS,//)                                 A  92
13     FORMAT (3X,7HT,YEARS,4X,1H2,11X,1H5,10X,2H10,10X,2H20,10X,2H50,9X,   A  93
       13H100,/)                                                           A  94
14     FORMAT (3X,1HX,3X,6E12.5,/,4X,1HT)                                   A  95
15     FORMAT (3X,1HS,3X,6E12.5,/,4X,1HT,//)                                A  96
16     FORMAT (3X,77HNOTE - FOR GOOD USE OF THIS DISTRIBUTION SKEW OF LOG   A  97
       1S SHOULD BE CLOSE TO ZERO,/)                                       A  98
17     FORMAT (20X,15HMEAN OF LN(X)   ,10X,E12.5)                          A  99
18     FORMAT (20X,19HVARIANCE OF LN(X)  ,6X,E12.5)                        A 100
19     FORMAT (20X,15HSKEW OF LN(X)   ,10X,E12.5,/)                        A 101
       END                                                                 A 102

10 ' 
20 ' Program LN2
30 ' Copyright G.W. Kite 1986
40 '
50 ' Compute method of moments and maximum likelihood estimates
60 ' for T year events and standard errors
70 ' for a 2-parameter lognormal distribution.
80 '
90 DIM SND#(6),X#(250),XT#(6),SX#(6)
100 FORM$="  ##.###^^^^"
110 VV$=CHR$(179)
120 DATA 0.0#,0.8416#,1.2816#,1.6449#,2.0538#,2.3264#
130 FOR I%=1 TO 6
140 READ SND#(I%)
150 NEXT I%
160 T$="Two Parameter Lognormal Distribution"
170 IERR%=0
180 GOSUB 1460
190 IF IERR% = 1 GOTO 1440
200 FLAG%=0:IF YN$ = "Y" OR YN$ = "y" THEN FLAG%=1
210 NREC%=0
220 OPEN FILE$ FOR INPUT AS #1
230 LINE INPUT#1,TITLE$
240 IF EOF(1) GOTO 290
250 NREC%=NREC%+1
260 LINE INPUT#1,I$
270 X#(NREC%)=VAL(MID$(I$,5,12))
280 GOTO 240
290 CLOSE#1
300 NREC#=NREC%
310 A%=LEN(TITLE$)
320 B%=(80-A%)/2
330 U$=STRING$(A%,205)
340 M$="Method of Moments"
350 E%=LEN(M$)
360 F%=(80-E%)/2
370 W$=STRING$(E%,205)
380 CLS
390 PRINT TAB(B%) TITLE$:PRINT TAB(B%) U$
400 PRINT TAB(D%) T$:PRINT TAB(D%) V$
410 PRINT TAB(F%) M$:PRINT TAB(F%) W$
420 IF FLAG% <> 1 GOTO 470
430 LPRINT CHR$(12):LPRINT:LPRINT
440 LPRINT TAB(B%) TITLE$:LPRINT TAB(B%) U$
450 LPRINT:LPRINT TAB(D%) T$:LPRINT TAB(D%) V$
460 LPRINT:LPRINT TAB(F%) M$:LPRINT TAB(F%) W$
470 A#=0#:B#=0#:C#=0#
480 FOR I%=1 TO NREC%
490 A#=A#+X#(I%)
500 B#=B#+X#(I%)*X#(I%)
510 C#=C#+X#(I%)*X#(I%)*X#(I%)
520 NEXT I%
530 M1#=A#/NREC#
540 M2#=(B#/NREC#)-M1#*M1#
550 M3#=(C#/NREC#)+2#*M1#*M1#*M1#-3#*M1#*(B#/NREC#)
560 M2#=M2#*NREC#/(NREC#-1#)
570 G#=M3#/(M2#^1.5#)
580 PRINT
590 PRINT TAB(20) "Mean is                 ";:PRINT USING FORM$;M1#
600 PRINT TAB(20) "Variance is             ";:PRINT USING FORM$;M2#
```

```
610 PRINT TAB(20) "Coefficient of skew is ";:PRINT USING FORM$;G#
620 IF FLAG% <> 1 GOTO 670
630 LPRINT
640 LPRINT TAB(20) "Mean is                    ";:LPRINT USING FORM$;M1#
650 LPRINT TAB(20) "Variance is                ";:LPRINT USING FORM$;M2#
660 LPRINT TAB(20) "Coefficient of skew is ";:LPRINT USING FORM$;G#
670 Z#=SQR(M2#)/M1#
680 A#=LOG(1#+Z#*Z#)
690 FOR J%=1 TO 6
700 T#=SND#(J%)
710 K#=(EXP(SQR(A#)*T#-A#/2#)-1#)/Z#
720 XT#(J%)=M1#+K#*SQR(M2#)
730 DELTA#=SQR(1#+((Z#*Z#*Z#+3#*Z#)*K#)+((Z#^8#+6#*Z#^6#+15#*Z#^4#+16#*Z#*Z#+2#)
*K#*K#)/4#)
740 SX#(J%)=DELTA#*SQR(M2#/NREC#)
750 NEXT J%
760 ROW$=CHR$(218)+STRING$(72,196)
770 PRINT:PRINT TAB(2) "T, years  2          5         10        20
    50          100":PRINT TAB(5) ROW$
780 PRINT TAB(2) "X  " VV$;:FOR J%=1 TO 6:PRINT USING FORM$;XT#(J%);:NEXT J%
790 PRINT TAB(2) " t " VV$
800 PRINT TAB(2) "S  " VV$;:FOR J%=1 TO 6:PRINT USING FORM$;SX#(J%);:NEXT J%
810 PRINT TAB(2) " t " VV$;
820 IF FLAG% <> 1 GOTO 880
830 LPRINT:LPRINT TAB(2) "T, years  2          5         10        20
    50          100":LPRINT TAB(5) ROW$
840 LPRINT TAB(2) "X  " VV$;:FOR J%=1 TO 6:LPRINT USING FORM$;XT#(J%);:NEXT J%
850 LPRINT TAB(2) " t " VV$
860 LPRINT TAB(2) "S  " VV$;:FOR J%=1 TO 6:LPRINT USING FORM$;SX#(J%);:NEXT J%
870 LPRINT TAB(2) " t " VV$
880 FOR J%=1 TO 6
890 T#=SND#(J%)
900 K#=(EXP(SQR(A#)*T#-A#/2#)-1#)/Z#
910 DELTA#=SQR((A#*((1#+K#*Z#)^2#)*(1#+(T#*T#)/2#))/(Z#*Z#))
920 SX#(J%)=DELTA#*SQR(M2#/NREC#)
930 NEXT J%
940 A#=0#:B#=0#:C#=0#
950 FOR I%=1 TO NREC%
960 X#(I%)=LOG(X#(I%))
970 A#=A#+X#(I%)
980 B#=B#+X#(I%)*X#(I%)
990 C#=C#+X#(I%)*X#(I%)*X#(I%)
1000 NEXT I%
1010 M1#=A#/NREC#
1020 M2#=(B#/NREC#)-M1#*M1#
1030 M3#=(C#/NREC#)+2#*M1#*M1#*M1#-3#*M1#*(B#/NREC#)
1040 M2#=M2#*NREC#/(NREC#-1#)
1050 G#=M3#/(M2#^1.5#)
1060 M$="Method of Maximum Likelihood"
1070 E%=LEN(M$)
1080 F%=(80-E%)/2
1090 W$=STRING$(E%,205)
1100 PRINT:PRINT"Press any key to continue"
1110 Q$=INKEY$:IF Q$ ="" GOTO 1110
1120 CLS
1130 PRINT TAB(B%) TITLE$:PRINT TAB(B%) U$
1140 PRINT TAB(D%) T$:PRINT TAB(D%) V$
1150 PRINT TAB(F%) M$:PRINT TAB(F%) W$
1160 IF FLAG% <> 1 GOTO 1180
1170 LPRINT:LPRINT:LPRINT TAB(F%) M$:LPRINT TAB(F%) W$
1180 PRINT
1190 PRINT TAB(20) "Mean of logs is            ";:PRINT USING FORM$;M1#
1200 PRINT TAB(20) "Variance of logs is        ";:PRINT USING FORM$;M2#
1210 PRINT TAB(20) "Coefficient of skew of logs is ";:PRINT USING FORM$;G#
1220 PRINT:PRINT TAB(2) "NOTE: For good use of this distribution the coeff. of s
kew of the logs"
1230 PRINT TAB(2) "should be close to zero"
1240 PRINT:PRINT TAB(2) "T, years  2          5         10        20
    50          100":PRINT TAB(5) ROW$
1250 PRINT TAB(2) "X  " VV$;:FOR J%=1 TO 6:PRINT USING FORM$;XT#(J%);:NEXT J%
1260 PRINT TAB(2) " t " VV$
1270 PRINT TAB(2) "S  " VV$;:FOR J%=1 TO 6:PRINT USING FORM$;SX#(J%);:NEXT J%
1280 PRINT TAB(2) " t " VV$
1290 IF FLAG% <> 1 GOTO 1420
1300 LPRINT
1310 LPRINT TAB(20) "Mean of logs is            ";:LPRINT USING FORM$;M1#
1320 LPRINT TAB(20) "Variance of logs is        ";:LPRINT USING FORM$;M2#
1330 LPRINT TAB(20) "Coefficient of skew of logs is ";:LPRINT USING FORM$;G#
1340 LPRINT:LPRINT TAB(2) "NOTE: For good use of this distribution the coeff. of
 skew of the logs"
```

```
1350 LPRINT TAB(2) "should be close to zero"
1360 LPRINT:LPRINT TAB(2) "T, years 2          5          10          20
      50          100":LPRINT TAB(5) ROW$
1370 LPRINT TAB(2) "X  " VV$;:FOR J%=1 TO 6:LPRINT USING FORM$;XT#(J%);:NEXT J%
1380 LPRINT TAB(2) " t " VV$
1390 LPRINT TAB(2) "S  " VV$;:FOR J%=1 TO 6:LPRINT USING FORM$;SX#(J%);:NEXT J%
1400 LPRINT TAB(2) " t " VV$
1410 LPRINT CHR$(12)
1420 PRINT:PRINT "Press any key to continue"
1430 Q$=INKEY$:IF Q$="" GOTO 1430
1440 CLS
1450 SYSTEM
1460 '
1470 ' Subroutine for standard screen format
1480 '
1490 CLS:PRINT:PRINT
1500 ROW$=STRING$(78,205)
1510 BOXTOP$=CHR$(201)+ROW$+CHR$(187)
1520 PRINT BOXTOP$;
1530 PRINT CHR$(186) TAB(80) CHR$(186);
1540 PRINT CHR$(186) TAB(80) CHR$(186);
1550 C%=LEN(T$)
1560 D%=(80-C%)/2
1570 V$=STRING$(C%,205)
1580 PRINT CHR$(186) TAB(D%) T$ TAB(80) CHR$(186);
1590 PRINT CHR$(186) TAB(D%) V$ TAB(80) CHR$(186);
1600 PRINT CHR$(186) TAB(80) CHR$(186);
1610 PRINT CHR$(186) TAB(80) CHR$(186);
1620 PRINT CHR$(186) "    What is the name of your data file?" TAB(80) CHR$(186);

1630 PRINT CHR$(186) TAB(80) CHR$(186);
1640 PRINT CHR$(186) "    Do you want printer output (Y/N)?" TAB(80) CHR$(186);
1650 PRINT CHR$(186) "    (press Alt Q to quit at this stage)" TAB(80) CHR$(186);

1660 PRINT CHR$(186) TAB(80) CHR$(186);
1670 PRINT CHR$(186) TAB(80) CHR$(186);
1680 BOXBOT$=CHR$(200)+ROW$+CHR$(188)
1690 PRINT BOXBOT$;
1700 LOCATE 24,1,0,0,0
1710 PRINT "G Kite";
1720 LOCATE 10,41,1,0,13
1730 FILE$=""
1740 I$=INPUT$(1)
1750 IF I$ = CHR$(13) GOTO 1850                         ' ENTER key
1760 IF I$ = CHR$(8)  GOTO 1800                         ' BACKSPACE key
1770 PRINT I$;
1780 FILE$=FILE$+I$
1790 GOTO 1740
1800 H%=POS(0)
1810 LOCATE 10,H%-1,1,0,13
1820 L%=LEN(FILE$)
1830 FILE$=LEFT$(FILE$,L%-1)
1840 GOTO 1740
1850 LOCATE 12,39,1,0,13
1860 YN$=""
1870 I$=INKEY$:IF I$ = "" GOTO 1870
1880 IF LEN(I$) = 1 GOTO 1920
1890 I$=RIGHT$(I$,1)
1900 IF I$ = CHR$(16) THEN IERR%=1:GOTO 1990            ' ALT Q key
1910 GOTO 1850
1920 IF I$ = CHR$(13) GOTO 1960                         ' ENTER key
1930 PRINT I$;
1940 YN$=I$
1950 GOTO 1870
1960 IF YN$ = "Y" OR YN$ = "y" OR YN$ = "N" OR YN$ = "n" GOTO 1980
1970 GOTO 1850
1980 LOCATE ,,0,13,13
1990 RETURN
```

## Two Parameter Lognormal Distribution

### Method of Moments

| | |
|---|---|
| Mean is | 4.132D+02 |
| Variance is | 2.067D+04 |
| Coefficient of skew is | 1.195D+00 |

| T, years | 2 | 5 | 10 | 20 | 50 | 100 |
|---|---|---|---|---|---|---|
| $X_t$ | 3.902D+02 | 5.186D+02 | 6.018D+02 | 6.805D+02 | 7.813D+02 | 8.568D+02 |
| $s_t$ | 1.565D+01 | 2.601D+01 | 3.480D+01 | 4.359D+01 | 5.520D+01 | 6.400D+01 |

### Method of Maximum Likelihood

| | |
|---|---|
| Mean of logs is | 5.969D+00 |
| Variance of logs is | 1.088D-01 |
| Coefficient of skew of logs is | 1.662D-01 |

NOTE: For good use of this distribution the coeff. of skew of the logs should be close to zero

| T, years | 2 | 5 | 10 | 20 | 50 | 100 |
|---|---|---|---|---|---|---|
| $X_t$ | 3.902D+02 | 5.186D+02 | 6.018D+02 | 6.805D+02 | 7.813D+02 | 8.568D+02 |
| $s_t$ | 1.555D+01 | 2.404D+01 | 3.236D+01 | 4.158D+01 | 5.489D+01 | 6.571D+01 |

# CHAPTER 7
## THREE-PARAMETER LOGNORMAL DISTRIBUTION

Introduction

    Just as the lognormal distribution represents the normal distribution of the logarithms of the variable x, so the 3- parameter lognormal represents the normal distribution of the logarithms of the reduced variable (x-a) where a is a lower boundary. The probability density distribution is then given by:

$$p(x) = \frac{1}{(x-a)\ \sigma_y\ \sqrt{2\pi}}\ e^{-\frac{[\ln(x-a)-\mu_y]^2}{2\sigma_y^2}} \qquad (7\text{-}1)$$

where $\mu_y$ and $\sigma_y^2$ are the form and scale parameters, shown later to be the mean and variance of the logarithms of (x-a).

Estimation of Parameters

    If the lower boundary, a, is known then the reduced variable (x-a) can be used together with the procedures described for the 2-parameter lognormal distribution.  If a is unknown the following methods are among those available.

    *Method of Moments* - Applying the general moment generating equation for moments about a, the lower bound, to the pdf for the 3-parameter lognormal distribution (Equation 7-1)

$$\mu_r(a) = \int_0^\infty \frac{(x-a)^r}{(x-a)\ \sigma_y\ \sqrt{2\pi}}\ \exp\left\{\left[-\ln(x-a)-\mu_y\right]^2/2\sigma_y^2\right\}\ dx \qquad (7\text{-}2)$$

Substitute $y = \ln(x-a)$, $(x-a)^r = e^{ry}$, $dx = (x-a)\ dy$ so that

$$\mu_r(a) = \frac{1}{\sigma_y\ \sqrt{2\pi}} \int_{-\infty}^\infty e^{ry}\ e^{-(y-\mu_y)^2/2\sigma_y^2}\ dx \qquad (7\text{-}3)$$

Complete the square within the exponents as

$$-\left\{(y-(\mu_y+r\sigma_y^2))^2 - r^2\sigma_y^4 - 2r\mu_y\sigma_y^2\right\}/2\sigma_y^2 \qquad (7\text{-}4)$$

then
$$\mu_r(a) = e^{r\mu_y + r^2\sigma_y^2/2}\ \frac{1}{\sigma_y\ \sqrt{2\pi}} \int_{-\infty}^\infty e^{-[y-(\mu_y+r\sigma_y^2)]^2/2\sigma_y^2}\ dy \qquad (7\text{-}5)$$

Since this integral is unity

$$\mu_r(a) = e^{r\mu_y + r^2\sigma_y^2/2} \tag{7-6}$$

The moments about the lower bound, a, can now be converted to moments about the origin by using the equation

$$\mu_r' = \sum_{j=0}^{r} \binom{r}{j} \mu_{r-j}(a) \cdot a^j \tag{7-7}$$

so that

$$\mu_1' = \mu = a + e^{\mu_y + \sigma_y^2/2} \tag{7-8}$$

Finally, converting to central moments leads to

$$\mu_2 = \sigma^2 = \left(e^{\sigma_y^2} - 1\right) e^{2\mu_y + \sigma_y^2} \tag{7-9}$$

etc. If $z_1$ and $z_2$ represent the coefficients of variation of the distributions x and x-a then

$$z_1 = \sigma/\mu \tag{7-10}$$

and

$$z_2 = \sigma/(\mu - a) \tag{7-11}$$

from which

$$a = \mu(1 - z_1/z_2) = \mu - \sigma/z_2 \tag{7-12}$$

The value of $z_1$ can be computed directly from the observed events. Since the second and third moments of the distribution (x-a) do not contain terms in a, the value of $z_2$ can be obtained from the relationship

$$\gamma_1 = 3z_2 + z_2^3 \tag{7-13}$$

where $\gamma_1$ is the coefficient of skew of the distribution x. The solution of this equation is

$$z_2 = \frac{1 - \omega^{2/3}}{\omega^{1/3}} \tag{7-14}$$

where

$$\omega = \frac{-\gamma_1 + (\gamma_1^2 + 4)^{\frac{1}{2}}}{2} \tag{7-15}$$

The procedure is then to compute the mean, $\mu$, standard deviation, $\sigma$, and coefficient of skew, $\gamma_1$, of the observed events, x; compute $z_1$ directly and $z_2$ from 7-14; compute a from Equation 7-12

73

and then

$$\sigma_y = [\ln (z_2^2 + 1)]^{\frac{1}{2}} \qquad (7\text{-}16)$$

and

$$\mu_y = \ln (\sigma/z_2) - \tfrac{1}{2} \ln (z_2^2 + 1) \qquad (7\text{-}17)$$

*Maximum Likelihood* - Taking the product of Equation 7-1 for $x_1 \to x_n$ and taking logarithms yields the logarithmic likelihood function:

$$\ln L = - \sum_{i=1}^{n} \ln (x_i - a) - \frac{n}{2} \ln (2\pi) - n \ln \sigma_y$$

$$- \frac{\sum_{i=1}^{n} [\ln (x_i - a) - \mu_y]^2}{2\sigma_y^2} \qquad (7\text{-}18)$$

Differentiating with respect to $\mu_y$, $\sigma_y^2$ and a and equating to zero leads to the three expressions:

$$\mu_y = \sum_{i=1}^{n} \ln (x_i - a)/n \qquad (7\text{-}19)$$

$$\sigma_y^2 = \sum_{i=1}^{n} [\ln (x_i - a) - \mu_y]^2/n \qquad (7\text{-}20)$$

and

$$\sum_{i=1}^{n} (x_i - a)^{-1} (\mu_y - \sigma_y^2) = \sum_{i=1}^{n} [(1/(x_i - a)) \ln (x_i - a)] \qquad (7\text{-}21)$$

By substituting for $\mu_y$ and $\sigma_y^2$ in Equation 7-21 an equation in a only is found which can be solved numerically.

*Other Methods* - Computation of the coefficient of skew from small samples is sometimes subject to error. Sangal and Biswas (3) derived a method of estimating the parameter a using only the mean, median and standard deviation of the original data. Their solution is:

$$a = \delta - \frac{\sigma^2}{2(\mu - \delta)} \qquad (7\text{-}22)$$

where $\delta$ is the median of x, determined as the mean of the middle 1/5th of the data. The determination of a is very sensitive to the difference between the mean and median, $(\mu - \delta)$. When this difference becomes small, then a takes a large negative value.

The assumption in this case is that the original data are symmetrically, but not necessarily normally, distributed.  This
follows from the empirical relationship for moderately skewed
distributions that (Mean - Mode) = 3(Mean - Median) so that if
the median equals the mean then the mode also equals the mean,
which only occurs in symmetrical distributions.  In such a case
if the median is set to 0.99 $\mu$ then equation 7-22 can be modified
to

$$a = \mu\ (0.99 - 50\ z^2) \qquad\qquad (7\text{-}23)$$

where z is the coefficient of variation.  Burges et al. (1) compared the performance of the methods of moments and median and
concluded that moments was better.

Condie (2) has described a simple graphical method of determining the parameter a which is applicable provided that:

(a)  At least one of the graph scales, vertical or horizontal, is logarithmic.
(b)  The curvature of the best fitting line is gradually
decreasing.

The other scale can be linear, logarithmic, normally probabilistic, time scale, etc.

Referring to Figure 7-1; $(x_1, y_1)$, $(x_2, y_2)$, $(x_3, y_3)$ are three

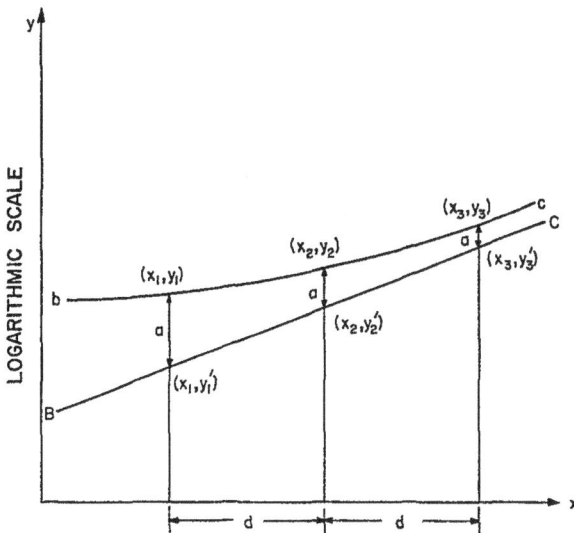

Figure 7-1.  Graphical technique of estimating parameter a in
3-parameter lognormal distribution.

points on the best-fitting line through the plotted event magnitudes (b-c) such that

$$x_3 - x_1 = x_2 - x_3 \qquad (7\text{-}24)$$

The object is then to find a constant a, which will move points $(x_1,y_1)$, $(x_2,y_2)$, $(x_3,y_3)$ to points $(x_1,y_1')$, $(x_2,y_2')$, $(x_3,y_3')$ all on a straight line (B-C). The straight line is easier to extrapolate than the original arbitrarily curved line.

Considering the slope of the logarithmically straight line B-C:

$$\frac{\ln(y_3-a) - \ln(y_1-a)}{x_3 - x_1} = \frac{\ln(y_2-a) - \ln(y_3-a)}{x_2 - x_1} \qquad (7\text{-}25)$$

but from Equation 7-24 this reduces to

$$\ln \frac{y_3-a}{y_1-a} = \ln \frac{y_2-a}{y_3-a} \qquad (7\text{-}26)$$

from which

$$a = \frac{y_1 y_2 - y_3^2}{y_1 + y_2 - 2y_3} \qquad (7\text{-}27)$$

The method is simple and can be applied to any distribution with a logarithmic scale such as lognormal, log-Gumbel, log-Pearson type III.

### Frequency Factor

Using the standard normal deviate as frequency factor the first expression obtained is

$$y_T = \mu_y + t\sigma_y \qquad (7\text{-}28)$$

where $\mu_y$ and $\sigma_y$ are the mean and standard deviation of the series $\ln (x-a)$ so that the T-year event, $x_T$, is

$$x_T = a + e^{\mu_y + t\sigma_y} \qquad (7\text{-}29)$$

Secondly, using the standard frequency equation

$$x_T = \mu + K\sigma \qquad (7\text{-}30)$$

the frequency factor K can be found by substituting for $x_T$, $\mu$ and $\sigma$ as

$$K = \frac{\exp (y_T - \mu_y - \sigma_y^2/2) - 1}{(e^{\sigma_y^2} - 1)^{\frac{1}{2}}} \qquad (7-31)$$

Now, substituting for $y_T$ and $\mu_y$ in terms of $\sigma_y$ from 7-28 and substituting for $\sigma_y$ in terms of $z_2$ from 7-16 yields:

$$K = \frac{\exp \{[\ln (1 + z_2^2)]^{\frac{1}{2}}t - [\ln (1 + z_2^2)]/2\} - 1.0}{z_2} \qquad (7-32)$$

But from Equations 7-14 and 7-15, $z_2$ is a function of $\gamma_1$, the sample coefficient of skew. Table 7-1 lists values of K for some typical values of $\gamma_1$. Note that at $\gamma_1 = 0.0$ the frequency factor K is zero for all return periods. The 3-parameter lognormal (together with the 2-parameter lognormal) is thus not a suitable distribution for use with data samples having skews approaching zero.

Standard Error

*Method of Moments* - As with the two-parameter lognormal distribution, 2 options are available to compute the standard error. The first method, using the frequency equation

Table 7-1

Frequency Factor for use in 3-Parameter Lognormal Distribution

| Coefficient of skew $\gamma_1$ | Cumulative Probability, P, % | | | | | |
|---|---|---|---|---|---|---|
| | 50 | 80 | 90 | 95 | 98 | 99 |
| | Corresponding Return Period, T, Years | | | | | |
| | 2 | 5 | 10 | 20 | 50 | 100 |
| -2.00 | .2366 | -.6144 | -1.2437 | -1.8916 | -2.7943 | -3.5196 |
| -1.80 | .2240 | -.6395 | -1.2621 | -1.8928 | -2.7578 | -3.4433 |
| -1.60 | .2092 | -.6654 | -1.2792 | -1.8901 | -2.7138 | -3.3570 |
| -1.40 | .1920 | -.6920 | -1.2943 | -1.8827 | -2.6615 | -3.2601 |
| -1.20 | .1722 | -.7186 | -1.3067 | -1.8696 | -2.6002 | -3.1521 |
| -1.00 | .1495 | -.7449 | -1.3156 | -1.8501 | -2.5294 | -3.0333 |
| -.80 | .1241 | -.7700 | -1.3201 | -1.8235 | -2.4492 | -2.9043 |
| -.60 | .0959 | -.7930 | -1.3194 | -1.7894 | -2.3600 | -2.7665 |
| -.40 | .0654 | -.8131 | -1.3128 | -1.7478 | -2.2631 | -2.6223 |
| -.20 | .0332 | -.8296 | -1.3002 | -1.6993 | -2.1602 | -2.4745 |
| .00 | 0.0000 | 0.0000 | 0.0000 | 0.0000 | 0.0000 | 0.0000 |
| .20 | -.0332 | .8296 | 1.3002 | 1.6993 | 2.1602 | 2.4745 |
| .40 | -.0654 | .8131 | 1.3128 | 1.7478 | 2.2631 | 2.6223 |
| .60 | -.0959 | .7930 | 1.3194 | 1.7894 | 2.3600 | 2.7665 |
| .80 | -.1241 | .7700 | 1.3201 | 1.8235 | 2.4492 | 2.9043 |
| 1.00 | -.1495 | .7449 | 1.3156 | 1.8501 | 2.5294 | 3.0333 |
| 1.20 | -.1722 | .7186 | 1.3067 | 1.8696 | 2.6002 | 3.1521 |
| 1.40 | -.1920 | .6920 | 1.2943 | 1.8827 | 2.6615 | 3.2601 |
| 1.60 | -.2092 | .6654 | 1.2792 | 1.8901 | 2.7138 | 3.3570 |
| 1.80 | -.2240 | .6395 | 1.2621 | 1.8928 | 2.7578 | 3.4433 |
| 2.00 | -.2366 | .6144 | 1.2437 | 1.8916 | 2.7943 | 3.5196 |

$$y_T = \ln (x_T - a) = \mu_y + t\sigma_y \qquad (7\text{-}33)$$

gives

$$S_T^2 = \frac{\sigma_y^2}{n} \left\{ 1 + t^2/2 \right\} \qquad (7\text{-}34)$$

but $S_T^2$ is now in logarithmic units and refers to $\ln (x_T - a)$ not to $x_T$.

Secondly we can use the frequency equation

$$x_T = \mu + K\sigma \qquad (7\text{-}35)$$

where $\mu$ and $\sigma$ are the mean and standard deviation of $x_i$, $i = 1$, $n$ and $K$ is the frequency factor from Section 7-3.

Since $K$ is a function of $\gamma_1$ and $T$, Equation 7-35 can be rewritten as

$$x_T = f(\mu_1', \mu_2, \mu_3, T) \qquad (7\text{-}36)$$

and since the return period, $T$, is invariable

$$\text{var } x_T = \left( \frac{\partial x_T}{\partial \mu_1'} \right)^2 \text{var } \mu_1' + \left( \frac{\partial x_T}{\partial \mu_2} \right)^2 \text{var } \mu_2 + \left( \frac{\partial x_T}{\partial \mu_3} \right)^2 \text{var } \mu_3$$

$$+ 2 \frac{\partial x_T}{\partial \mu_1'} \frac{\partial x_T}{\partial \mu_2} \text{cov } (\mu_1', \mu_2) + 2 \frac{\partial x_T}{\partial \mu_1'} \frac{\partial x_T}{\partial \mu_3} \text{cov } (\mu_1', \mu_3)$$

$$+ 2 \frac{\partial x_T}{\partial \mu_2} \frac{\partial x_T}{\partial \mu_3} \text{cov } (\mu_2, \mu_3) \qquad (7\text{-}37)$$

where, from Equation 7-35

$$\frac{\partial x_T}{\partial \mu_1'} = 1 \qquad (7\text{-}38)$$

$$\frac{\partial x_T}{\partial \mu_2} = \frac{1}{2\sigma} \left[ K - 3\gamma_1 \frac{\partial K}{\partial \gamma_1} \right] \qquad (7\text{-}39)$$

$$\frac{\partial x_T}{\partial \mu_3} = \frac{1}{\sigma^2} \frac{\partial K}{\partial \gamma_1} \qquad (7\text{-}40)$$

From Equations 7-14, 7-15 and 7-32

$$\frac{\partial K}{\partial \gamma_1} = \frac{\partial K}{\partial z_2} \cdot \frac{\partial z_2}{\partial \omega} \cdot \frac{\partial \omega}{\partial \gamma_1} \qquad (7\text{-}41)$$

with

$$\frac{\partial K}{\partial z_2} = \left\{\frac{2z_2}{1 + z_2^2}\right\}\left\{\left[\exp\left\{[\ln(1 + z_2^2)]^{\frac{1}{2}}t - [\ln(1 + z_2^2)]/2\right\}\right].$$

$$\left[\frac{t}{2z_2[\ln(1 + z_2^2)]^{\frac{1}{2}}} - \frac{1}{z_2} - \frac{1}{2z_2^3}\right] + \frac{1}{2z_2^3} + \frac{1}{2z_2}\right\} \qquad (7\text{-}42)$$

$$\frac{\partial z_2}{\partial \omega} = -\frac{1}{3\omega^{2/3}}\left[1 + \frac{1}{\omega^{2/3}}\right] \qquad (7\text{-}43)$$

$$\frac{\partial \omega}{\partial \gamma_1} = -\frac{1}{2} + \frac{\gamma_1}{2(\gamma_1^2 + 4)^{\frac{1}{2}}} \qquad (7\text{-}44)$$

Also, from Standard Error by Moments in Chapter 3,

$$\text{var } \mu_1' = \mu_2/n \qquad (7\text{-}45)$$

$$\text{var } \mu_2 = \frac{1}{n}\left\{\mu_4 - \mu_2^2\right\} \qquad (7\text{-}46)$$

$$\text{var } \mu_3 = \frac{1}{n}\left\{\mu_6 - \mu_3^2 - 6\mu_4\mu_2 + 9\mu_2^3\right\} \qquad (7\text{-}47)$$

$$\text{cov } (\mu_1', \mu_2) = \mu_3/n \qquad (7\text{-}48)$$

$$\text{cov } (\mu_1', \mu_3) = \frac{1}{n}\left\{\mu_4 - 3\mu_2^2\right\} \qquad (7\text{-}49)$$

$$\text{cov } (\mu_2, \mu_3) = \frac{1}{n}\left\{\mu_5 - 4\mu_3\mu_2\right\} \qquad (7\text{-}50)$$

and central moments may be obtained from Method of Moments in this chapter as

$$\mu_2 = e^{\sigma_y^2 + 2\mu_y}\left(e^{\sigma_y^2} - 1\right) \qquad (7\text{-}51)$$

$$\mu_3 = e^{3\sigma_y^2/2 + 3\mu_y}\left(e^{\sigma_y^2} - 1\right)^2\left(e^{\sigma_y^2} + 2\right) \qquad (7\text{-}52)$$

$$\mu_4 = e^{2\sigma_y^2 + 4\mu_y} \left(e^{\sigma_y^2} - 1\right)^2 \left(e^{4\sigma_y^2} + 2e^{3\sigma_y^2} + 3e^{2\sigma_y^2} - 3\right) \quad (7\text{-}53)$$

$$\mu_5 = e^{5\sigma_y^2/2 + 5\mu_y} \left(e^{10\sigma_y^2} - 5e^{6\sigma_y^2} + 10e^{3\sigma_y^2} - 10e^{\sigma_y^2} + 4\right) \quad (7\text{-}54)$$

$$\mu_6 = e^{3\sigma_y^2 + 6\mu_y} \left(e^{15\sigma_y^2} - 6e^{10\sigma_y^2} + 15e^{6\sigma_y^2} - 20e^{3\sigma_y^2} + 15e^{\sigma_y^2} - 5\right) \quad (7\text{-}55)$$

Substitution of these results into Equation 7-37 ultimately yields an expression for the standard error of estimate.

The initial computations are to obtain the mean, variance and coefficient of skew, $\mu_1'$, $\mu_2$ and $\gamma_1$, from the observed events; calculate $\mu_y$ and $\sigma_y$ and the moments $\mu_4 - \mu_6$. The variances and covariances of the moments, Equation 7-45 - 7-50, and the derivatives $\partial x_T/\partial \mu_1'$, $\partial x_T/\mu_2$ and $\partial x_T/\partial \mu_3$ can then be calculated and the variance of the estimate obtained from 7-37.

*Maximum Likelihood* - For the 3-parameter lognormal distribution with pdf

$$f(x) = \frac{1}{(x-a)\,\sigma_y\,\sqrt{2\pi}} \ \exp\left\{-[(\ln x\text{-}a) - \mu_y]^2/2\sigma_y^2\right\} \quad (7\text{-}56)$$

the standard error of estimate by maximum likelihood is given by:

$$S_T^2 = \left(\frac{\partial x}{\partial a}\right)^2 \text{var } a + \left(\frac{\partial x}{\partial \sigma_y^2}\right)^2 \text{var } \sigma_y^2 + \left(\frac{\partial x}{\partial \mu_y}\right)^2 \text{var } \mu_y$$

$$+ \ 2\ \frac{\partial x}{\partial a}\ \frac{\partial x}{\partial \sigma_y^2}\ \text{cov } (a, \sigma_y^2) + 2\ \frac{\partial x}{\partial a}\ \frac{\partial x}{\partial \mu_y}\ \text{cov } (a, \mu_y)$$

$$+ \ 2\ \frac{\partial x}{\partial \sigma_y^2}\ \frac{\partial x}{\partial \mu_y}\ \text{cov } (\sigma_y^2, \mu_y) \quad (7\text{-}57)$$

From Equation 7-29

$$x_T = a + e^{\mu_y + t\sigma_y} \quad (7\text{-}58)$$

so that

$$\frac{\partial x}{\partial a} = 1 \quad (7\text{-}59)$$

$$\frac{\partial x}{\partial \sigma_y^2} = \frac{te^{\mu_y + t\sigma_y}}{2\sigma_y} \tag{7-60}$$

$$\frac{\partial x}{\partial \mu_y} = e^{\mu_y + t\sigma_y} \tag{7-61}$$

so that, writing $\omega$ for $e^{\mu_y + t\sigma_y}$ Equation 7-57 becomes

$$S_T^2 = \text{var } a + \frac{t^2\omega^2}{4\sigma_y^2} \text{ var } \sigma_y^2 + \omega^2 \text{ var } \mu_y + \frac{t\omega}{\sigma_y} \text{ cov } (a, \sigma_y^2)$$

$$+ 2\omega \text{ cov } (a, \mu_y) + \frac{t\omega^2}{\sigma_y} \text{ cov } (\sigma_y^2, \mu_y) \tag{7-62}$$

From Equation 7-56 the logarithm of the likelihood function is

$$\ln L = - \sum_{i=1}^{n} \ln (x_i - a) - \frac{n}{2} \ln \sigma_y^2 - \frac{n}{2} \ln (2\pi)$$

$$- \sum_{i=1}^{n} [\ln (x_i - a) - \mu_y]^2 / 2\sigma_y^2 \tag{7-63}$$

from which

$$\frac{\partial^2 \ln L}{\partial a^2} = \frac{1}{\sigma_y^2} \left\{ (\sigma_y^2 - \mu_y - 1) \sum_{i=1}^{n} (x_i - a)^{-2} + \sum_{i=1}^{n} \left[\ln (x-a)\right] \left[x_i - a\right]^{-2} \right\} \tag{7-64}$$

$$\frac{\partial^2 \ln L}{\partial \mu_y^2} = - \frac{n}{\sigma_y^2} \tag{7-65}$$

$$\frac{\partial^2 \ln L}{\partial (\sigma_y^2)^2} = \frac{n}{2\sigma_y^4} - \frac{\sum_{i=1}^{n} [\ln (x_i - a) - \mu_y]^2}{2\sigma_y^6} = \frac{n}{2\sigma_y^4} \tag{7-66}$$

$$\frac{\partial^2 \ln L}{\partial a \partial \mu_y} = - \frac{1}{\sigma_y^2} \sum_{i=1}^{n} (x_i - a)^{-1} \tag{7-67}$$

$$\frac{\partial^2 \ln L}{\partial a \partial (\sigma_y^2)} = - \frac{\Sigma [\ln (x-a) - \mu_y] \cdot (x-a)^{-1}}{\sigma_y^4} \qquad (7\text{-}68)$$

$$\frac{\partial^2 \ln L}{\partial \mu_y \partial (\sigma_y^2)} = - \frac{\sum\limits_{i=1}^{n} [\ln (x-a) - \mu_y]}{\sigma_y^4} = 0 \qquad (7\text{-}69)$$

To evaluate $\sum\limits_{i=1}^{n} (x_i - a)^r$, say M, introduce this factor into the pdf

$$M = \sum\limits_{i=1}^{n} (x_i - a)^r = \frac{n}{\sigma_y \sqrt{2\pi}} \int_a^\infty \frac{(x_i - a)^r}{(x_i - a)}$$

$$\exp - \left\{ \left[ \ln (x_i - a) - \mu_y \right]^2 / 2\sigma_y^2 \right\} \, dx \qquad (7\text{-}70)$$

substitute $y = \ln (x_i - a)$

$$M = \frac{n}{\sigma_y \sqrt{2\pi}} \int_{-\infty}^{\infty} \exp \left\{ (- (y - \mu_y)^2 + 2ry\sigma_y^2)/2\sigma_y^2 \right\} \, dy \qquad (7\text{-}71)$$

completing the square within the exponent

$$M = e^{r\mu_y + r\sigma_y^2/2} \cdot \frac{n}{\sigma_y \sqrt{2\pi}} \int_{-\infty}^{\infty} \exp \left\{ - \left[ y - (\mu_y + r\sigma_y^2) \right]^2 / 2\sigma_y^2 \right\} \, dy$$

$$(7\text{-}72)$$

so that

$$\sum\limits_{i=1}^{n} (x_i - a)^r = ne^{r\mu_y + r^2\sigma_y^2/2} \qquad (7\text{-}73)$$

Similarly, to evaluate $\sum\limits_{i=1}^{n} \left\{ \left[ \ln (x_i - a) \right] \left[ x_i - a \right]^r \right\}$, say N,

$$N = \frac{n}{\sigma_y \sqrt{2\pi}} \int_a^\infty \frac{\ln(x_i-a)(x_i-a)^r}{(x_i - a)} \exp - \left\{ \left[ \ln (x_i-a) - \mu_y \right]^2 / 2\sigma_y^2 \right\} \, dx$$

Put $y = \ln (x_i-a)$, $(x_i-a)^r = e^{ry}$, $dy = \frac{dx}{(x_i-a)}$, then $\qquad (7\text{-}74)$

$$N = \frac{n}{\sigma_y \sqrt{2\pi}} \int_{-\infty}^{\infty} y \exp -\left\{\left[(y - \mu_y)^2 - 2ry\sigma_y^2\right]/2\sigma_y^2\right\} \, dy \qquad (7\text{-}75)$$

completing the square within the exponent

$$N = e^{r\mu_y + r^2\sigma_y^2/2} \frac{n}{\sigma_y \sqrt{2\pi}} \int_{-\infty}^{\infty} y \, e^{-\left[y - (\mu_y + r\sigma_y^2)\right]/2\sigma_y^2} \, dy \qquad (7\text{-}76)$$

so that

$$\sum_{i=1}^{n} \left\{\left[\ln (x_i - a)\right]\left[x_i - a\right]^r\right\} = n(\mu + r\sigma_y^2) \, e^{r\mu_y + r^2\sigma_y^2/2} \, ) \qquad (7\text{-}77)$$

Utilizing these expressions the derivatives of the likelihood equation are simplified to

$$\frac{\partial^2 \ln L}{\partial a^2} = -\frac{n}{\sigma_y^2} \left\{(1 + \sigma_y^2) \, e^{2\sigma_y^2 - 2\mu_y}\right\} \qquad (7\text{-}78)$$

$$\frac{\partial^2 \ln L}{\partial \mu_y^2} = -\frac{n}{\sigma_y^2} \qquad (7\text{-}79)$$

$$\frac{\partial^2 \ln L}{\partial (\sigma_y^2)^2} = -\frac{n}{2\sigma_y^4} \qquad (7\text{-}80)$$

$$\frac{\partial^2 \ln L}{\partial a \partial \mu_y} = -\frac{n}{\sigma_y^2} \, e^{\sigma_y^2/2 - \mu_y} \qquad (7\text{-}81)$$

$$\frac{\partial^2 \ln L}{\partial a \partial (\sigma_y^2)} = \frac{n}{\sigma_y^2} \, e^{\sigma_y^2/2 - \mu_y} \qquad (7\text{-}82)$$

From these expressions, the likelihood or information matrix is:

$$I = \frac{n}{\sigma_y^2} \begin{bmatrix} (\sigma_y^2 + 1)\, e^{2(\sigma_y^2 - \mu_y)} & e^{\sigma_y^2/2 - \mu_y} & - e^{\sigma_y^2/2 - \mu_y} \\[2em] e^{\sigma_y^2/2 - \mu_y} & 1 & 0 \\[2em] - e^{\sigma_y^2/2 - \mu_y} & 0 & \dfrac{1}{2\sigma_y^2} \end{bmatrix}$$

$$(7\text{-}83)$$

The determinant, D, of this matrix is:

$$D = \frac{(\sigma_y^2 + 1)\, e^{2(\sigma_y^2 - \mu_y)}}{2\sigma_y^2} - \frac{(2\sigma_y^2 + 1)\, e^{\sigma_y^2 - 2\mu_y}}{2\sigma_y^2} \qquad (7\text{-}84)$$

and, by inverting the matrix, the variances and covariances can be obtained as

$$\text{var } a = \frac{1}{2nD} \qquad (7\text{-}85)$$

$$\text{var } \mu_y = \frac{\sigma_y^2}{nD} \left[ \frac{(\sigma_y^2 + 1)}{2\sigma_y^2}\, e^{2(\sigma_y^2 - \mu_y)} - e^{\sigma_y^2 - 2\mu_y} \right] \qquad (7\text{-}86)$$

$$\text{var } \sigma_y^2 = \frac{\sigma_y^2}{nD} \left[ (\sigma_y^2 + 1)\, e^{2(\sigma_y^2 - \mu_y)} - e^{\sigma_y^2 - 2\mu_y} \right] \qquad (7\text{-}87)$$

$$\text{cov } (a, \mu_y) = - \frac{e^{\sigma_y^2/2 - \mu_y}}{2nD} \qquad (7\text{-}88)$$

$$\text{cov } (a, \sigma_y^2) = \frac{\sigma_y^2}{nD}\, e^{\sigma_y^2/2 - \mu_y} \qquad (7\text{-}89)$$

$$\text{cov} \left( \mu_y, \sigma_y^2 \right) = - \frac{\sigma_y^2}{nD} e^{\sigma_y^2 - 2\mu_y} \qquad (7\text{-}90)$$

Substitution into 7-62 then yields the variance of the T-year event.

Table 7-2 shows that, as would be expected, the standard error of estimate by method of moments is larger than by maximum likelihood.

References

1.  Burges, S. J., Lettenmaier, D. P., and C. L. Bates, 1975, Properties of the Three-Parameter Lognormal Distribution, Wat. Res. Res., Vol. 11, No. 2, pp. 229-235.

2.  Condie, R., 1973, Unpublished Notes, Environment Canada, Ottawa.

3.  Sangal, B. P. and A. K. Biswas, 1970, The 3-Parameter Lognormal Distribution and its Applications in Hydrology, Wat. Res. Res., Vol. 6, No. 2, pp. 505-515.

```
      PROGRAM LN3(INPUT,OUTPUT,TAPE5=INPUT,TAPE6=OUTPUT)            A    1
C                                                                   A    2
C                                                                   A    3
C     COMPUTES METHOD OF MOMENTS AND MAXIMUM LIKELIHOOD ESTIMATES FOR  A    4
C     T YEAR EVENTS AND STANDARD ERRORS FOR 3 PARAMETER LOGNORMAL   A    5
C     DISTRIBUTION                                                  A    6
C     INPUT                                                         A    7
C     TITLE                                                         A    8
C     N NUMBER OF ANNUAL MAXIMUM EVENTS                             A    9
C     X SERIES OF EVENTS                                            A   10
C                                                                   A   11
C                                                                   A   12
      REAL M1,M2,M3,M4,M5,M6,MY,K,MU                                A   13
      DIMENSION SND(6), X(100)                                      A   14
      DIMENSION XT(6), SX(6), TITLE(80)                             A   15
      SND(1)=0.0                                                    A   16
      SND(2)=0.8416                                                 A   17
      SND(3)=1.2816                                                 A   18
      SND(4)=1.6449                                                 A   19
      SND(5)=2.0538                                                 A   20
      SND(6)=2.3264                                                 A   21
      READ (5,16) TITLE                                             A   22
      READ (5,17) N                                                 A   23
      XN=N                                                          A   24
      READ (5,18) (X(I),I=1,N)                                      A   25
      WRITE (6,19) TITLE                                            A   26
      WRITE (6,20)                                                  A   27
      A=0.0                                                         A   28
      B=0.0                                                         A   29
      C=0.0                                                         A   30
      DO 1 I=1,N                                                    A   31
      A=A+X(I)                                                      A   32
      B=B+X(I)**2                                                   A   33
      C=C+X(I)**3                                                   A   34
    1 CONTINUE                                                      A   35
      M1=A/XN                                                       A   36
      M2=(B/XN)-(A/XN)**2                                           A   37
      M3=(C/XN)+2.0*M1**3-3.0*M1*(B/XN)                             A   38
      G=M3/(M2**1.5)                                                A   39
      M2=M2*XN/(XN-1.0)                                             A   40
      WRITE (6,11) M1                                               A   41
      WRITE (6,12) M2                                               A   42
      WRITE (6,13) G                                                A   43
      IF (G.LT.0.0) GO TO 3                                         A   44
      W=(-G+((G**2)+4.0)**0.5)/2.0                                  A   45
      Z2=(1.0-W**(2./3.))/(W**(1./3.))                             A   46
      AMO=M1-(M2**0.5)/Z2                                           A   47
      WRITE (6,21) AMO                                              A   48
      SY=(ALOG(Z2**2+1.0))**0.5                                     A   49
      SY2=SY**2                                                     A   50
      MY=ALOG((M2**0.5)/Z2)-0.5*ALOG(Z2**2+1.0)                     A   51
      E=EXP(SY2)                                                    A   52
      EA=EXP(2.0*SY2)                                               A   53
      EB=EXP(2.5*SY2)                                               A   54
      EC=EXP(3.0*SY2)                                               A   55
      ED=EXP(4.0*SY2)                                               A   56
      EF=EXP(6.0*SY2)                                               A   57
      EG=EXP(10.0*SY2)                                              A   58
      EH=EXP(15.0*SY2)                                              A   59
      EI=EXP(4.0*MY)                                                A   60
      EJ=EXP(5.0*MY)                                                A   61
      EK=EXP(6.0*MY)                                                A   62
      EL=(EXP(SY2)-1.0)**2                                          A   63
      M4=EA*EI*EL*(ED+2.0*EC+3.0*EA-3.0)                            A   64
      M5=EB*EJ*(EG-5.0*EF+10.0*EC-10.0*E+4.0)                       A   65
      M6=EC*EK*(EH-6.0*EG+15.0*EF-20.0*EC+15.0*E-5.0)               A   66
      VM1=M2/XN                                                     A   67
      VM2=(M4-M2**2)/XN                                             A   68
      VM3=(M6-M3**2-6.0*M4*M2+9.0*M2**3)/XN                         A   69
      CM1M2=M3/XN                                                   A   70
      CM1M3=(M4-3.0*M2**2)/XN                                       A   71
      CM2M3=(M5-4.0*M3*M2)/XN                                       A   72
      DO 2 J=1,6                                                    A   73
      T=SND(J)                                                      A   74
      DXDM1=1.0                                                     A   75
      DWDG=-0.5*G/(2.0*(G**2+4.0)**0.5)                             A   76
      DZ2DW=(-1./3.)*(W**(-4./3.)+W**(-2./3.))                      A   77
      D1=ALOG(Z2**2+1.0)                                            A   78
      D2=EXP((SQRT(D1))*T-D1/2.)                                    A   79
      D3=(2.0*Z2)/(1.0+Z2**2)                                       A   80
```

86

```
      D4=T/(2.0*Z2*SQRT(D1))                                    A  81
      D5=1.0/Z2                                                 A  82
      D6=1.0/(2.0*Z2**3)                                        A  83
      DKDZ2=D3*(D2*(D4-D5-D6)+D6+D5/2.)                         A  84
      K=(D2-1.0)/Z2                                             A  85
      DKDG=DKDZ2*DZ2DW*DWDG                                     A  86
      DXDM2=(1.0/(2.0*SQRT(M2)))*(K-3.0*G*DKDG)                 A  87
      DXDM3=DKDG/M2                                             A  88
      SX(J)=SQRT((DXDM1**2)*VM1+(DXDM2**2)*VM2+(DXDM3**2)*VM3+2.0*DXDM1*  A  89
     1DXDM2*CM1M2+2.0*DXDM1*DXDM3*CM1M3+2.0*DXDM2*DXDM3*CM2M3)  A  90
2     XT(J)=M1+K*M2**0.5                                        A  91
      WRITE (6,24)                                              A  92
      WRITE (6,25) (XT(J),J=1,6)                                A  93
      WRITE (6,26) (SX(J),J=1,6)                                A  94
      GO TO 4                                                   A  95
3     WRITE (6,14)                                              A  96
4     WRITE (6,27)                                              A  97
      WRITE (6,28)                                              A  98
      XMIN=10000000.                                            A  99
      DO 5 I=1,N                                                A 100
5     IF (X(I).LT.XMIN) XMIN=X(I)                               A 101
      AML=XMIN*0.80                                             A 102
      ICOUNT=0                                                  A 103
6     ICOUNT=ICOUNT+1                                           A 104
      A=0.0                                                     A 105
      B=0.0                                                     A 106
      C=0.0                                                     A 107
      D=0.0                                                     A 108
      E=0.0                                                     A 109
      F=0.0                                                     A 110
      P=0.0                                                     A 111
      DO 7 I=1,N                                                A 112
      A=A+ALOG(X(I)-AML)                                        A 113
      B=B+(ALOG(X(I)-AML))**2                                   A 114
      P=P+(ALOG(X(I)-AML))**3                                   A 115
      C=C+1.0/((X(I)-AML))                                      A 116
      D=D+1.0/((X(I)-AML)**2)                                   A 117
      E=E+(1.0/((X(I)-AML)))*ALOG(X(I)-AML)                     A 118
7     F=F+(1.0/((X(I)-AML)**2))*ALOG(X(I)-AML)                  A 119
      G=(B/XN)-(A/XN)**2-(A/XN)                                 A 120
      H=(-2.0*E/XN)+(2.0*A/XN)*(C/XN)+(C/XN)                    A 121
      FCN=C*G+E                                                 A 122
      FPN=C*H+D*G+F-D                                           A 123
      AS=AML-(FCN/FPN)                                          A 124
      WRITE (6,29) ICOUNT,AS,FCN                                A 125
      DELTA=ABS(0.00001*AS)                                     A 126
      IF (ABS(AS-AML).LT.DELTA) GO TO 8                         A 127
      IF (ICOUNT.GT.25) GO TO 10                                A 128
      AML=AS                                                    A 129
      GO TO 6                                                   A 130
8     CONTINUE                                                  A 131
      AML=AS                                                    A 132
      MU=A/XN                                                   A 133
      VAR=(B/XN)-(A/XN)**2                                      A 134
      VAR=VAR*XN/(XN-1)                                         A 135
      SKEW=(P/XN)+2.0*MU**3-3.0*MU*(B/XN)                       A 136
      SD=SQRT(VAR)                                              A 137
      A=EXP(VAR-2.0*MU)                                         A 138
      B=EXP(2.0*VAR-2.0*MU)                                     A 139
      C=EXP(VAR/2.0-MU)                                         A 140
      D1=(VAR+1.0)/(2.0*VAR)                                    A 141
      D2=1.0/(2.0*VAR)                                          A 142
      D=D1*B-D2*A-A                                             A 143
      E=1.0/(N*D)                                               A 144
      VA=E*0.5                                                  A 145
      VMU=(VAR*E)*(D1*B-A)                                      A 146
      VVAR=VAR*E*((VAR+1.0)*B-A)                                A 147
      CAMU=C*E/2.0                                              A 148
      CAVAR=VAR*E*C                                             A 149
      CMUVAR=VAR*E*A                                            A 150
      CAMU=-CAMU                                                A 151
      CMUVAR=-CMUVAR                                            A 152
      DO 9 J=1,6                                                A 153
      T=SND(J)                                                  A 154
      Z=EXP(MU+T*SD)                                            A 155
      VX=VA+(VMU*Z**2)+(T*Z*CAVAR/SD)+(2.0*Z*CAMU)+(T*Z**2*CMUVAR/SD)+(T  A 156
     1**2*Z**2*VVAR/(4.0*VAR))                                  A 157
      XT(J)=AML+Z                                               A 158
      SX(J)=SQRT(VX)                                            A 159
9     CONTINUE                                                  A 160
```

87

```
       WRITE (6,30)                                                    A 161
       WRITE (6,32) AML                                                A 162
       WRITE (6,22) MU                                                 A 163
       WRITE (6,23) VAR                                                A 164
       WRITE (6,15) SKEW                                               A 165
       WRITE (6,33)                                                    A 166
       WRITE (6,24)                                                    A 166
       WRITE (6,25) (XT(J),J=1,6)                                      A 167
       WRITE (6,26) (SX(J),J=1,6)                                      A 168
       WRITE (6,31)                                                    A 169
10     CONTINUE                                                        A 170
       STOP                                                            A 171
C                                                                      A 172
C                                                                      A 173
C                                                                      A 174
11     FORMAT (20X,9HMEAN OF X,16X,E12.5)                              A 175
12     FORMAT (20X,13HVARIANCE OF X,12X,E12.5)                         A 176
13     FORMAT (20X,9HSKEW OF X,16X,E12.5)                              A 177
14     FORMAT (/,3X,52H NO MOMENTS SOLUTION IS POSSIBLE BECAUSE OF -VE SK A 178
      1EW,/)                                                           A 179
15     FORMAT (20X,15HSKEW OF LN(X-A),10X,E12.5)                       A 180
16     FORMAT (80A1)                                                   A 181
17     FORMAT (I5)                                                     A 182
18     FORMAT (8F10.0)                                                 A 183
19     FORMAT (1H1,/,80A1,//,21X,38HTHREE PARAMETER LOGNORMAL DISTRIBUTIO A 184
      1N,/)                                                            A 185
20     FORMAT (31X,17HMETHOD OF MOMENTS,//)                            A 186
21     FORMAT (20X,1HA,24X,E12.5,/)                                    A 187
22     FORMAT (20X,15HMEAN OF LN(X-A),10X,E12.5)                       A 188
23     FORMAT (20X,19HVARIANCE OF LN(X-A),6X,E12.5)                    A 189
24     FORMAT (3X,7HT,YEARS,4X,1HZ,11X,1H5,10X,2H10,10X,2H20,10X,2H50,9X, A 190
      13H100,/)                                                        A 191
25     FORMAT (3X,1HX,3X,6E12.5,/,4X,1HT)                              A 192
26     FORMAT (3X,1HS,3X,6E12.5,/,4X,1HT,//)                           A 193
27     FORMAT (25X,28HMAXIMUM LIKELIHOOD PROCEDURE,//)                 A 194
28     FORMAT (21X,5HTRIAL,11X,1HA,11X,4HF(A),/)                       A 195
29     FORMAT (22X,I2,8X,E12.5,1X,E12.5)                               A 196
30     FORMAT (//)                                                     A 197
31     FORMAT (/1H1)                                                   A 198
32     FORMAT (20X,1HA,24X,E12.5)                                      A 199
33     FORMAT(/,3X,73HFOR GOOD USE OF THIS DISTRIBUTION SKEW OF LN(X-A) S A 200
      1HOULD BE CLOSE TO ZERO,/)                                       A 201
       END                                                            A 202
```

```
10 '
20 ' Program LN3
30 ' Copyright G.W. Kite 1986
40 '
50 ' Compute method of moments and maximum likelihood estimates
60 ' for T year events and standard errors
70 ' for a 3-parameter lognormal distribution.
80 '
90 DIM SND#(6),X#(250),XT#(6),SX#(6)
91 FORM$="  ##.###^^^^"
92 VV$=CHR$(179)
100 DATA 0.0#,0.8416#,1.2816#,1.6449#,2.0538#,2.3264#
110 FOR I%=1 TO 6
120 READ SND#(I%)
130 NEXT I%
140 T$="Three Parameter Lognormal Distribution"
141 IERR%=0
150 GOSUB 3000
160 IF IERR% = 1 GOTO 2410
230 FLAG%=0:IF YN$ = "Y" OR YN$ = "y" THEN FLAG%=1
240 NREC%=0
250 OPEN FILE$ FOR INPUT AS #1
260 LINE INPUT#1,TITLE$
270 IF EOF(1) GOTO 310
280 NREC%=NREC%+1
290 LINE INPUT#1,I$
291 X#(NREC%)=VAL(MID$(I$,5,12))
300 GOTO 270
310 CLOSE#1
311 NREC#=NREC%
320 A%=LEN(TITLE$)
330 B%=(80-A%)/2
340 U$=STRING$(A%,205)
350 M$="Method of Moments"
360 E%=LEN(M$)
370 F%=(80-E%)/2
380 W$=STRING$(E%,205)
390 CLS
400 PRINT TAB(B%) TITLE$:PRINT TAB(B%) U$
410 PRINT TAB(D%) T$:PRINT TAB(D%) V$
420 PRINT TAB(F%) M$:PRINT TAB(F%) W$
430 IF FLAG% <> 1 GOTO 480
440 LPRINT CHR$(12):LPRINT:LPRINT
450 LPRINT TAB(B%) TITLE$:LPRINT TAB(B%) U$
460 LPRINT:LPRINT TAB(D%) T$:LPRINT TAB(D%) V$
470 LPRINT:LPRINT TAB(F%) M$:LPRINT TAB(F%) W$
480 A#=0#:B#=0#:C#=0#
490 FOR I%=1 TO NREC%
500 A#=A#+X#(I%)
510 B#=B#+X#(I%)*X#(I%)
520 C#=C#+X#(I%)*X#(I%)*X#(I%)
530 NEXT I%
540 M1#=A#/NREC#
550 M2#=(B#/NREC#)-M1#*M1#
560 M3#=(C#/NREC#)+2#*M1#*M1#*M1#-3#*M1#*(B#/NREC#)
570 G#=M3#/(M2#^1.5#)
580 M2#=M2#*NREC#/(NREC#-1#)
590 IF G# < 0# GOTO 1290
600 W#=(-G#+SQR(G#*G#+4#))/2#
610 Z2#=(1#-W#^(2#/3#))/(W#^(1#/3#))
620 AM0#=M1#-SQR(M2#)/Z2#
630 PRINT
640 PRINT TAB(20) "Mean is                ";:PRINT USING FORM$;M1#
650 PRINT TAB(20) "Variance is            ";:PRINT USING FORM$;M2#
660 PRINT TAB(20) "Coefficient of skew is ";:PRINT USING FORM$;G#
670 PRINT TAB(20) "Parameter A is         ";:PRINT USING FORM$;AM0#
680 IF FLAG% <> 1 GOTO 740
690 LPRINT
700 LPRINT TAB(20) "Mean is                ";:LPRINT USING FORM$;M1#
710 LPRINT TAB(20) "Variance is            ";:LPRINT USING FORM$;M2#
720 LPRINT TAB(20) "Coefficient of skew is ";:LPRINT USING FORM$;G#
730 LPRINT TAB(20) "Parameter A is         ";:LPRINT USING FORM$;AM0#
740 SY2#=LOG(Z2#*Z2#+1#)
750 SY#=SQR(SY2#)
760 MY#=LOG(SQR(M2#)/Z2#)-.5#*SY2#
770 E#=EXP(SY2#)
780 EA#=EXP(2#*SY2#)
790 EB#=EXP(2.5#*SY2#)
800 EC#=EXP(3#*SY2#)
810 ED#=EXP(4#*SY2#)
```

```
820 EF#=EXP(6#*SY2#)
830 EG#=EXP(10#*SY2#)
840 EH#=EXP(15#*SY2#)
850 EI#=EXP(4#*MY#)
860 EJ#=EXP(5#*MY#)
870 EK#=EXP(6#*MY#)
880 EL#=(EXP(SY2#)-1#)*(EXP(SY2#)-1#)
890 M4#=EA#*EI#*EL#*(ED#+2#*EC#+3#*EA#-3#)
900 M5=EB#*EJ#*(EG#-5#*EF#+10#*EC#-10#*E#+4#)
910 M6#=EC#*EK#*(EH#-6#*EG#+15#*EF#-20#*EC#+15#*E#-5#)
920 VM1#=M2#/NREC#
930 VM2#=(M4#-M2#*M2#)/NREC#
940 VM3#=(M6#-M3#*M3#-6#*M4#*M2#+9#*M2#*M2#*M2#)/NREC#
950 CM1M2#=M3#/NREC#
960 CM1M3#=(M4#-3#*M2#*M2#)/NREC#
970 CM2M3#=(M5#-4#*M3#*M2#)/NREC#
980 FOR J%=1 TO 6
990 T#=SND#(J%)
1000 DXDM1#=1#
1010 DWDG#=-.5#+G#/(2#*SQR(G#*G#+4#))
1020 DZ2DW#=(-1#/3#)*(W#^(-4#/3#)+W#^(-2#/3#))
1030 D1#=LOG(Z2#*Z2#+1#)
1040 D2#=EXP((SQR(D1#))*T#-D1#/2#)
1050 D3#=(2#*Z2#)/(1#+Z2#*Z2#)
1060 D4#=T#/(2#*Z2#*SQR(D1#))
1070 D5#=1#/Z2#
1080 D6#=1#/(2#*Z2#*Z2#*Z2#)
1090 DKDZ2#=D3#*(D2#*(D4#-D5#-D6#)+D6#+D5#/2#)
1100 K#=(D2#-1#)/Z2#
1110 DKDG#=DKDZ2#*DZ2DW#*DWDG#
1120 DXDM2#=(1#/(2#*SQR(M2#)))*(K#-3#*G#*DKDG#)
1130 DXDM3#=DKDG#/M2#
1140 SX#(J%)=SQR((DXDM1#*DXDM1#)*VM1#+(DXDM2#*DXDM2#)*VM2#+(DXDM3#*DXDM3#)*VM3#+
2#*DXDM1#*DXDM2#*CM1M2#+2#*DXDM1#*DXDM3#*CM1M3#+2#*DXDM2#*DXDM3#*CM2M3#)
1150 XT#(J%)=M1#+K#*SQR(M2#)
1160 NEXT J%
1161 ROW$=CHR$(218)+STRING$(72,196)
1170 PRINT:PRINT TAB(2) "T, years  2           5           10          20
    50          100":PRINT TAB(5) ROW$
1180 PRINT TAB(2) "X  " VV$;:FOR J%=1 TO 6:PRINT USING FORM$;XT#(J%);:NEXT J%
1190 PRINT TAB(2) " t " VV$
1200 PRINT TAB(2) "S  " VV$;:FOR J%=1 TO 6:PRINT USING FORM$;SX#(J%);:NEXT J%
1210 PRINT TAB(2) " t " VV$
1220 IF FLAG% <> 1 GOTO 1300
1230 LPRINT:LPRINT TAB(2) "T, years  2           5           10          20
    50          100":LPRINT TAB(5) ROW$
1240 LPRINT TAB(2) "X  " VV$;:FOR J%=1 TO 6:LPRINT USING FORM$;XT#(J%);:NEXT J%
1250 LPRINT TAB(2) " t " VV$
1260 LPRINT TAB(2) "S  " VV$;:FOR J%=1 TO 6:LPRINT USING FORM$;SX#(J%);:NEXT J%
1270 LPRINT TAB(2) " t " VV$
1280 GOTO 1300
1290 PRINT:PRINT "No moments solution is possible because of the -ve coeff. of s
kew":PRINT
1300 M$="Method of Maximum Likelihood"
1310 E%=LEN(M$)
1320 F%=(80-E%)/2
1330 W$=STRING$(E%,205)
1340 PRINT:PRINT"Press any key to continue"
1350 Q$=INKEY$:IF Q$ ="" GOTO 1350
1360 CLS
1370 PRINT:PRINT TAB(F%) M$:PRINT TAB(F%) W$
1380 PRINT:PRINT TAB(19) "Iteration" TAB(37) "A" TAB(53) "F(A)"
1390 PRINT TAB(19) "---------" TAB(37) "-" TAB(53) "----"
1400 IF FLAG% <> 1 GOTO 1440
1410 LPRINT:LPRINT TAB(F%) M$:LPRINT TAB(F%) W$
1420 LPRINT:LPRINT TAB(19) "Iteration" TAB(37) "A" TAB(53) "F(A)"
1430 LPRINT TAB(19)"---------" TAB(37) "-" TAB(53) "----"
1440 XMIN#=1000000000#
1450 FOR I%=1 TO NREC%
1460 IF X#(I%) < XMIN# THEN XMIN#=X#(I%)
1470 NEXT I%
1480 AML#=XMIN#*.8#
1490 COUNT%=0
1500 COUNT%=COUNT%+1
1510 A#=0#
1520 B#=0#
1530 C#=0#
1540 D#=0#
1550 E#=0#
```

90

```
1560 F#=0#
1570 P#=0#
1580 FOR I%=1 TO NREC%
1590 TEMP#=LOG(X#(I%)-AML#)
1600 A#=A#+TEMP#
1610 B#=B#+TEMP#*TEMP#
1620 P#=P#+TEMP#*TEMP#*TEMP#
1630 TEMP2#=X#(I%)-AML#
1640 C#=C#+1#/TEMP2#
1650 D#=D#+1#/(TEMP2#*TEMP2#)
1660 E#=E#+(1#/TEMP2#)*TEMP#
1670 F#=F#+(1#/(TEMP2#*TEMP2#))*TEMP#
1680 NEXT I%
1690 TEMP3#=A#/NREC#
1700 G#=(B#/NREC#)-TEMP3#*TEMP3#-TEMP3#
1710 H#=(-2#*E#/NREC#)+(2#*A#/NREC#)*(C#/NREC#)+(C#/NREC#)
1720 FCN#=C#*G#+E#
1730 FPN#=C#*H#+D#*G#+F#-D#
1740 AAS#=AML#-(FCN#/FPN#)
1750 PRINT TAB(22);:PRINT USING "##";COUNT%;:PRINT USING "         ##.####^^^^";AA
S#,FCN#
1760 IF FLAG% = 1 THEN LPRINT TAB(22);:LPRINT USING "##";COUNT%;:LPRINT USING "
       ##.####^^^^";AAS#,FCN#
1770 DELTA#=ABS(.00001#*AAS#)
1780 IF ABS(AAS#-AML#) < DELTA# GOTO 1820
1790 IF COUNT% > 25 THEN PRINT:PRINT "Procedure does not converge":PRINT:GOTO 23
90
1800 AML#=AAS#
1810 GOTO 1500
1820 AML#=AAS#
1830 MU#=TEMP3#
1840 VAR#=(B#/NREC#)-TEMP3#*TEMP3#
1850 VAR#=VAR#*NREC#/(NREC#-1)
1860 SKEW#=(P#/NREC#)+2#*MU#*MU#*MU#-3#*MU#*(B#/NREC#)
1870 SD#=SQR(VAR#)
1880 A#=EXP(VAR#-2#*MU#)
1890 B#=EXP(VAR#*2-2#*MU#)
1900 D1#=(VAR#+1#)/(2#*VAR#)
1910 C#=EXP(VAR#/2#-MU#)
1920 D2#=1#/(VAR#*2)
1930 D#=D1#*B#-D2#*A#-A#
1940 E#=1#/(NREC#*D#)
1950 VA#=E#*.5#
1960 VMU#=(VAR#*E#)*(D1#*B#-A#)
1970 VVAR#=VAR#*E#*((VAR#+1#)*B#-A#)
1980 CAMU#=C#*E#/2#
1990 CAVAR#=VAR#*E#*C#
2000 CMUVAR#=VAR#*E#*A#
2010 CAMU#=-CAMU#
2020 CMUVAR#=-CMUVAR#
2030 FOR J% = 1 TO 6
2040 T#=SND#(J%)
2050 Z#=EXP(MU#+T#*SD#)
2060 VX#=VA#+(VMU#*Z#*Z#)+(T#*Z#*CAVAR#/SD#)+(2*Z#*CAMU#)+(T#*Z#*Z#*CMUVAR#/SD#)
+(T#*T#*Z#*Z#*VVAR#/(4*VAR#))
2070 XT#(J%)=AML#+Z#
2080 SX#(J%)=SQR(VX#)
2090 NEXT J%
2100 PRINT
2110 CLS
2120 PRINT:PRINT TAB(B%) TITLE$:PRINT TAB(B%) U$
2130 PRINT TAB(D%) T$:PRINT TAB(D%) V$
2140 PRINT TAB(F%) M$:PRINT TAB(F%) W$
2150 PRINT TAB(19) "Parameter A is              ";:PRINT USING FORM$;AML#
2160 PRINT TAB(19) "Mean of ln(x-A) is          ";:PRINT USING FORM$;MU#
2170 PRINT TAB(19) "Variance of ln(x-A) is      ";:PRINT USING FORM$;VAR#
2180 PRINT TAB(19) "Coeff. of skew of ln(x-A) is ";:PRINT USING FORM$;SKEW#
2190 PRINT:PRINT TAB(2) "NOTE: For good use of this distribution the coeff. of s
kew of ln(x-A) should be close to zero"
2200 IF FLAG% <> 1 GOTO 2280
2210 LPRINT
2220 LPRINT TAB(19) "Parameter A is              ";:LPRINT USING FORM$;AML#
2230 LPRINT TAB(19) "Mean of ln(x-A) is          ";:LPRINT USING FORM$;MU#
2240 LPRINT TAB(19) "Variance of ln(x-A) is      ";:LPRINT USING FORM$;VAR#
2250 LPRINT TAB(19) "Coeff. of skew of ln(x-A) is ";:LPRINT USING FORM$;SKEW#
2260 LPRINT:LPRINT TAB(2) "NOTE: For good use of this distribution the coeff. of
 skew of ln(x-A) should"
2270 LPRINT TAB(2) "be close to zero"
2280 PRINT:PRINT TAB(2) "T, years 2         5         10        20
   50        100":PRINT TAB(5) ROW$
```

```
2290 PRINT TAB(2) "X  " VV$;:FOR J%=1 TO 6:PRINT USING FORM$;XT#(J%);:NEXT J%
2300 PRINT TAB(2) " t " VV$
2310 PRINT TAB(2) "S  " VV$;:FOR J%=1 TO 6:PRINT USING FORM$;SX#(J%);:NEXT J%
2320 PRINT TAB(2) " t " VV$
2330 IF FLAG% <> 1 GOTO 2390
2340 LPRINT:LPRINT TAB(2) "T, years  2        5        10        20
        50       100":LPRINT TAB(5) ROW$
2350 LPRINT TAB(2) "X  " VV$;:FOR J%=1 TO 6:LPRINT USING FORM$;XT#(J%);:NEXT J%
2360 LPRINT TAB(2) " t " VV$
2370 LPRINT TAB(2) "S  " VV$;:FOR J%=1 TO 6:LPRINT USING FORM$;SX#(J%);:NEXT J%
2380 LPRINT TAB(2) " t " VV$
2390 PRINT:PRINT"Press any key to continue"
2400 Q$=INKEY$:IF Q$ ="" GOTO 2400
2410 CLS
2420 SYSTEM
3000 '
3010 ' Subroutine for standard screen format
3020 '
3030 CLS:PRINT:PRINT
3040 ROW$=STRING$(78,205)
3050 BOXTOP$=CHR$(201)+ROW$+CHR$(187)
3060 PRINT BOXTOP$;
3070 PRINT CHR$(186) TAB(80) CHR$(186);
3080 PRINT CHR$(186) TAB(80) CHR$(186);
3090 C%=LEN(T$)
3100 D%=(80-C%)/2
3110 V$=STRING$(C%,205)
3140 PRINT CHR$(186) TAB(D%) T$ TAB(80) CHR$(186);
3150 PRINT CHR$(186) TAB(D%) V$ TAB(80) CHR$(186);
3160 PRINT CHR$(186) TAB(80) CHR$(186);
3170 PRINT CHR$(186) TAB(80) CHR$(186);
3180 PRINT CHR$(186) "   What is the name of your data file?" TAB(80) CHR$(186);

3190 PRINT CHR$(186) TAB(80) CHR$(186);
3200 PRINT CHR$(186) "   Do you want printer output (Y/N)?" TAB(80) CHR$(186);
3201 PRINT CHR$(186) "   (press Alt Q to quit at this stage)" TAB(80) CHR$(186);

3210 PRINT CHR$(186) TAB(80) CHR$(186);
3220 PRINT CHR$(186) TAB(80) CHR$(186);
3230 BOXBOT$=CHR$(200)+ROW$+CHR$(188)
3240 PRINT BOXBOT$;
3241 LOCATE 24,1,0,0,0
3242 PRINT "G Kite";
3250 LOCATE 10,41,1,0,13
3260 FILE$=""
3270 I$=INPUT$(1)
3280 IF I$ = CHR$(13) GOTO 3380                              ' ENTER key
3290 IF I$ = CHR$(8)  GOTO 3330                              ' BACKSPACE key
3300 PRINT I$;
3310 FILE$=FILE$+I$
3320 GOTO 3270
3330 H%=POS(0)
3340 LOCATE 10,H%-1,1,0,13
3350 L%=LEN(FILE$)
3360 FILE$=LEFT$(FILE$,L%-1)
3370 GOTO 3270
3380 LOCATE 12,39,1,0,13
3381 YN$=""
3390 I$=INKEY$:IF I$ = "" GOTO 3390
3400 IF LEN(I$) = 1 GOTO 3410
3401 I$=RIGHT$(I$,1)
3402 IF I$ = CHR$(16) THEN IERR%=1:GOTO 3490                 ' ALT Q key
3403 GOTO 3380
3410 IF I$ = CHR$(13) GOTO 3440                              ' ENTER key
3420 PRINT I$;
3421 YN$=I$
3430 GOTO 3390
3440 IF YN$ = "Y" OR YN$ = "y" OR YN$ = "N" OR YN$ = "n" GOTO 3460
3450 GOTO 3380
3460 LOCATE ,,0,13,13
3490 RETURN
```

## Three Parameter Lognormal Distribution

### Method of Moments

| | |
|---|---|
| Mean is | 4.132D+02 |
| Variance is | 2.067D+04 |
| Coefficient of skew is | 1.221D+00 |
| Parameter A is | 4.221D+01 |

| T, years | 2 | 5 | 10 | 20 | 50 | 100 |
|---|---|---|---|---|---|---|
| $X_t$ | 3.881D+02 | 5.161D+02 | 6.009D+02 | 6.822D+02 | 7.879D+02 | 8.680D+02 |
| $S_t$ | 1.955D+01 | 2.568D+01 | 3.415D+01 | 5.067D+01 | 8.321D+01 | 1.144D+02 |

### Method of Maximum Likelihood

| Iteration | A | F(A) |
|---|---|---|
| 1 | 1.340D+02 | -2.379D-02 |
| 2 | 1.131D+02 | -9.503D-03 |
| 3 | 9.224D+01 | -3.543D-03 |
| 4 | 7.605D+01 | -1.175D-03 |
| 5 | 6.829D+01 | -3.019D-04 |
| 6 | 6.690D+01 | -4.045D-05 |
| 7 | 6.686D+01 | -1.057D-06 |
| 8 | 6.686D+01 | -7.749D-10 |

| | |
|---|---|
| Parameter A is | 6.686D+01 |
| Mean of ln(x-A) is | 5.768D+00 |
| Variance of ln(x-A) is | 1.617D-01 |
| Coeff. of skew of ln(x-A) is | -1.496D-03 |

NOTE: For good use of this distribution the coeff. of skew of ln(x-A) should be close to zero

| T, years | 2 | 5 | 10 | 20 | 50 | 100 |
|---|---|---|---|---|---|---|
| $X_t$ | 3.867D+02 | 5.156D+02 | 6.024D+02 | 6.867D+02 | 7.974D+02 | 8.821D+02 |
| $S_t$ | 1.621D+01 | 2.475D+01 | 3.522D+01 | 4.916D+01 | 7.230D+01 | 9.307D+01 |

# CHAPTER 8
## TYPE I EXTREMAL DISTRIBUTION

## Introduction

Suppose that from N samples each containing n events the maximum or minimum event in each sample is selected. As n increases, the distribution of the N maxima or minima approaches a limiting or assymptotic form. The type of the limiting form depends on the type of the initial distribution of the Nn values. The distribution of the maxima or minima is given by the functional equation

$$p^n(x) = P(a_n x + b_n) \qquad (8-1)$$

where $a_n$ and $b_n$ are functions of n.

Fisher and Tippett (quoted in (15)) have shown that there are three possible solutions to the functional equation. These are known, logically enough, as types I, II and III extremal distributions. The type I distribution is unbounded, the type II has a lower limit and the type III has an upper limit.

The type I distribution (or Gumbel (3)) is often used for maximum type events and results from any initial unlimited distribution of exponential type which converges to an exponential function. Examples of this type of distribution include the normal and lognormal distributions. The derivation of the type I distribution for a simple exponential function can be described (13) as follows:

(a) Let $\varepsilon_1$, $\varepsilon_2$,..., $\varepsilon_n$ be a series of independent random variables with cumulative probability distribution given by:

$$P(x) = P(\varepsilon_v \leq y) \qquad (8-2)$$

(b) Define $X_n$ as the maximum value of $\varepsilon$ in a sample of length n i.e. $\max_{1 \leq v \leq n} \varepsilon_v$ so that

$$P(X_n \leq y) = P(\varepsilon_1 \leq y, \varepsilon_2 \leq y,..., \varepsilon_n \leq y) \qquad (8-3)$$

or

$$P(X_n \leq y) = [P(y)]^n \qquad (8-4)$$

(c) Now assume that the tail of the distribution P(y) is exponential such that

$$P(y) = 1 - \alpha e^{-y} \qquad (8-5)$$

(d)  From Equation 8-4 if ln $(\alpha n)$ is a normalizing constant

$$P(X_n \leq y + \ln (\alpha n)) = [P(y + \ln (\alpha n))]^n \qquad (8\text{-}6)$$

and from Equation 8-5

$$P(y + \ln (\alpha n)) = 1 - \alpha e^{-(y + \ln (\alpha n))} \qquad (8\text{-}7)$$

so that

$$P(X_n \leq y + \ln (\alpha n)) = [1 - \alpha e^{-(y + \ln (\alpha n))}]^n \qquad (8\text{-}8)$$

or

$$P(X_n \leq y + \ln (\alpha n)) = [1 - e^{-y}/n]^n \qquad (8\text{-}9)$$

(e)  If $n \to \infty$, then:

$$\lim_{n \to \infty} P(X_n \leq y + \ln (\alpha n)) = \lim_{n \to \infty} [1 - e^{-y}/n]^n \qquad (8\text{-}10)$$

or

$$\lim_{n \to \infty} P(X_n \leq y + \ln (\alpha n)) = e^{-e^{-y}} \qquad (8\text{-}11)$$

This is the reduced form of the cumulative probability distribution.  Substituting the expression

$$y = \alpha(x-\beta) \qquad (8\text{-}12)$$

where $\alpha$ is a concentration parameter and $\beta$ is a measure of central tendency, the cumulative probability of the type I extremal distribution becomes

$$P(x) = e^{-e^{-\alpha(x-\beta)}} \qquad (8\text{-}13)$$

and the probability density becomes

$$p(x) = \alpha e^{\{-\alpha(x-\beta)-e^{-\alpha(x-\beta)}\}} \qquad (8\text{-}14)$$

## Estimation of Parameters

*Method of Moments* - Applying the standard equation for generation of origin moments to the pdf of the reduced type I extremal distribution gives

$$\mu'_{y,r} = \int_{-\infty}^{\infty} y^r e^{-y - e^{-y}} \, dy \qquad (8\text{-}15)$$

Substituting z for $e^{-y}$ results in

$$\mu'_{y,r} = \int_0^{\infty} (-\ln z)^r \cdot z \cdot e^{-z} \cdot -\frac{1}{z} \, dz \qquad (8\text{-}16)$$

The first moment about the origin is then

**95**

$$\mu'_{y,1} = \int_0^\infty \ln z \cdot z \cdot e^{-z} \cdot \frac{dz}{z} \qquad (8\text{-}17)$$

but $\int_0^\infty z\, e^{-z}\, dz$ is $\Gamma(1)$ so that, from Kendall and Stuart (6),

$$\mu'_{y,1} = -\psi(1) = \gamma_E \qquad (8\text{-}18)$$

where $\gamma_E$ is Eulers constant, approximately 0.5772157. Reconverting to the original variate x as $x = y/\alpha + \beta$

$$\mu'_1 = \beta + \gamma_E/\alpha \qquad (8\text{-}19)$$

Similarly, Gumbel (3) has shown that the second moment about the mean, $\mu_2$, is given by

$$\mu_2 = \pi^2/(6\alpha^2) \qquad (8\text{-}20)$$

Rearranging gives expressions for the parameters $\alpha$ and $\beta$

$$\alpha = 1.2825/\sigma \qquad (8\text{-}21)$$

and

$$\beta = \mu - 0.4500\ \sigma \qquad (8\text{-}22)$$

where $\mu$ and $\sigma$ are the mean and standard deviation from the sample.

The coefficients of skew and kurtosis for the type I extremal distribution are constants at approximately 1.14 and 5.40 respectively.

*Maximum Likelihood* - The maximum likelihood solution of the type I extremal distribution was first proposed by Kimball (7) in 1946 (quoted in Gumbel (3)) but was not practical until the advent of computers.

The maximum likelihood method of estimating $\alpha$ and $\beta$ postulates that $\alpha$ and $\beta$ should be such that the probability of n individual maximum events $x_1, \ldots, x_n$ actually being observed as n annual peaks should be a maximum. The probability that $x_1$ occurs as an annual peak event is:

$$p(x_1) = \alpha e^{\{-\alpha(x_1-\beta)\ -e^{-\alpha(x_1-\beta)}\}} \qquad (8\text{-}23)$$

and for $x_2$:

$$p(x_2) = \alpha e^{\{-\alpha(x_2-\beta)\ -e^{-\alpha(x_2-\beta)}\}} \qquad (8\text{-}24)$$

now

$$p(x_1,\ldots,x_n) = p(x_1)\, p(x_2),\ldots,p(x_n) \qquad (8\text{-}25)$$

so that:

$$P(x_1,\ldots,x_n) = \alpha^n\, e^{\{-\alpha \sum_{i=1}^{n} (x_i-\beta) - \sum_{i=1}^{n} e^{-\alpha(x_i-\beta)}\}} \qquad (8\text{-}26)$$

The method of maximum likelihood then takes the logarithm of Equation 8-26, partially differentiates with respect to $\alpha$ and $\beta$ and equates to zero:

$$\ln L = n \ln \alpha - \alpha \sum_{i=1}^{n} (x_i-\beta) - \sum_{i=1}^{n} e^{-\alpha(x_i-\beta)} \qquad (8\text{-}27)$$

$$\frac{\partial \ln L}{\partial \alpha} = \frac{n}{\alpha} - \sum_{i=1}^{n} (x_i-\beta) + \sum_{i=1}^{n} (x_i-\beta)\, e^{-\alpha(x_i-\beta)} \qquad (8\text{-}28)$$

$$\frac{\partial \ln L}{\partial \beta} = n\alpha - \alpha \sum_{i=1}^{n} e^{-\alpha(x_i-\beta)} \qquad (8\text{-}29)$$

Setting Equation 8-29 equal to zero:

$$\sum_{i=1}^{n} e^{-\alpha(x_i-\beta)} = n \qquad (8\text{-}30)$$

so that:

$$e^{\alpha\beta} = n/ \sum_{i=1}^{n} e^{-\alpha x_i} \qquad (8\text{-}31)$$

or

$$\beta = \frac{1}{\alpha} \ln [n/ \sum_{i=1}^{n} e^{-\alpha x_i}] \qquad (8\text{-}32)$$

If the arithmetic mean of the series $x_1,\ldots,x_n$ is denoted by $\mu$, then Equation 8-28 can be written as:

$$\frac{\partial \ln L}{\partial \alpha} = \frac{n}{\alpha} - n(\mu-\beta) + e^{\alpha\beta} \sum_{i=1}^{n} (x_i-\beta)\, e^{-\alpha x_i} \qquad (8\text{-}33)$$

Substituting for $e^{\alpha\beta}$ from Equation 8-31

$$\frac{\partial \ln L}{\partial \alpha} = \frac{n}{\alpha} - n(\mu-\beta) + \frac{n \sum\limits_{i=1}^{n} (x_i-\beta) e^{-\alpha x_i}}{\sum\limits_{i=1}^{n} e^{-\alpha x_i}} \qquad (8-34)$$

Equating this to zero and simplifying:

$$F(\alpha) = \sum_{i=1}^{n} x_i e^{-\alpha x_i} - (\mu-1/\alpha) \sum_{i=1}^{n} e^{-\alpha x_i} = 0 \qquad (8-35)$$

Equation 8-35 cannot be solved for $\alpha$ analytically and so a Taylor's expansion has been used by Panchang (12).

$$F(\alpha_{j+1}) = F(\alpha_j + h_j) \qquad (8-36)$$

$$F(\alpha_{j+1}) = F(\alpha_j) + h_j F'(\alpha_j) \qquad (8-37)$$

where $F'(\alpha_j)$ is the first order derivative of $F(\alpha)$ with respect to $\alpha$

$$F'(\alpha) = - \sum_{i=1}^{n} x_i^2 e^{-\alpha x_i} + (\mu-1/\alpha) \sum_{i=1}^{n} x_i e^{-\alpha x_i} - \frac{1}{\alpha^2} \sum_{i=1}^{n} e^{-\alpha x_i}$$

$$(8-38)$$

and $\alpha_j$ and $\alpha_{j+1}$ are successive approximations to $\alpha$. The procedure adopted by Panchang is to estimate $\alpha_1$ from the method of moments (18) (see earlier discussion). By evaluating $F(\alpha_1)$ and $F'(\alpha_1)$ from Equations 8-35 and 8-38 then:

$$h_1 = - F(\alpha_1)/F'(\alpha_1) \qquad (8-39)$$

and

$$\alpha_2 = \alpha_1 + h_1 \qquad (8-40)$$

This procedure is repeated until sufficiently small value of $F(\alpha_j)$ is obtained when $\beta$ can be obtained from Equation 8-32.

In most cases only 3 or 4 steps will be required.

Samuelsson (13) has described a similar procedure using different first estimates of $\alpha$ and $\beta$ and suggests that the true parameter values can be estimated to within 1% in only 3 iterations.

Other methods of deriving the maximum likelihood parameter estimates are available (3), (17) but these are generally more complex. Leese (9) has described the modifications to the maximum likelihood equations which are needed to accomodate missing data and historic flood records.

*Other Methods* - A simple approximation has been used by
Verma and Advani (15) to estimate the parameters $\alpha$ and $\beta$. If
$x_{max}$ is the largest event in a series of n maxima and $x_{min}$ is
the smallest event, then the reduced events $y_{max}$ and $y_{min}$ are
defined as

$$y_{max} = \alpha(x_{max} - \beta) \qquad (8-41)$$

and

$$y_{min} = \alpha(x_{min} - \beta) \qquad (8-42)$$

By taking the probability of exceedence of the largest event,
$x_{max}$, as 1/n and the probability of exceedence of the smallest
event, $x_{min}$, as 1/1.01, then the following expressions apply

$$y_{max} = - \ln (- \ln (1-1/n)) \qquad (8-43)$$

and

$$y_{min} = - \ln (- \ln (1-1/1.01)) \qquad (8-44)$$

Evaluation of Equations 8-43 and 8-44 and substitution into
Equations 8-41 and 8-42 will, by simultaneous solution, produce
rough estimates of $\alpha$ and $\beta$. By using the expansion of ln and
neglecting terms above second order, Verma and Advani have
produced expressions for $\alpha$ and $\beta$.

Yevjevich (18) has described a process by which estimates
of $\alpha$ and $\beta$ can be determined graphically. If, in Equation 8-13,
$x = \beta$, then

$$P(\beta) = e^{-1} = 0.368 \qquad (8-45)$$

Plot the event magnitudes, x, versus return period, $T = (n + 1)/m$,
on graph paper with a double exponential scale and fit a straight
line through the plotted points. By entering the graph at
$T = 1/P(\beta) = 2.717$ the value of $\beta$ is determined. The slope of
the best fitting straight line is then equal to $1/\alpha$. This method
is very easy to apply but its accuracy is not to be compared with
the method of maximum likelihood.

Lowery and Nash (10) have compared various methods of estimating the parameters of the type I extremal distribution. They
recognized the greater efficiency of the maximum likelihood technique but recommended moments because of the method's simplicity
and lack of bias.

## Frequency Factor

From the cumulative probability distribution (Equation 8-11)
the expression relating the reduced variable, y, to return period,
T, is

$$y_T = - \ln \left( - \ln \left( (T-1)/T \right) \right) \tag{8-46}$$

For convenience, Table 8-1 gives values of the reduced variable $y_T$ for some common return periods.

Table 8-1

Values of the Reduced Variable, $y_T$, of the Type I Extremal Distribution for Some Commonly Used Return Periods, T

| Return Period T | Reduced Variable $y_T$ |
|---|---|
| 2 | 0.3665 |
| 5 | 1.4999 |
| 10 | 2.2504 |
| 20 | 2.9702 |
| 50 | 3.9019 |
| 100 | 4.6001 |

If the n recorded events are placed in order of magnitude so that m = 1 for the largest event and m = n for the smallest event then $T = (n+1)/m$ and Equation 8-46 can be written as

$$y_m = - \ln \left[ - \ln \left\{ (n+1-m)/(n+1) \right\} \right] \tag{8-47}$$

If the mean, $\mu_y$, and the variance, $\sigma_y^2$, of the series $y_m$, m = 1, 2,...,n, are computed from the reduced sample as:

$$\mu_y = \sum_{m=1}^{n} y_m/n \tag{8-48}$$

and

$$\sigma_y^2 = \sum_{m=1}^{n} (y_m - \mu_y)^2/n \tag{8-49}$$

and if $\mu$ and $\sigma^2$ are the mean and variance of the recorded events, then the parameters $\alpha$ and $\beta$ can be defined (5) as:

$$\alpha = \sigma_y/\sigma \tag{8-50}$$

and

$$\beta = \mu - \mu_y/\alpha \tag{8-51}$$

Now, introducing these relationships into the equation for the reduced variate y

$$y_m = \alpha(x-\beta) \tag{8-52}$$

and rearranging for x, gives:

$$x = \mu + (y_m - \mu_y) \sigma/\sigma_y \tag{8-53}$$

Comparing Equation 8-53 with the general frequency equation (e.g. Equation 3-30) it is apparent that for the type I extremal distribution the frequency factor, K, is defined as:

$$K = \frac{y_m - \mu_y}{\sigma_y} \tag{8-54}$$

## Table 8-2

### Mean and Standard Deviation of Order Statistics, m/(n+1), for Various Sample Sizes, n

| Sample Size n | Mean $\mu_y$ | Standard Deviation $\sigma_y$ |
|---|---|---|
| 10 | 0.4952 | 0.9496 |
| 15 | 0.5128 | 1.0206 |
| 20 | 0.5236 | 1.0628 |
| 25 | 0.5309 | 1.0914 |
| 30 | 0.5362 | 1.1124 |
| 35 | 0.5403 | 1.1285 |
| 40 | 0.5436 | 1.1413 |
| 45 | 0.5463 | 1.1518 |
| 50 | 0.5485 | 1.1607 |
| 55 | 0.5504 | 1.1682 |
| 60 | 0.5521 | 1.1747 |
| 65 | 0.5535 | 1.1803 |
| 70 | 0.5548 | 1.1854 |
| 75 | 0.5559 | 1.1898 |
| 80 | 0.5569 | 1.1938 |
| 85 | 0.5578 | 1.1974 |
| 90 | 0.5586 | 1.2007 |
| 95 | 0.5593 | 1.2037 |
| 100 | 0.5600 | 1.2065 |

Since $\mu_y$ and $\sigma_y$ are functions of the sample size only, they can be tabulated. Table 8-2 shows values of $\mu_y$ and $\sigma_y$ for some typical values of n. Alternatively, for a predetermined set of return periods and samples sizes the frequency factor, K, can be tabulated (5) as in Table 8-3.

As an example, for a sample size of 55, Table 8-2 gives values of $\mu_y$ and $\sigma_y$ of 0.5504 and 1.1681 respectively. For a 100 year return period, Table 8-1 or Equation 8-46 gives the reduced variable, $y_m$, as 4.6001 so that, from Equation 8-54, the frequency factor, K, is 3.4670. Alternatively this figure can be found directly from Table 8-3. To determine the 100 year event magnitude it is then only necessary to estimate the population mean, $\mu$ and standard deviation, $\sigma$ from the 55 recorded events and substitute, $\mu$, K and $\sigma$ in the general frequency equation.

Weiss (16) has devised a convenient nomogram for performing graphically the solution of Equation 8-53 given the sample mean and standard deviation. Shown here as Figure 8-1 this nomogram is entered on the left hand side with the required return period, T. From the intersection of the horizontal line through T with the slanting line through the appropriate sample size, n, draw a vertical line to intersect the sloping line corresponding to the sample standard deviation, $\sigma$. From this second intersection draw a horizontal line to cut the right hand edge of the diagram at the value of $K\sigma$, the numerical value to be added to the mean, $\mu$, to give the required event magnitude, x. Examples of the use of this nomogram are given in Weiss (16) and Kendall(5).

Chow (1) has considered the type I extremal distribution as a special case of the lognormal distribution for which the coefficient of skew, $\gamma_1$, is a constant at 1.1396. From the cumulative probability distribution (Equation 8-13) the event magnitude $x_T$ is related to return period, T, by:

$$x_T = \beta - \frac{1}{\alpha} \ln (- \ln (1-1/T)) \qquad (8\text{-}55)$$

where T is $1/[1-P(x)]$. Substituting the moments solutions for $\alpha$ and $\beta$ leads to:

$$x_T = \mu - \{0.45 + 0.7797 \ln (- \ln [1-1/T])\} \sigma \qquad (8\text{-}56)$$

Comparison with the standard frequency equation shows that the frequency factor for the type I extremal distribution by this method is:

$$K = - \{0.45 + 0.7797 \ln (- \ln [1-1/T])\} \qquad (8\text{-}57)$$

The last line of Table 8-3 gives the value of the frequency factor determined from this equation. These values correspond to the asymptotic results of using Equation 8-54 as $n \to \infty$.

## Standard Error

*Method of Moments* - The general equation for the standard error of estimate of a 2-parameter distribution has been given earlier as:

$$s_T^2 = \frac{\mu_2}{n} \{1 + K\gamma_1 + \frac{K^2}{4} [\gamma_2 - 1]\} \qquad (8\text{-}58)$$

Table 8-3

Frequency Factor for Type I Extremal Distribution

| Sample Size n | Cumulative Probability, P, % | | | | | |
|---|---|---|---|---|---|---|
| | 50 | 80 | 90 | 95 | 98 | 99 |
| | Corresponding Return Period, T, Years | | | | | |
| | 2 | 5 | 10 | 20 | 50 | 100 |
| 10 | -0.1355 | 1.0580 | 1.8483 | 2.6063 | 3.5874 | 4.3227 |
| 15 | -0.1434 | 0.9672 | 1.7025 | 2.4078 | 3.3208 | 4.0049 |
| 20 | -0.1478 | 0.9187 | 1.6248 | 2.3020 | 3.1787 | 3.8356 |
| 25 | -0.1506 | 0.8879 | 1.5754 | 2.2350 | 3.0886 | 3.7284 |
| 30 | -0.1526 | 0.8664 | 1.5410 | 2.1881 | 3.0257 | 3.6534 |
| 35 | -0.1540 | 0.8504 | 1.5154 | 2.1532 | 2.9789 | 3.5976 |
| 40 | -0.1552 | 0.8379 | 1.4954 | 2.1261 | 2.9425 | 3.5543 |
| 45 | -0.1561 | 0.8279 | 1.4794 | 2.1044 | 2.9133 | 3.5195 |
| 50 | -0.1568 | 0.8197 | 1.4663 | 2.0865 | 2.8892 | 3.4908 |
| 55 | -0.1574 | 0.8128 | 1.4552 | 2.0714 | 2.8690 | 3.4667 |
| 60 | -0.1580 | 0.8069 | 1.4458 | 2.0586 | 2.8518 | 3.4461 |
| 65 | -0.1584 | 0.8018 | 1.4376 | 2.0475 | 2.8368 | 3.4284 |
| 70 | -0.1588 | 0.7974 | 1.4305 | 2.0377 | 2.8238 | 3.4128 |
| 75 | -0.1592 | 0.7934 | 1.4242 | 2.0291 | 2.8122 | 3.3991 |
| 80 | -0.1595 | 0.7900 | 1.4185 | 2.0215 | 2.8020 | 3.3868 |
| 85 | -0.1597 | 0.7868 | 1.4135 | 2.0146 | 2.7928 | 3.3758 |
| 90 | -0.1600 | 0.7840 | 1.4090 | 2.0084 | 2.7844 | 3.3659 |
| 95 | -0.1602 | 0.7814 | 1.4048 | 2.0028 | 2.7769 | 3.3569 |
| 100 | -0.1604 | 0.7791 | 1.4011 | 1.9977 | 2.7700 | 3.3487 |

Corresponding Frequency Factors from Equation 8.57

| | | | | | | |
|---|---|---|---|---|---|---|
| | -0.1643 | 0.7194 | 1.3046 | 1.8658 | 2.5923 | 3.1367 |

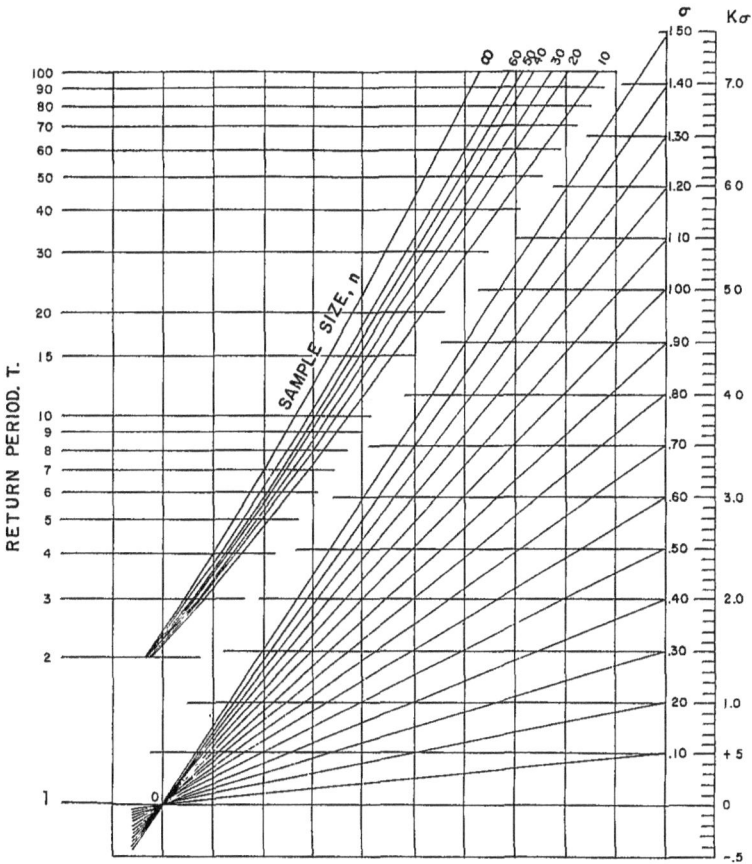

Figure 8-1.   Nomogram for use with type I extremal distribution.
(after Weiss (16)).

From the method of moments,

$$\gamma_1 = 1.1396 \tag{8-59}$$

$$\gamma_2 = \mu_4/\mu_2^2 = 5.4002 \tag{8-60}$$

So that:

$$S_T^2 = \frac{\sigma^2}{n} [1 + 1.1396K + 1.1000K^2] \tag{8-61}$$

Taking the square root and simplifying to:

$$S_T = \delta \sqrt{\frac{\sigma^2}{n}} \tag{8-62}$$

values of $\delta$ depend only on K and can thus be predetermined and tabulated for typical values of return period, T, and sample size, n (see Table 8-4). Kaczmarek (4) and Kendall (5) have given similar tables, although as pointed out by Lowery and Nash (10) there is a computational error in the table of Kaczmarek.

Using the same example as before, the 100 year return period event computed from a 55 year sample will have an $\delta$ value of 4.265. Knowing the standard deviation of the recorded events, $\sigma$, the standard error, $S_T$, is computed from Equation 8-62 and the 95% confidence limits are applied as $x_T \pm 1.96\ S_T$.

*Maximum Likelihood* - Since the coefficient of skew is a constant for the type I extremal distribution the maximum likelihood estimate of the standard error of estimate is

$$S_T^2 = \left(\frac{\partial x}{\partial \alpha}\right)^2 \text{var } \alpha + \left(\frac{\partial x}{\partial \beta}\right)^2 \text{var } \beta + 2 \frac{\partial x}{\partial \alpha} \frac{\partial x}{\partial \beta} \text{cov } (\alpha,\beta) \qquad (8\text{-}63)$$

and since from 8-52⁻

Table 8-4

Paramater $\delta$ for Use in Standard Error
of Type I Extremal Distribution

| Sample Size n | Cumulative Probability, P, % | | | | | |
|---|---|---|---|---|---|---|
| | 50 | 80 | 90 | 95 | 98 | 99 |
| | Corresponding Return Period, T, Years | | | | | |
| | 2 | 5 | 10 | 20 | 50 | 100 |
| 10 | 0.9305 | 1.8539 | 2.6199 | 3.3826 | 4.3869 | 5.1459 |
| 15 | 0.9269 | 1.7695 | 2.4756 | 3.1814 | 4.1127 | 4.8174 |
| 20 | 0.9250 | 1.7249 | 2.3990 | 3.0745 | 3.9670 | 4.6427 |
| 25 | 0.9238 | 1.6968 | 2.3506 | 3.0069 | 3.8747 | 4.5320 |
| 30 | 0.9229 | 1.6772 | 2.3169 | 2.9597 | 3.8103 | 4.4548 |
| 35 | 0.9223 | 1.6627 | 2.2919 | 2.9247 | 3.7624 | 4.3974 |
| 40 | 0.9218 | 1.6514 | 2.2725 | 2.8975 | 3.7252 | 4.3527 |
| 45 | 0.9214 | 1.6424 | 2.2569 | 2.8756 | 3.6954 | 4.3169 |
| 50 | 0.9211 | 1.6350 | 2.2441 | 2.8577 | 3.6708 | 4.2874 |
| 55 | 0.9208 | 1.6288 | 2.2333 | 2.8426 | 3.6502 | 4.2627 |
| 60 | 0.9206 | 1.6235 | 2.2241 | 2.8297 | 3.6326 | 4.2415 |
| 65 | 0.9204 | 1.6189 | 2.2162 | 2.8186 | 3.6173 | 4.2232 |
| 70 | 0.9202 | 1.6149 | 2.2093 | 2.8089 | 3.6040 | 4.2073 |
| 75 | 0.9201 | 1.6114 | 2.2032 | 2.8003 | 3.5923 | 4.1931 |
| 80 | 0.9199 | 1.6083 | 2.1977 | 2.7926 | 3.5818 | 4.1806 |
| 85 | 0.9198 | 1.6055 | 2.1928 | 2.7858 | 3.5724 | 4.1693 |
| 90 | 0.9197 | 1.6030 | 2.1884 | 2.7796 | 3.5639 | 4.1591 |
| 95 | 0.9196 | 1.6007 | 2.1844 | 2.7739 | 3.5562 | 4.1498 |
| 100 | 0.9195 | 1.5986 | 2.1808 | 2.7688 | 3.5491 | 4.1414 |

$$x_T = \beta + y_T/\alpha \qquad (8\text{-}64)$$

$$\frac{\partial X}{\partial \alpha} = - y_T/\alpha^2 \qquad (8\text{-}65)$$

$$\frac{\partial X}{\partial \beta} = 1 \qquad (8\text{-}66)$$

then

$$S_T^2 = \frac{y_T^2}{\alpha^4} \text{ var } \alpha + \text{var } \beta - 2 \frac{y_T}{\alpha^2} \text{ cov } (\alpha,\beta) \qquad (8\text{-}67)$$

From the maximum likelihood equation (Equation 8-27) the partial derivatives are as follows:

$$\frac{\partial^2 \ln L}{\partial \alpha^2} = - \frac{n}{\alpha^2} - \sum_{i=1}^{n} (x_i-\beta)^2 \, e^{-\alpha(x_i-\beta)} \qquad (8\text{-}68)$$

$$\frac{\partial^2 \ln L}{\partial \beta^2} = - \alpha^2 \sum_{i=1}^{n} e^{-\alpha(x_i-\beta)} \qquad (8\text{-}69)$$

$$\frac{\partial^2 \ln L}{\partial \alpha \partial \beta} = n + \alpha \sum_{i=1}^{n} (x_i-\beta) \, e^{-\alpha(x_i-\beta)} \qquad (8\text{-}70)$$

Substituting from Equation 8-30

$$\frac{\partial^2 \ln L}{\partial \beta^2} = - n\alpha^2 \qquad (8\text{-}71)$$

and from Kimball (8)

$$\frac{\partial^2 \ln L}{\partial \alpha^2} = - \frac{1.8237 \, n}{\alpha^2} \qquad (8\text{-}72)$$

$$\frac{\partial^2 \ln L}{\partial \alpha \partial \beta} = - .4228 \, n \qquad (8\text{-}73)$$

Inverting the likelihood or information matrix leads to the expressions for variances and covariance:

$$\text{var } \alpha = \frac{0.6079 \, \alpha^2}{n} \qquad (8\text{-}74)$$

$$\text{var } \beta = \frac{0.2570}{n} \qquad (8\text{-}75)$$

and

$$\text{cov } (\alpha,\beta) = \frac{1.1086}{n\alpha^2} \qquad (8\text{-}76)$$

Substituting these values in Equation 8-67 leads to:

$$S_T^2 = \frac{1}{n\alpha^2} [1.1086 + 0.5140\ y_T + 0.6079\ y_T^2] \qquad (8\text{-}77)$$

which is equivalent to the expression given by Kimball (8):

$$S_T^2 = \frac{1}{n\alpha^2} [1 + (1 - \gamma_E + y_T)^2 / (\pi^2/6)] \qquad (8\text{-}78)$$

where $\gamma_E$ is the Euler constant of approximately 0.5772. Simplifying Equations 8-77 and 8-78 to

$$S_T = \delta_\ell / \sqrt{n\alpha^2} \qquad (8\text{-}79)$$

Table 8-5 gives values of $\delta_\ell$ for some values of T. For a comparison with moment estimates, Equation 8-21 was used to replace $\alpha$ with the sample standard deviation. The last line in Table 8-5 shows the resulting values of $\delta_m$ in Equation 8-80.

$$S_T = \delta_m \sqrt{\frac{\sigma^2}{n}} \qquad (8\text{-}80)$$

Table 8-5

| | | Cumulative Probability, P, % | | | |
|---|---|---|---|---|---|
| 50 | 80 | 90 | 95 | 98 | 99 |
| | | Corresponding Return Period, T, Years | | | |
| 2 | 5 | 10 | 20 | 50 | 100 |
| | | Value of $\delta_\ell$ in Equation 8-79 | | | |
| 1.1742 | 1.8020 | 2.3118 | 2.8282 | 3.5171 | 4.0420 |
| | | Value of $\delta_m$ in Equation 8-80 | | | |
| 0.7139 | 1.0956 | 1.4055 | 1.7195 | 2.1383 | 2.4574 |

Comparison with Table 8-4 shows that the values of $\delta_\ell$, the moments solution, are larger than $\delta_m$, the likelihood solution.

*Other Methods* - Using the relationship

$$\mathrm{cov}\,(\mu,\sigma) = \rho(\mathrm{var}\,\mu)^{\frac{1}{2}} (\mathrm{var}\,\sigma)^{\frac{1}{2}} \qquad (8\text{-}81)$$

where $\rho$ is the simple linear correlation coefficient between $\mu$ and $\sigma$, Nash and Amorocho (11) found that $\rho$ has a mean value of 0.56 and a standard deviation of 0.02. Substitution in the standard error relationship (Equation 8-58) yields

$$S_T^2 = \frac{\sigma^2}{n} (1 + 1.18K + 1.1K^2) \qquad (8\text{-}82)$$

which approaches the theoretical relationship of Equation 8-61.

Dalrymple (2) has given the following equation for the standard error of the reduced variate, y, of the type I extremal distribution

$$S' = \frac{e^y}{\sqrt{n}} \sqrt{\frac{1}{T - 1}} \qquad (8\text{-}83)$$

The confidence interval for the reduced variable is then computed as

$$y_T \pm t\, S' \qquad (8\text{-}84)$$

Dalrymple used this confidence region to determine the homogeneous hydrologic region in his index flood method of regional flood frequency analysis (see Chapter 12).

## References

1.  Chow, V. T., 1954, The Log-Probability Law and its Engineering Applications, Proc. ASCE, Vol. 80, pp. 1-25.

2.  Dalrymple, T., 1960, Flood Frequency Analysis, USGS Water Supply Paper 1543-A.

3.  Gumbel, E. J., 1958, Statistics of Extremes, Columbia University Press.

4.  Kaczmarek, Z., 1958, Efficiency of the Estimation of Floods with a Given Return Period, Proc. AISH General Assembly of Toronto, 1957, Vol. III, 144-159.

5.  Kendall, G. R., 1959, Statistical Analysis of Extreme Values, Proc. Hydrology Symp. No. 1, Spillway Design Floods, NRC, Ottawa, pp. 54-78.

6.  Kendall, M. G. and A. Stuart, 1963, The Advanced Theory of Statistics, Vol. I, Griffin, London.

7.  Kimball, B. F., 1946, Sufficient Statistical Estimation Functions for the Parameters of the Distribution of Maximum Values, Ann. Math. Stats., Vol. 17, pp. 299-309.

8.  Kimball, B. F., 1949, An Approximation to the Sampling Variances of an Estimated Maximum Value of Given Frequency Based on fit of Doubly Exponential Distribution of Maximum Values, Ann. Math. Stats., Vol. 20, pp. 110-113.

9.  Leese, M. N., 1973, The Use of Censored Data in Estimating T-Year Floods, Proceedings of the UNESCO/WMO/IAHS

Symposium on the Design of Water Resources Projects
with Inadequate Data, Madrid, Vol. 1, pp. 235-247.

10. Lowery, M. D. and J. E. Nash, 1970, A Comparison of Methods
    of Fitting the Double Exponential Distribution, Journal
    of Hydrology, Vol. 10, No. 3, pp. 259-275.

11. Nash, J. E. and J. Amorocho, 1966, The Accuracy of the Pre-
    diction of Floods of High Return Period, Wat. Res.
    Res., Vol. 2, No. 2, pp. 191-198.

12. Panchang, G. M., 1967, Improved Precision of Future High
    Floods, Proc. Symp. on Floods and their Computations,
    Leningrad, pp. 51-59.

13. Samuelsson, B., 1972, Statistical Interpretation of Hydro-
    meteorological Extreme Values, Nordic Hydrology, Vol.
    3, No. 4, pp. 199-214.

14. Todorovic, P. and J. Rousselle, 1971, Some Problems of
    Flood Analysis, Wat. Res. Res., Vol. 7, No. 5, pp.
    1144-1150.

15. Verma, R. D. and R. M. Advani, 1973, Flood Control Planning
    with Inadequate Hydrologic Data, Proceedings of Second
    International Symposium in Hydrology, Water Resources
    Publications, Fort Collins, Colorado 80521, pp. 259-268.

16. Weiss, L. L., 1955, A Nomogram Based on the Theory of
    Extreme Values for Determining Values for Various
    Return Periods, Monthly Weather Review, Vol. 83, No. 3,
    pp. 69-71.

17. WMO, 1969, Estimation of Maximum Floods, Technical Note No.
    98, WMO No. 233.TP.126, Geneva.

18. Yevjevich, V., 1972, Probability and Statistics in Hydrology,
    Water Resources Publications, Fort Collins, Colorado.

```
      PROGRAM T1E(INPUT,OUTPUT,TAPE5=INPUT,TAPE6=OUTPUT)             A   1
C                                                                    A   2
C                                                                    A   3
C     COMPUTES METHOD OF MOMENTS AND MAXIMUM LIKELIHOOD ESTIMATES FOR A   4
C     T YEAR EVENTS AND STANDARD ERRORS FOR TYPE 1 EXTREMAL          A   5
C     DISTRIBUTION                                                   A   6
C     INPUT                                                          A   7
C     TITLE                                                          A   8
C     N NUMBER OF ANNUAL MAXIMUM EVENTS                              A   9
C     X SERIES OF EVENTS                                             A  10
C                                                                    A  11
C                                                                    A  12
      REAL M1,M2,K                                                   A  13
      DIMENSION T(6), X(100)                                         A  14
      DIMENSION XT(6), SX(6), TITLE(80)                              A  15
      T(1)=2.                                                        A  16
      T(2)=5.                                                        A  17
      T(3)=10.                                                       A  18
      T(4)=20.                                                       A  19
      T(5)=50.                                                       A  20
      T(6)=100.                                                      A  21
      READ (5,9) TITLE                                               A  22
      READ (5,10) N                                                  A  23
      XN=N                                                           A  24
      READ (5,11) (X(I),I=1,N)                                       A  25
      A=0.0                                                          A  26
      B=0.0                                                          A  27
      C=0.0                                                          A  28
      DO 1 I=1,N                                                     A  29
      A=A+X(I)                                                       A  30
      B=B+X(I)**2                                                    A  31
      C=C+X(I)**3                                                    A  32
1     CONTINUE                                                       A  33
      M1=A/XN                                                        A  34
      M2=(B/XN)-(A/XN)**2                                            A  35
      M2=M2*XN/(XN-1.0)                                              A  36
      M3=(C/XN)+2.0*M1**3-3.0*M1*(B/XN)                              A  37
      SKEW=M3/(M2**1.5)                                              A  38
      ALPHA=1.2825/(SQRT(M2))                                        A  39
      BETA=M1-0.45*SQRT(M2)                                          A  40
      A=0.0                                                          A  41
      B=0.0                                                          A  42
      DO 2 I=1,N                                                     A  43
      XI=I                                                           A  44
      XN=N                                                           A  45
      Y=-ALOG(-ALOG((XN+1.0-XI)/(XN+1.0)))                           A  46
      A=A+Y                                                          A  47
      B=B+Y**2                                                       A  48
2     CONTINUE                                                       A  49
      YBAR=A/XN                                                      A  50
      YSTD=SQRT((B/XN)-YBAR**2)                                      A  51
      DO 3 J=1,6                                                     A  52
      YM=-ALOG(-ALOG((T(J)-1.0)/T(J)))                               A  53
      K=(YM-YBAR)/YSTD                                               A  54
      XT(J)=M1+K*SQRT(M2)                                            A  55
      DELTA=1.0+1.139547093*K+1.100000027*K**2                       A  56
      SX(J)=SQRT(M2*DELTA/XN)                                        A  57
3     CONTINUE                                                       A  58
      WRITE (6,12) TITLE                                             A  59
      WRITE (6,13)                                                   A  60
      WRITE (6,20) ALPHA,M1                                          A  61
      WRITE (6,21) BETA,M2                                           A  62
      WRITE (6,22) SKEW                                              A  63
      WRITE (6,25)                                                   A  64
      WRITE (6,14)                                                   A  65
      WRITE (6,15) (XT(J),J=1,6)                                     A  66
      WRITE (6,16) (SX(J),J=1,6)                                     A  67
      WRITE (6,17)                                                   A  68
      WRITE (6,18)                                                   A  69
      ICOUNT=0                                                       A  70
      AML=ALPHA                                                      A  71
4     ICOUNT=ICOUNT+1                                                A  72
      A=1.0/(AML**2)                                                 A  73
      B=M1-1.0/AML                                                   A  74
      C=0.0                                                          A  75
      D=0.0                                                          A  76
      E=0.0                                                          A  77
      DO 5 I=1,N                                                     A  78
      TEMP=EXP(-AML*X(I))                                            A  79
      C=C+TEMP                                                       A  80
```

109

```
                  D=D+TEMP*X(I)                                               A  81
                  E=E+TEMP*X(I)**2                                            A  82
    5             CONTINUE                                                    A  83
                  FCN=D-B*C                                                   A  84
                  FPN=B*D-E-A*C                                               A  85
                  AS=AML-(FCN/FPN)                                            A  86
                  WRITE (6,19) ICOUNT,AS,FCN                                  A  87
                  DELTA=ABS(0.0000001*AS)                                     A  88
                  IF (ABS(AS-AML).LT.DELTA) GO TO 6                           A  89
                  IF (ICOUNT.GT.25) GO TO 8                                   A  90
                  AML=AS                                                      A  91
                  GO TO 4                                                     A  92
    6             CONTINUE                                                    A  93
                  ALPHA=AS                                                    A  94
                  BETA=(1.0/ALPHA)*ALOG(XN/C)                                 A  95
                  M2=1.2825/ALPHA                                            A  96
                  MI=BETA+0.45*M2                                            A  97
                  M2=M2**2                                                    A  98
                  DO 7 J=1,6                                                  A  99
                  YM=-ALOG(-ALOG(1.0-1.0/T(J)))                               A 100
                  XT(J)=BETA-YM/ALPHA                                         A 101
                  SX(J)=SQRT((1.1086+0.5140*YM+0.6079*YM**2)/(XN*ALPHA**2))   A 102
    7             CONTINUE                                                    A 103
                  WRITE (6,23)                                                A 104
                  WRITE (6,20) ALPHA,MI                                       A 105
                  WRITE (6,24) BETA,M2                                        A 106
                  WRITE (6,14)                                                A 107
                  WRITE (6,15) (XT(J),J=1,6)                                  A 108
                  WRITE (6,16) (SX(J),J=1,6)                                  A 109
    8             CONTINUE                                                    A 110
                  STOP                                                        A 111
    C                                                                         A 112
    9             FORMAT (80A1)                                               A 113
   10             FORMAT (I5)                                                 A 114
   11             FORMAT (8F10.0)                                             A 115
   12             FORMAT (1H1,/,80A1,//,26X,28HTYPE 1 EXTREMAL DISTRIBUTION,/) A 116
   13             FORMAT (31X,17HMETHOD OF MOMENTS,//)                        A 117
   14             FORMAT (3X,7HT,YEARS,4X,1H2,11X,1H5,10X,2H10,10X,2H20,10X,2H50,9X, A 118
              1   3H100,/)                                                     A 119
   15             FORMAT (3X,1HX,3X,6E12.5,/,4X,1HT)                          A 120
   16             FORMAT (3X,1HS,3X,6E12.5,/,4X,1HT,//)                       A 121
   17             FORMAT (25X,28HMAXIMUM LIKELIHOOD PROCEDURE,//)             A 122
   18             FORMAT (21X,5HTRIAL,11X,1HA,11X,4HF(A),/)                   A 123
   19             FORMAT (22X,I2,8X,E12.5,1X,E12.5)                           A 124
   20             FORMAT (9X,5HALPHA,5X,E12.5,14X,4HM1  ,6X,E12.5)            A 125
   21             FORMAT (9X,5HBETA ,5X,E12.5,14X,4HM2  ,6X,E12.5)            A 126
   22             FORMAT (45X,4HSKEW,6X,E12.5,/)                              A 127
   23             FORMAT (//)                                                 A 128
   24             FORMAT (9X,5HBETA ,5X,E12.5,14X,4HM2  ,6X,E12.5,/)          A 129
   25             FORMAT (3X,67HNOTE - FOR GOOD USE OF THIS DISTRIBUTION SKEW SHOULD A 130
              1   BE AROUND 1.13,/)                                           A 131
                  END                                                         A 132
```

```
10 '
20 ' Program T1E
30 ' Copyright G.W. Kite 1986
40 '
50 ' Compute method of moments and maximum likelihood estimates
60 ' for T year events and standard errors
70 ' for a Type I Extremal Distribution.
80 '
90 DIM T#(6),X#(250),XT#(6),SX#(6)
100 FORM$="   ##.###^^^^"
110 DATA 2#,5#,10#,20#,50#,100#
120 FOR I%=1 TO 6
130 READ T#(I%)
140 NEXT I%
150 T$="Type I Extremal Distribution"
160 IERR%=0
170 GOSUB 1800
180 IF IERR% = 1 GOTO 1780
190 FLAG%=0:IF YN$ = "Y" OR YN$ = "y" THEN FLAG%=1
200 NREC%=0
210 OPEN FILE$ FOR INPUT AS #1
220 LINE INPUT#1,TITLE$
230 IF EOF(1) GOTO 280
240 NREC%=NREC%+1
250 LINE INPUT#1,I$
260 X#(NREC%)=VAL(MID$(I$,5,12))
270 GOTO 230
280 CLOSE#1
290 NREC#=NREC%
300 A%=LEN(TITLE$)
310 B%=(80-A%)/2
320 U$=STRING$(A%,205)
330 M$="Method of Moments"
340 E%=LEN(M$)
350 F%=(80-E%)/2
360 W$=STRING$(E%,205)
370 CLS
380 PRINT TAB(B%) TITLE$:PRINT TAB(B%) U$
390 PRINT TAB(D%) T$:PRINT TAB(D%) V$
400 PRINT TAB(F%) M$:PRINT TAB(F%) W$
410 IF FLAG% <> 1 GOTO 460
420 LPRINT CHR$(12):LPRINT:LPRINT
430 LPRINT TAB(B%) TITLE$:LPRINT TAB(B%) U$
440 LPRINT:LPRINT TAB(D%) T$:LPRINT TAB(D%) V$
450 LPRINT:LPRINT TAB(F%) M$:LPRINT TAB(F%) W$
460 A#=0#:B#=0#:C#=0#
470 FOR I%=1 TO NREC%
480 A#=A#+X#(I%)
490 B#=B#+X#(I%)*X#(I%)
500 C#=C#+X#(I%)*X#(I%)*X#(I%)
510 NEXT I%
520 M1#=A#/NREC#
530 M2#=(B#/NREC#)-M1#*M1#
540 M2#=M2#*NREC#/(NREC#-1#)
550 M3#=(C#/NREC#)+2#*M1#*M1#*M1#-3#*M1#*(B#/NREC#)
560 SKEW#=M3#/(M2#^1.5#)
570 ALPHA#=1.2825#/(SQR(M2#))
580 BETA#=M1#-.45#*SQR(M2#)
590 A#=0#:B#=0#
600 FOR I% = 1 TO NREC%
610 I#=I%
620 Y#=-LOG(-LOG((NREC#+1#-I#)/(NREC#+1#)))
630 A#=A#+Y#
640 B#=B#+Y#*Y#
650 NEXT I%
660 YBAR#=A#/NREC#
670 YSTD#=SQR((B#/NREC#)-YBAR#*YBAR#)
680 PRINT
690 PRINT TAB(20) "Mean is                  ";:PRINT USING FORM$;M1#
700 PRINT TAB(20) "Variance is              ";:PRINT USING FORM$;M2#
710 PRINT TAB(20) "Coefficient of skew is ";:PRINT USING FORM$;SKEW#
720 PRINT TAB(20) "Parameter Alpha is     ";:PRINT USING FORM$;ALPHA#
730 PRINT TAB(20) "Parameter Beta is      ";:PRINT USING FORM$;BETA#
740 PRINT TAB(2) "NOTE: For good use of the T1E the coeff. of skew should be clo
se to 1.13"
750 IF FLAG% <> 1 GOTO 840
760 LPRINT
770 LPRINT TAB(20) "Mean is                  ";:LPRINT USING FORM$;M1#
780 LPRINT TAB(20) "Variance is              ";:LPRINT USING FORM$;M2#
790 LPRINT TAB(20) "Coefficient of skew is ";:LPRINT USING FORM$;SKEW#
```

```
800 LPRINT TAB(20) "Parameter Alpha is    ";:LPRINT USING FORM$;ALPHA#
810 LPRINT TAB(20) "Parameter Beta is     ";:LPRINT USING FORM$;BETA#
820 LPRINT
830 LPRINT TAB(2) "NOTE: For good use of the T1E the coeff. of skew should be cl
ose to 1.13"
840 FOR J%=1 TO 6
850 YM#=-LOG(-LOG((T#(J%)-1#)/T#(J%)))
860 K#=(YM#-YBAR#)/YSTD#
870 XT#(J%)=M1#+K#*SQR(M2#)
880 DELTA#=1#+1.139547093#*K#+1.100000027#*K#*K#
890 SX#(J%)=SQR(M2#*DELTA#/NREC#)
900 NEXT J%
910 ROW$=CHR$(218)+STRING$(72,196)
920 VV$=CHR$(179)
930 PRINT:PRINT TAB(2) "T, years  2           5          10          20
   50        100":PRINT TAB(5) ROW$
940 PRINT TAB(2) "X " VV$;:FOR J%=1 TO 6:PRINT USING FORM$;XT#(J%);:NEXT J%
950 PRINT TAB(2) " t " VV$
960 PRINT TAB(2) "S " VV$;:FOR J%=1 TO 6:PRINT USING FORM$;SX#(J%);:NEXT J%
970 PRINT TAB(2) " t " VV$
980 IF FLAG% <> 1 GOTO 1040
990 LPRINT:LPRINT TAB(2) "T, years  2           5          10          20
   50        100":LPRINT TAB(5) ROW$
1000 LPRINT TAB(2) "X " VV$;:FOR J%=1 TO 6:LPRINT USING FORM$;XT#(J%);:NEXT J%
1010 LPRINT TAB(2) " t " VV$
1020 LPRINT TAB(2) "S " VV$;:FOR J%=1 TO 6:LPRINT USING FORM$;SX#(J%);:NEXT J%
1030 LPRINT TAB(2) " t " VV$
1040 M$="Method of Maximum Likelihood"
1050 E%=LEN(M$)
1060 F%=(80-E%)/2
1070 W$=STRING$(E%,205)
1080 PRINT:PRINT"Press any key to continue"
1090 Q$=INKEY$:IF Q$ ="" GOTO 1090
1100 CLS
1110 PRINT:PRINT TAB(F%) M$:PRINT TAB(F%) W$
1120 PRINT:PRINT TAB(19) "Iteration" TAB(37) "A" TAB(53) "f(A)"
1130 PRINT TAB(19) "---------" TAB(37) "-" TAB(53) "----"
1140 IF FLAG% <> 1 GOTO 1180
1150 LPRINT:LPRINT TAB(F%) M$:LPRINT TAB(F%) W$
1160 LPRINT:LPRINT TAB(19) "Iteration" TAB(37) "A" TAB(53) "f(A)"
1170 LPRINT TAB(19)"---------" TAB(37) "-" TAB(53) "----"
1180 COUNT%=0
1190 AML#=ALPHA#
1200 COUNT%=COUNT%+1
1210 A#=1#/(AML#*AML#)
1220 B#=M1#-1#/AML#
1230 C#=0#:D#=0#:E#=0#
1240 FOR I%=1 TO NREC%
1250 TEMP#=EXP(-AML#*X#(I%))
1260 C#=C#+TEMP#
1270 D#=D#+TEMP#*X#(I%)
1280 E#=E#+TEMP#*X#(I%)*X#(I%)
1290 NEXT I%
1300 FCN#=D#-B#*C#
1310 FPN#=B#*D#-E#-A#*C#
1320 AAS#=AML#-FCN#/FPN#
1330 PRINT TAB(22);:PRINT USING "##";COUNT%;:PRINT USING "        ##.###^^^^";AA
S#,FCN#
1340 IF FLAG% = 1 THEN LPRINT TAB(22);:PRINT USING "##";COUNT%;:LPRINT USING "
      ##.###^^^^";AAS#,FCN#
1350 DELTA#=ABS(.0000001#*AAS#)
1360 IF ABS(AAS#-AML#) < DELTA# GOTO 1400
1370 IF COUNT% > 24 THEN PRINT:PRINT "Procedure does not converge":PRINT:GOTO 17
60
1380 AML#=AAS#
1390 GOTO 1200
1400 ALPHA#=AAS#
1410 BETA#=(1#/ALPHA#)*LOG(NREC#/C#)
1420 M2#=1.285#/ALPHA#
1430 M1#=BETA#+.45#*M2#
1440 M2#=M2#*M2#
1450 FOR J% = 1 TO 6
1460 YM#=-LOG(-LOG(1#-1#/T#(J%)))
1470 XT#(J%)=BETA#+YM#/ALPHA#
1480 SX#(J%)=SQR((1.1086#+.514#*YM#+.6079#*YM#*YM#)/(NREC#*ALPHA#*ALPHA#))
1490 NEXT J%
1500 PRINT
1510 CLS
1520 PRINT:PRINT TAB(B%) TITLE$:PRINT TAB(B%) U$
1530 PRINT TAB(D%) T$:PRINT TAB(D%) V$
```

112

```
1540 PRINT TAB(F%) M$:PRINT TAB(F%) W$
1550 PRINT TAB(19) "Mean is                    ";:PRINT USING FORM$;M1#
1560 PRINT TAB(19) "Variance is                ";:PRINT USING FORM$;M2#
1570 PRINT TAB(19) "Parameter Alpha is         ";:PRINT USING FORM$;ALPHA#
1580 PRINT TAB(19) "Parameter Beta is          ";:PRINT USING FORM$;BETA#
1590 IF FLAG% <> 1 GOTO 1650
1600 LPRINT
1610 LPRINT TAB(19) "Mean is                    ";:LPRINT USING FORM$;M1#
1620 LPRINT TAB(19) "Variance is                ";:LPRINT USING FORM$;M2#
1630 LPRINT TAB(19) "Parameter Alpha is         ";:LPRINT USING FORM$;ALPHA#

1640 LPRINT TAB(19) "Parameter Beta is          ";:LPRINT USING FORM$;BETA#
1650 PRINT:PRINT TAB(2) "T, years  2           5           10          20
  50          100":PRINT TAB(5) ROW$
1660 PRINT TAB(2) "X  " VV$;:FOR J%=1 TO 6:PRINT USING FORM$;XT#(J%);:NEXT J%
1670 PRINT TAB(2) " t " VV$
1680 PRINT TAB(2) "S  " VV$;:FOR J%=1 TO 6:PRINT USING FORM$;SX#(J%);:NEXT J%
1690 PRINT TAB(2) " t " VV$
1700 IF FLAG% <> 1 GOTO 1760
1710 LPRINT:LPRINT TAB(2) "T, years  2           5           10          20
  50          100":LPRINT TAB(5) ROW$
1720 LPRINT TAB(2) "X  " VV$;:FOR J%=1 TO 6:LPRINT USING FORM$;XT#(J%);:NEXT J%
1730 LPRINT TAB(2) " t " VV$
1740 LPRINT TAB(2) "S  " VV$;:FOR J%=1 TO 6:LPRINT USING FORM$;SX#(J%);:NEXT J%
1750 LPRINT TAB(2) " t " VV$
1760 PRINT:PRINT"Press any key to continue"
1770 Q$=INKEY$:IF Q$ ="" GOTO 1770
1780 CLS
1790 SYSTEM
1800 '
1810 ' Subroutine for standard screen format
1820 '
1830 CLS:PRINT:PRINT
1840 ROW$=STRING$(78,205)
1850 BOXTOP$=CHR$(201)+ROW$+CHR$(187)
1860 PRINT BOXTOP$;
1870 PRINT CHR$(186) TAB(80) CHR$(186);
1880 PRINT CHR$(186) TAB(80) CHR$(186);
1890 C%=LEN(T$)
1900 D%=(80-C%)/2
1910 V$=STRING$(C%,205)
1920 PRINT CHR$(186) TAB(D%) T$ TAB(80) CHR$(186);
1930 PRINT CHR$(186) TAB(D%) V$ TAB(80) CHR$(186);
1940 PRINT CHR$(186) TAB(80) CHR$(186);
1950 PRINT CHR$(186) TAB(80) CHR$(186);
1960 PRINT CHR$(186) "   What is the name of your data file?" TAB(80) CHR$(186);

1970 PRINT CHR$(186) TAB(80) CHR$(186);
1980 PRINT CHR$(186) "   Do you want printer output (Y/N)?" TAB(80) CHR$(186);
1990 PRINT CHR$(186) "   (press Alt Q to quit at this stage)" TAB(80) CHR$(186);

2000 PRINT CHR$(186) TAB(80) CHR$(186);
2010 PRINT CHR$(186) TAB(80) CHR$(186);
2020 BOXBOT$=CHR$(200)+ROW$+CHR$(188)
2030 PRINT BOXBOT$;
2040 LOCATE 24,1,0,0,0
2050 PRINT "G Kite";
2060 LOCATE 10,41,1,0,13
2070 FILE$=""
2080 I$=INPUT$(1)
2090 IF I$ = CHR$(13) GOTO 2190                 ' ENTER key
2100 IF I$ = CHR$(8)  GOTO 2140                 ' BACKSPACE key
2110 PRINT I$;
2120 FILE$=FILE$+I$
2130 GOTO 2080
2140 H%=POS(0)
2150 LOCATE 10,H%-1,1,0,13
2160 L%=LEN(FILE$)
2170 FILE$=LEFT$(FILE$,L%-1)
2180 GOTO 2080
2190 LOCATE 12,39,1,0,13
2200 YN$=""
2210 I$=INKEY$:IF I$ = "" GOTO 2210
2220 IF LEN(I$) = 1 GOTO 2260
2230 I$=RIGHT$(I$,1)
2240 IF I$ = CHR$(16) THEN IERR%=1:GOTO 2330    ' ALT Q key
2250 GOTO 2190
2260 IF I$ = CHR$(13) GOTO 2300                 ' ENTER key
2270 PRINT I$;
2280 YN$=I$
```

113

```
2290 GOTO 2210
2300 IF YN$ = "Y" OR YN$ = "y" OR YN$ = "N" OR YN$ = "n" GOTO 2320
2310 GOTO 2190
2320 LOCATE ,,0,13,13
2330 RETURN
```

St. Mary's River at Stillwater, Nova Scotia. Station No. 01E0001. m**3/s.

### Type I Extremal Distribution

#### Method of Moments

| | |
|---|---|
| Mean is | 4.132D+02 |
| Variance is | 2.067D+04 |
| Coefficient of skew is | 1.195D+00 |
| Parameter Alpha is | 8.921D-03 |
| Parameter Beta is | 3.485D+02 |

NOTE: For good use of the T1E the coeff. of skew should be close to 1.13

| T, years | 2 | 5 | 10 | 20 | 50 | 100 |
|---|---|---|---|---|---|---|
| X$_t$ | 3.903D+02 | 5.276D+02 | 6.184D+02 | 7.056D+02 | 8.184D+02 | 9.030D+02 |
| S$_t$ | 1.559D+01 | 2.734D+01 | 3.739D+01 | 4.753D+01 | 6.098D+01 | 7.118D+01 |

#### Method of Maximum Likelihood

| Iteration | A | f(A) |
|---|---|---|
| 1 | 9.196D-03 | 1.869D+01 |
| 2 | 9.229D-03 | 1.817D+00 |
| 3 | 9.230D-03 | 2.301D-02 |
| 4 | 9.230D-03 | 3.826D-06 |

| | |
|---|---|
| Mean is | 4.122D+02 |
| Variance is | 1.938D+04 |
| Parameter Alpha is | 9.230D-03 |
| Parameter Beta is | 3.496D+02 |

| T, years | 2 | 5 | 10 | 20 | 50 | 100 |
|---|---|---|---|---|---|---|
| X$_t$ | 3.893D+02 | 5.121D+02 | 5.934D+02 | 6.714D+02 | 7.723D+02 | 8.480D+02 |
| S$_t$ | 1.499D+01 | 2.301D+01 | 2.952D+01 | 3.611D+01 | 4.491D+01 | 5.161D+01 |

# CHAPTER 9
## PEARSON TYPE III DISTRIBUTION

Introduction

The probability density distribution of the Pearson type III distribution is of the form

$$p(x) = \frac{1}{\alpha\Gamma(\beta)} \{\frac{x-\gamma}{\alpha}\}^{\beta-1} e^{-\{\frac{x-\gamma}{\alpha}\}} \qquad (9\text{-}1)$$

where $\alpha$, $\beta$ and $\gamma$ are parameters to be defined and $\Gamma(\beta)$ is the gamma function.

If the substitution $y = (x-\gamma)/\alpha$ is made, then Equation 9-1 simplifies to

$$p(y) = \frac{y^{\beta-1} e^{-y}}{\Gamma(\beta)} \qquad (9\text{-}2)$$

which is a one parameter gamma distribution described in many statistics texts.

Estimation of Parameters

*Method of Moments* - Substituting the probability density function of the Pearson type III distribution (Equation 9-1) into the general equation for moments about the origin yields

$$\mu'_r = \int_0^\infty x^r \frac{1}{\alpha\Gamma(\beta)} (\frac{x-\gamma}{\alpha})^{\beta-1} e^{-(\frac{x-\gamma}{\alpha})} dx \qquad (9\text{-}3)$$

Substituting $y$ for $(x-\gamma)/\alpha$ gives

$$\mu'_r = \frac{1}{\Gamma(\beta)} \int_0^\infty (\alpha y+\gamma)^r y^{\beta-1} e^{-y} dy \qquad (9\text{-}4)$$

which can be evaluated by noting that

$$\int_0^\infty y^{\beta-1} e^{-y} dy = \Gamma(\beta) \qquad (9\text{-}5)$$

the gamma function. As an example, for r=1, the first moment about the origin, $\mu'_1$, is given by

$$\mu'_1 = \frac{1}{\Gamma(\beta)} \int_0^\infty (\alpha y^\beta e^{-y} + \gamma y^{\beta-1} e^{-y}) dy \qquad (9\text{-}6)$$

$$\mu'_1 = \frac{\alpha\Gamma(\beta+1)}{\Gamma(\beta)} + \gamma \frac{\Gamma(\beta)}{\Gamma(\beta)} = \alpha\beta + \gamma \qquad (9\text{-}7)$$

For r = 2,

$$\mu_2' = \alpha^2 \beta(\beta+1) + 2\alpha\beta\gamma + \gamma^2 \qquad (9\text{-}8)$$

and since

$$\mu_2 = \mu_2' - \mu_1'^2 \qquad (9\text{-}9)$$

the second central moment is given by

$$\mu_2 = \sigma^2 = \alpha^2\beta \qquad (9\text{-}10)$$

Similarly, higher order central moments can be calculated as

$$\mu_3 = 2\alpha^3\beta \qquad (9\text{-}11)$$

$$\mu_4 = 3\alpha^4\beta(\beta+2) \qquad (9\text{-}12)$$

$$\mu_5 = 4\alpha^5\beta(5\beta+6) \qquad (9\text{-}13)$$

$$\mu_6 = 5\alpha^6\beta(3\beta^2+26\beta+24) \qquad (9\text{-}14)$$

If moment ratios $\gamma_1 \rightarrow \gamma_4$ are defined as

$$\gamma_1 = \mu_3/\mu_2^{3/2} \qquad (9\text{-}15)$$

$$\gamma_2 = \mu_4/\mu_2^2 \qquad (9\text{-}16)$$

$$\gamma_3 = \mu_5/\mu_2^{5/2} \qquad (9\text{-}17)$$

$$\gamma_4 = \mu_6/\mu_2^3 \qquad (9\text{-}18)$$

(note that $\gamma_1$ and $\gamma_2$ are the coefficients of skew and kurtosis respectively), then, for the Pearson type III distribution, $\gamma_2 - \gamma_4$ can be defined in terms of $\gamma_1$ as

$$\gamma_2 = 3(1+\gamma_1^2/2) \qquad (9\text{-}19)$$

$$\gamma_3 = \gamma_1(10+3\gamma_1^2) \qquad (9\text{-}20)$$

$$\gamma_4 = 5(3+13\gamma_1^2/2 + 3\gamma_1^4/2) \qquad (9\text{-}21)$$

From the sample mean, $\mu$, standard deviation, $\sigma$, and coefficient of skew, $\gamma_1$, the parameters $\alpha$, $\beta$ and $\gamma$ can be determined. Bobée and Robitaille (4) have shown that the sample coefficient of skew, $\gamma_1$, from Equation 9-15 should be corrected for bias as

$$\hat{\gamma}_1 = \hat{\gamma}_1 \frac{\sqrt{N(N-1)}}{N-2} (1 + \frac{8.5}{N})$$ (9-22)

where N is the sample size. Then, from 9-10, 9-11 and 9-15

$$\beta = (2/\gamma_1)^2$$ (9-23)

from 9-7 and 9-10

$$\alpha = \sigma/\sqrt{\beta}$$ (9-24)

and

$$\gamma = \mu - \sigma\sqrt{\beta}$$ (9-25)

*Maximum Likelihood* - The likelihood function is set up, as usual, as the logarithm of the product of the probability density distributions:

$$\ln L = - n \ln \Gamma(\beta) - \frac{1}{\alpha} \sum_{i=1}^{n} (x_i - \gamma) + (\beta-1) \sum_{i=1}^{n} \ln (x_i - \gamma) - n\beta \ln \alpha$$ (9-26)

Differentiating with respect to $\alpha$, $\beta$ and $\gamma$ and equating to zero gives the following three equations (9):

$$\frac{\partial \ln L}{\partial \alpha} = \frac{1}{\alpha^2} \sum_{i=1}^{n} (x_i - \gamma) - \frac{n\beta}{\alpha} = 0$$ (9-27)

$$\frac{\partial \ln L}{\partial \beta} = - n \Gamma'(\beta)/\Gamma(\beta) + \sum_{i=1}^{n} \ln (x_i - \gamma) - n \ln \alpha = 0$$ (9-28)

$$\frac{\partial \ln L}{\partial \gamma} = \frac{n}{\alpha} - (\beta-1) \sum_{i=1}^{n} (1/(x_i - \gamma)) = 0$$ (9-29)

The maximum likelihood solution then depends on a simultaneous solution of these three equations. The psi or digamma function, $\psi(\beta)$ or $\Gamma'(\beta)/\Gamma(\beta)$ in Equation 9-28 can be calculated using an asymptotic equation (1) of the form

$$\psi(\beta) = \ln (\beta) - \frac{1}{2\beta} - \frac{1}{12\beta^2} + \frac{1}{120\beta^4} - \frac{1}{252\beta^6}$$ (9-30)

Condie and Nix (6) found that to preserve accuracy at low values of $\beta$ it was necessary to include a recurrence equation so that

$$\psi(\beta) = \ln (\beta+2) - \frac{1}{2(\beta+2)} - \frac{1}{12(\beta+2)^2} + \frac{1}{120(\beta+2)^4}$$

$$- \frac{1}{252(\beta+2)^6} - \frac{1}{(\beta+1)} - \frac{1}{\beta}$$ (9-31)

Eliminating $\alpha$ from Equations 9-27 and 9-29 yields an expression for $\beta$

$$\beta = 1/\left[1 - \frac{n^2}{\sum\limits_{i=1}^{n}(x_i-\gamma) \cdot \sum\limits_{i=1}^{n}(1/(x_i-\gamma))}\right] \tag{9-32}$$

Substituting for $\beta$ in 9-27 leads to

$$\alpha = \frac{\sum\limits_{i=1}^{n}(x_i-\gamma)}{n} - \frac{n}{\sum\limits_{i=1}^{n}(1/(x_i-\gamma))} \tag{9-33}$$

Further substitution of $\alpha$ and $\beta$ from these two expressions into Equation 9-28 yields an equation in $\gamma$ only which can be solved numerically at $\partial \ln L/\partial\beta = 0$. Using an iterative procedure such as Newton's method gives a closure although multiple roots sometimes occur.

If the value of $\gamma$ can be derived from a source other than the data then direct maximum likelihood estimates of $\alpha$ and $\beta$ are available. If the arithmetic mean of the reduced series x-$\gamma$ is given by

$$A = \frac{1}{n}\sum\limits_{i=1}^{n}(x-\gamma) \tag{9-34}$$

then Equation 9-27 reduces to

$$\alpha = A/\beta \tag{9-35}$$

Now if the geometric mean of the reduced series x-$\gamma$ is given by

$$B = \left[\prod\limits_{i=1}^{n}(x-\gamma)\right]^{1/n} \tag{9-36}$$

and if

$$C = \ln A - \ln B \tag{9-37}$$

then substitution of Equations 9-35 and 9-37 into Equation 9-28 and simplifying yields

$$\ln \beta - \Gamma'(\beta)/\Gamma(\beta) = C \tag{9-38}$$

Solution of Equation 9-38 gives an estimate of $\beta$ and substitution back into Equation 9-35 will result in an estimate of $\alpha$. Solution of this equation is made easier by the availability of tables of C versus $\beta C$. Thus calculation of C from the reduced data and interpolation from a table gives $\beta C$ from which $\beta$ can be quickly found. Greenwood and Durand (7) have also given the following polynomial approximations to estimate $\beta$ directly from C:

for $0 \le C \le 0.5772$

$$\hat{\beta} = (0.5000876 + 0.1648852C - 0.054427C^2)/C \qquad (9\text{-}39)$$

for which the maximum error is 0.0088% and

for $0.5772 \le C \le 17.0$

$$\hat{\beta} = \frac{8.898919 + 9.059950C + 0.9775373C^2}{C(17.79728 + 11.968477C + C^2)} \qquad (9\text{-}40)$$

for which the maximum error is 0.00554%.

The maximum likelihood methods described above may not always be applicable. Matalas and Wallis (10) have noted that for very small values of sample skew a solution may not be possible. Similarly if $\beta < 1$ a maximum likelihood solution is not possible. If $\beta$ must be greater than 1 then the sample coefficient of skew cannot exceed 2 (see Equation 9-23). Finally if the sample coefficient of skew is negative the Pearson type III distribution becomes bounded at the upper limit which is not suitable for analysis of maximum events. In general, it is not possible with any automatic procedure to guarantee finding the minimum variance solution each time. Each data set requires a careful investigation of the shape of the function.

Frequency Factor

The cumulative probability distribution of the Pearson type III can be expressed as:

$$P(x) = \frac{1}{\alpha\Gamma(\beta)} \int_{0}^{x_o} e^{-\{\frac{x-\gamma}{\alpha}\}} \{\frac{x-\gamma}{\alpha}\}^{\beta-1} dx \qquad (9\text{-}41)$$

If $\alpha$, $\beta$ and $\gamma$ are derived as in the previous section and the probability required for a given return period, T, is $P(x) = 1 - 1/T$ then, to define the event magnitude, Equation 9-41 must be solved for $x_o$.

Making the substitution $y = (x-\gamma)/\alpha$ the distribution is given by:

$$P(y) = \frac{1}{\Gamma(\beta)} \int_{0}^{y_o} y^{\beta-1} e^{-y} dy \qquad (9\text{-}42)$$

but from (1)

$$P(y) = P(\chi^2|v) \qquad (9\text{-}43)$$

where $P(\chi^2|v)$ is the chi-square distribution with $2\beta$ degrees of freedom and $\chi^2 = 2y$. So, from the tabulated value of $\chi^2$ for probability $1 - 1/T$ and $2\beta$ degrees of freedom the reduced event magnitude, $y_0$, is obtained as:

$$y_0 = \chi^2/2 \qquad (9-44)$$

and the expected event magnitude is given by:

$$x = \frac{\chi^2_\alpha}{2} + \gamma \qquad (9-45)$$

Tables of chi-square distribution are commonly given in statistical texts, and, for convenience, Table 9-1 provides

Table 9-1

## Percentage Points of the Chi-Square Distribution[1]

| Degrees of Freedom $v$ | Cumulative Probability, $P(\chi^2\|v)$, % | | | | | |
|---|---|---|---|---|---|---|
| | 50 | 75 | 90 | 95 | 97.5 | 99 |
| | | | Corresponding Return Period, T, Years | | | |
| | 2 | 4 | 10 | 20 | 40 | 100 |
| 1 | 0.46 | 1.32 | 2.71 | 3.84 | 5.02 | 6.63 |
| 2 | 1.39 | 2.77 | 4.61 | 5.99 | 7.38 | 9.21 |
| 3 | 2.37 | 4.11 | 6.25 | 7.81 | 9.35 | 11.34 |
| 4 | 3.36 | 5.39 | 7.78 | 9.49 | 11.14 | 13.28 |
| 5 | 4.35 | 6.63 | 9.24 | 11.07 | 12.83 | 15.09 |
| 6 | 5.35 | 7.84 | 10.64 | 12.59 | 14.45 | 16.81 |
| 7 | 6.35 | 9.04 | 12.02 | 14.07 | 16.01 | 18.48 |
| 8 | 7.34 | 10.22 | 13.36 | 15.51 | 17.53 | 20.09 |
| 9 | 8.34 | 11.39 | 14.68 | 16.92 | 19.02 | 21.67 |
| 10 | 9.34 | 12.55 | 15.99 | 18.31 | 20.48 | 23.21 |
| 11 | 10.34 | 13.70 | 17.28 | 19.68 | 21.92 | 24.73 |
| 12 | 11.34 | 14.84 | 18.55 | 21.03 | 23.34 | 26.32 |
| 13 | 12.34 | 15.98 | 19.81 | 22.36 | 24.74 | 27.69 |
| 14 | 13.34 | 17.11 | 21.06 | 23.68 | 26.12 | 29.14 |
| 15 | 14.34 | 18.25 | 22.31 | 25.00 | 27.49 | 30.58 |
| 16 | 15.34 | 19.37 | 23.54 | 26.30 | 28.85 | 32.00 |
| 17 | 16.34 | 20.48 | 24.77 | 27.59 | 30.19 | 33.41 |
| 18 | 17.34 | 21.61 | 25.99 | 29.87 | 31.53 | 34.81 |
| 19 | 18.34 | 22.72 | 27.20 | 30.14 | 32.85 | 36.19 |
| 20 | 19.34 | 23.83 | 28.41 | 31.41 | 34.17 | 37.57 |
| 21 | 20.34 | 24.93 | 29.62 | 32.67 | 35.48 | 38.93 |
| 22 | 21.34 | 26.04 | 30.81 | 33.92 | 36.78 | 40.29 |
| 23 | 22.34 | 27.14 | 32.01 | 35.17 | 38.08 | 41.64 |
| 24 | 23.34 | 28.24 | 33.20 | 36.42 | 39.36 | 42.98 |
| 25 | 24.34 | 29.34 | 34.38 | 37.65 | 40.65 | 44.31 |
| 26 | 25.34 | 30.43 | 35.56 | 38.89 | 41.92 | 45.64 |
| 27 | 26.34 | 31.53 | 36.74 | 40.11 | 43.19 | 46.96 |
| 28 | 27.34 | 32.62 | 37.92 | 41.34 | 44.46 | 48.28 |
| 29 | 28.34 | 33.71 | 39.09 | 42.56 | 45.72 | 49.59 |
| 30 | 29.34 | 34.80 | 40.26 | 43.77 | 46.98 | 50.89 |
| 40 | 39.34 | 45.62 | 51.81 | 55.76 | 59.34 | 63.69 |
| 50 | 49.33 | 56.33 | 63.17 | 67.50 | 71.42 | 76.17 |
| 60 | 59.33 | 66.98 | 74.40 | 79.08 | 83.30 | 88.38 |
| 70 | 69.33 | 77.57 | 85.53 | 90.53 | 95.02 | 100.42 |
| 80 | 79.33 | 88.13 | 96.58 | 101.88 | 106.63 | 112.33 |
| 90 | 89.33 | 98.65 | 107.56 | 113.14 | 118.14 | 124.12 |
| 100 | 99.33 | 109.14 | 118.49 | 124.34 | 129.56 | 135.81 |

[1] The function, $\chi^2_\alpha$, tabulated is that value of $\chi^2$ with v degrees of freedom beyond which $(1-P)$% of the distribution lies.

values of $x^2$ for some commonly used probabilities and for various degrees of freedom. Note that in Table 9-1 the probabilities are arranged so that larger event magnitudes correspond to larger cumulative probabilities and smaller probabilities of exceedence (larger return periods). This table would be suitable for an analysis of maxima such as flood events. In the study of minima such as drought events, for which this distribution is sometimes used, (c.f. Chin (5)), these probabilities should be reversed.

It has been found (8) that the expression

$$\left\{ \left(\frac{x^2}{v}\right)^{1/3} + \frac{2}{9v} - 1 \right\} \left\{\frac{9v}{2}\right\}^{1/2} \qquad (9\text{-}46)$$

(known as the Wilson-Hilferty approximation) is approximately normally distributed with zero mean and unit variance for $v > 30$. Thus the value of $x^2$ for a particular probability level, P, and number of degrees of freedom, $v$, can be approximated by substituting the corresponding standard normal deviate, t, in the equation:

$$x^2 \approx v \left\{ 1 - \frac{2}{9v} + t.\sqrt{\frac{2}{9v}} \right\}^3 \qquad (9\text{-}47)$$

A refinement to the approximation may be made by substituting $(t - h_v)$ for t in Equation 9-47 where

$$h_v = \frac{60}{v} \, h_{60} \qquad (9\text{-}48)$$

and $h_{60}$ is tabulated in Abramowitz and Stegun (1).

Combining Equations 9-45 and 9-47 and substituting the degrees of freedom, $v = 2\beta$,

$$x \approx \alpha\beta \left\{ 1 - \frac{1}{9\beta} + t \sqrt{\frac{1}{9\beta}} \right\}^3 + \gamma \qquad (9\text{-}49)$$

Thus, knowing, $\alpha$, $\beta$ and $\gamma$ the value of x corresponding to any given probability level can be approximated.

Substituting for $\alpha$ and $\gamma$ from Equations 9-24 and 9-25 in Equation 9-45 an expression:

$$x \approx \mu + [\frac{x^2\gamma_1}{4} - \frac{2}{\gamma_1}]\sigma \qquad (9\text{-}50)$$

is obtained, where $\gamma_1$ is the coefficient of skew of the sample data.

Comparing Equation 9-50 with the general frequency equation it can be seen that, for the Pearson type III distribution, the frequency factor, K, can be approximated by

$$K \approx \frac{x^2 \gamma_1}{4} - \frac{2}{\gamma_1} \qquad (9\text{-}51)$$

For computation purposes and for later use in calculating standard errors it is more useful to express K directly in terms of $\gamma_1$. By substituting Equation 9-47 in Equation 9-51 and noting that degrees of freedom, v, is equal to $8/\gamma_1^2$ the expression becomes

$$K \approx t + (t^2-1) \frac{\gamma_1}{6} + \frac{1}{3} (t^3-6t) (\frac{\gamma_1}{6})^2$$
$$- (t^2-1) (\frac{\gamma_1}{6})^3 + t (\frac{\gamma_1}{6})^4 - \frac{1}{3} + \frac{1}{3} (\frac{\gamma_1}{6})^5 \qquad (9\text{-}52)$$

From this equation tables of K for different coefficients of skew, $\gamma_1$, and probability, P, can be calculated. Table 9-2 is an example. The Soil Conservation Service, U. S. Dept. of Agriculture (13) has prepared very comprehensive tables of frequence factors for the Pearson type III distribution. Chin (5) has given a table of frequency factors for probability levels suitable for analysis of minimum events such as droughts.

### Standard Error

*Method of Moments* - The general equation for the standard error of estimate of a 3-parameter distribution has been derived earlier as

$$S_T^2 = \frac{\mu_2}{n} \{1 + K\gamma_1 + \frac{K^2}{4} [\gamma_2 - 1] + \frac{\partial K}{\partial \gamma_1} [2\gamma_2 - 3\gamma_1^2 - 6$$

$$+ K(\gamma_3 - 6\gamma_1\gamma_2/4 - 10\gamma_1/4)] + (\frac{\partial K}{\partial \gamma_1})^2 [\gamma_4$$

$$- 3\gamma_3\gamma_1 - 6\gamma_2 + 9\gamma_1^2\gamma_2/4 + 35\gamma_1^2/4 + 9]\} \qquad (9\text{-}53)$$

Substituting the relationships between the moment ratios of the Pearson type III distribution (Equations 9-19 - 9-21) this expression simplifies to

Table 9-2

## Frequency Factor for Use in Pearson Type III Distribution

| Coefficient of Skew $\gamma_1$ | Cumulative Probability, P, % | | | | | |
|---|---|---|---|---|---|---|
| | 50 | 80 | 90 | 95 | 98 | 99 |
| | Corresponding Return Period, T, Years | | | | | |
| | 2 | 5 | 10 | 20 | 50 | 100 |
| 0.0 | 0.0000 | .8416 | 1.2816 | 1.6448 | 2.0537 | 2.3264 |
| 0.1 | -.0167 | .8363 | 1.2917 | 1.6728 | 2.1070 | 2.3997 |
| 0.2 | -.0333 | .8303 | 1.3009 | 1.6996 | 2.1595 | 2.4727 |
| 0.3 | -.0499 | .8234 | 1.3089 | 1.7254 | 2.2112 | 2.5453 |
| 0.4 | -.0664 | .8157 | 1.3159 | 1.7501 | 2.2619 | 2.6172 |
| 0.5 | -.0828 | .8072 | 1.3218 | 1.7735 | 2.3117 | 2.6884 |
| 0.6 | -.0990 | .7980 | 1.3267 | 1.7958 | 2.3603 | 2.7588 |
| 0.7 | -.1151 | .7880 | 1.3304 | 1.8168 | 2.4078 | 2.8283 |
| 0.8 | -.1310 | .7773 | 1.3330 | 1.8366 | 2.4541 | 2.8968 |
| 0.9 | -.1467 | .7659 | 1.3345 | 1.8551 | 2.4991 | 2.9641 |
| 1.0 | -.1621 | .7537 | 1.3349 | 1.8723 | 2.5428 | 3.0303 |
| 1.1 | -.1772 | .7409 | 1.3342 | 1.8881 | 2.5851 | 3.0952 |
| 1.2 | -.1921 | .7275 | 1.3324 | 1.9026 | 2.6260 | 3.1588 |
| 1.3 | -.2067 | .7134 | 1.3295 | 1.9157 | 2.6653 | 3.2209 |
| 1.4 | -.2209 | .6987 | 1.3255 | 1.9274 | 2.7031 | 3.2816 |
| 1.5 | -.2347 | .6834 | 1.3204 | 1.9378 | 2.7394 | 3.3406 |
| 1.6 | -.2482 | .6676 | 1.3143 | 1.9467 | 2.7740 | 3.3981 |
| 1.7 | -.2612 | .6513 | 1.3072 | 1.9543 | 2.8070 | 3.4538 |
| 1.8 | -.2738 | .6344 | 1.2990 | 1.9604 | 2.8383 | 3.5078 |
| 1.9 | -.2860 | .6171 | 1.2897 | 1.9651 | 2.8678 | 3.5600 |
| 2.0 | -.2977 | .5993 | 1.2795 | 1.9684 | 2.8956 | 3.6103 |

$$S_T^2 = \frac{\mu_2}{n} \{1 + K\gamma_1 + \frac{K^2}{2} [3\gamma_1^2/4 + 1] + 3K \frac{\partial K}{\partial \gamma_1} [\gamma_1 + \gamma_1^3/4]$$

$$+ 3 (\frac{\partial K}{\partial \gamma_1})^2 [2 + 3\gamma_1^2 + 5\gamma_1^4/8]\} \qquad (9\text{-}54)$$

and now from Equation 9-52 the partial derivative of K with respect to $\gamma_1$ is

$$\frac{\partial K}{\partial \gamma_1} \approx \frac{t^2-1}{6} + \frac{4(t^3-6t)}{6^3} \gamma_1 - \frac{3(t^2-1)}{6^3} \gamma_1^2 + \frac{4t}{6^4} \gamma_1^3 - \frac{10}{6^6} \gamma_1^4$$

$$(9\text{-}55)$$

where t is the standard normal deviate. Substituting Equation 9-55 into Equation 9-54 and simplifying

$$S_T \approx \delta \sqrt{\frac{\mu_2}{n}} \qquad (9\text{-}56)$$

where $\delta$ is now dependent only on the required return period, T, and the sample coefficient of skew, $\gamma_1$. Table 9-3 gives values of $\delta$ for some values of T and $\gamma_1$. Bobée (2) has derived Equation 9-54 in a slightly different form.

Table 9-3

Parameter δ for Use in Standard Error
of Pearson Type III Distribution

| Coefficient of skew $\gamma_1$ | Cumulative Probability, P, % | | | | | |
|---|---|---|---|---|---|---|
| | 50 | 80 | 90 | 95 | 98 | 99 |
| | Corresponding Return Period, T, Years | | | | | |
| | 2 | 5 | 10 | 20 | 50 | 100 |
| 0.0 | 1.0801 | 1.1698 | 1.3748 | 1.6845 | 2.1988 | 2.6363 |
| 0.1 | 1.0808 | 1.2006 | 1.4367 | 1.7810 | 2.3425 | 2.8168 |
| 0.2 | 1.0830 | 1.2309 | 1.4989 | 1.8815 | 2.4986 | 3.0175 |
| 0.3 | 1.0866 | 1.2609 | 1.5610 | 1.9852 | 2.6656 | 3.2365 |
| 0.4 | 1.0918 | 1.2905 | 1.6227 | 2.0915 | 2.8423 | 3.4724 |
| 0.5 | 1.0987 | 1.3199 | 1.6838 | 2.1998 | 3.0277 | 3.7238 |
| 0.6 | 1.1073 | 1.3492 | 1.7441 | 2.3094 | 3.2209 | 3.9895 |
| 0.7 | 1.1179 | 1.3785 | 1.8032 | 2.4198 | 3.4208 | 4.2684 |
| 0.8 | 1.1304 | 1.4082 | 1.8609 | 2.5303 | 3.6266 | 4.5595 |
| 0.9 | 1.1449 | 1.4385 | 1.9170 | 2.6403 | 3.8374 | 4.8618 |
| 1.0 | 1.1614 | 1.4699 | 1.9714 | 2.7492 | 4.0522 | 5.1741 |
| 1.1 | 1.1799 | 1.5030 | 2.0240 | 2.8564 | 4.2699 | 5.4952 |
| 1.2 | 1.2003 | 1.5382 | 2.0747 | 2.9613 | 4.4896 | 5.8240 |
| 1.3 | 1.2223 | 1.5764 | 2.1237 | 3.0631 | 4.7100 | 6.1592 |
| 1.4 | 1.2457 | 1.6181 | 2.1711 | 3.1615 | 4.9301 | 6.4992 |
| 1.5 | 1.2701 | 1.6643 | 2.2173 | 3.2557 | 5.1486 | 6.8427 |
| 1.6 | 1.2952 | 1.7157 | 2.2627 | 3.3455 | 5.3644 | 7.1881 |
| 1.7 | 1.3204 | 1.7732 | 2.3081 | 3.4303 | 5.5761 | 7.5339 |
| 1.8 | 1.3452 | 1.8374 | 2.3541 | 3.5100 | 5.7827 | 7.8783 |
| 1.9 | 1.3690 | 1.9091 | 2.4018 | 3.5844 | 5.9829 | 8.2196 |
| 2.0 | 1.3913 | 1.9888 | 2.4525 | 3.6536 | 6.1755 | 8.5562 |

*Maximum Likelihood* - As given earlier in Chapter 3 the variance of the T-year event, $S_T^2$, is expressed by maximum likelihood in terms of the parameter as

$$S_T^2 = (\frac{\partial x}{\partial \alpha})^2 \text{ var } \alpha + (\frac{\partial x}{\partial \beta})^2 \text{ var } \beta + (\frac{\partial x}{\partial \gamma})^2 \text{ var } \gamma$$

$$+ 2 \frac{\partial x}{\partial \alpha} \frac{\partial x}{\partial \beta} \text{ cov } (\alpha,\beta) + 2 \frac{\partial x}{\partial \alpha} \frac{\partial x}{\partial \gamma} \text{ cov } (\alpha,\gamma)$$

$$+ 2 \frac{\partial x}{\partial \beta} \frac{\partial x}{\partial \gamma} \text{ cov } (\beta,\gamma) \tag{9-57}$$

From the logarithm of the likelihood equation

$$\ln L = - n \ln \Gamma(\beta) - \frac{1}{\alpha} \sum_{i=1}^{n} (x_i - \gamma) + (\beta-1) \sum_{i=1}^{n} \ln (x_i - \gamma) - n\beta \ln \alpha$$

$$\tag{9-58}$$

the following derivatives are obtained:

$$\frac{\partial^2 \ln L}{\partial \alpha^2} = \frac{-2 \sum\limits_{i=1}^{n} (x_i - \gamma)}{\alpha^3} + \frac{n\beta}{\alpha^2} \qquad (9\text{-}59)$$

$$\frac{\partial^2 \ln L}{\partial \beta^2} = - n\psi'(\beta) \qquad (9\text{-}60)$$

where $\psi'(\beta)$ is the trigamma function.

The trigamma function, $\psi'(\beta)$, may be approximated by an asymptotic equation given by Abramowitz and Stegun (1). As with the psi function, this approximation is best combined with a recurrence equation to preserve accuracy at low $\beta$. A suitable equation is

$$\psi'(\beta) = \frac{1}{\beta+2} + \frac{1}{2(\beta+2)^2} + \frac{1}{6(\beta+2)^3} - \frac{1}{30(\beta+2)^5}$$

$$+ \frac{1}{42(\beta+2)^7} - \frac{1}{30(\beta+2)^9} + \frac{1}{(\beta+1)^2} + \frac{1}{\beta^2} \qquad (9\text{-}61)$$

The remaining derivatives are:

$$\frac{\partial^2 \ln L}{\partial \gamma^2} = - (\beta-1) \sum\limits_{i=1}^{n} (x_i-\gamma)^{-2} \qquad (9\text{-}62)$$

$$\frac{\partial^2 \ln L}{\partial \alpha \partial \beta} = - \frac{n}{\alpha} \qquad (9\text{-}63)$$

$$\frac{\partial^2 \ln L}{\partial \alpha \partial \gamma} = - \frac{n}{\alpha^2} \qquad (9\text{-}64)$$

$$\frac{\partial^2 \ln L}{\partial \beta \partial \gamma} = - \sum\limits_{i=1}^{n} (x_i-\gamma)^{-1} \qquad (9\text{-}65)$$

The expression $\sum\limits_{i=1}^{n} (x_i-\gamma)^r$ can be further evaluated from the pdf of the reduced variate, $y$,

$$f(x) = \frac{1}{\Gamma(\beta)} \int_{0}^{\infty} y^{\beta-1} e^{-y} dy \qquad (9\text{-}66)$$

where $y = (x-\gamma)/\alpha$. If $(x_i-\gamma)^r$ is now introduced into the expression for moments about $\gamma$:

$$\mu_r^\gamma = \frac{1}{\Gamma(\beta)} \int_{0}^{\infty} (x_i - \gamma)^r y^{\beta-1} e^{-y} dy \qquad (9\text{-}67)$$

$$= \frac{\alpha^r}{\Gamma(\beta)} \int_0^\infty y^{\beta+r-1} e^{-y} \, dy \qquad (9\text{-}68)$$

$$= \alpha^r \Gamma(\beta+r)/\Gamma(\beta) \qquad (9\text{-}69)$$

so that

$$\sum_{i=1}^n (x_i-\gamma)^r = n\mu_r^\gamma = n\alpha^r \Gamma(\beta+r)/\Gamma(\beta) \qquad (9\text{-}70)$$

Utilizing this expression for appropriate values of r the derivatives of the likelihood equation are reduced to

$$\frac{\partial^2 \ln L}{\partial \alpha^2} = 1 \frac{n\beta}{\alpha^2} \qquad (9\text{-}71)$$

$$\frac{\partial^2 \ln L}{\partial \beta^2} = - n\psi'(\beta) \qquad (9\text{-}72)$$

$$\frac{\partial^2 \ln L}{\partial \gamma^2} = - \frac{n}{\alpha^2(\beta-2)} \qquad (9\text{-}73)$$

$$\frac{\partial^2 \ln L}{\partial \alpha \partial \beta} = - \frac{n}{\alpha} \qquad (9\text{-}74)$$

$$\frac{\partial^2 \ln L}{\partial \alpha \partial \gamma} = - \frac{n}{\alpha^2} \qquad (9\text{-}75)$$

$$\frac{\partial^2 \ln L}{\partial \beta \partial \gamma} = \frac{-n}{\alpha(\beta-1)} \qquad (9\text{-}76)$$

The likelihood or information matrix is then

$$I = n \begin{bmatrix} \dfrac{\beta}{\alpha^2} & \dfrac{1}{\alpha} & \dfrac{1}{\alpha^2} \\[2em] & \psi'(\beta) & \dfrac{1}{\alpha(\beta-1)} \\[2em] & & \dfrac{1}{\alpha^2(\beta-2)} \end{bmatrix} \qquad (9\text{-}77)$$

The variance and covariances of the parameters are obtained as the inverse of this matrix as:

$$\text{var } \alpha = \frac{1}{n\alpha^2 D} \left\{ \frac{\psi'(\beta)}{(\beta-2)} - \frac{1}{(\beta-1)^2} \right\} \qquad (9\text{-}78)$$

$$\text{var } \beta = \frac{2}{nD\alpha^4 (\beta-2)} \qquad (9\text{-}79)$$

$$\text{var } \gamma = \frac{\beta\psi'(\beta) - 1}{n\alpha^2 D} \qquad (9\text{-}80)$$

$$\text{cov } (\alpha,\beta) = \frac{-1}{n\alpha^3 D} \left\{ \frac{1}{(\beta-2)} - \frac{1}{(\beta-1)} \right\} \qquad (9\text{-}81)$$

$$\text{cov } (\alpha,\gamma) = \frac{1}{n\alpha^2 D} \left[ \frac{1}{(\beta-1)} - \psi'(\beta) \right] \qquad (9\text{-}82)$$

$$\text{cov } (\beta,\gamma) = \frac{-1}{n\alpha^3 D} \left[ \frac{\beta}{(\beta-1)} - 1 \right] \qquad (9\text{-}83)$$

where D is the determinant of the matrix of likelihood derivatives

$$D = \frac{1}{(\beta-2)\alpha^4} \left[ 2\psi'(\beta) - \frac{(2\beta-3)}{(\beta-1)^2} \right] \qquad (9\text{-}84)$$

The partial derivatives needed can be obtained from 9-49 as:

$$\frac{\partial x}{\partial \alpha} = \left[ \beta^{1/3} - \frac{1}{9\beta^{2/3}} + \frac{t}{3\beta^{1/6}} \right]^3 \qquad (9\text{-}85)$$

$$\frac{\partial x}{\partial \beta} = 3\alpha \left[ \beta^{1/3} - \frac{1}{9\beta^{2/3}} + \frac{t}{3\beta^{1/6}} \right]^2 \left[ \frac{1}{3\beta^{2/3}} + \frac{2}{27\beta^{5/3}} - \frac{t}{18\beta^{7/6}} \right] \qquad (9\text{-}86)$$

$$\frac{\partial x}{\partial \gamma} = 1 \qquad (9\text{-}87)$$

so that expression 9-57 can now be evaluated.

The maximum likelihood standard error of estimate for the two-parameter gamma distribution has been described by Moran (11) and Santos (12).

127

*Other Methods* - Bobée and Morin (3) have used order statistics to derive confidence intervals for the Pearson type III distribution.  Tables and graphs show values of the standardized variable for given sample size, sample coefficient of skew, probability of exceedence (as an order statistic) and confidence level.  The equivalent to this type of table using the standard error of estimate would be a modification of Equation 9-54 yielding confidence limits as

$$CL_{1,2} \simeq \mu + K\sigma \pm \frac{t \delta \sigma}{\sqrt{n}} \qquad (9\text{-}88)$$

standardizing, this becomes

$$CL_{1,2} \simeq K \pm \frac{t \delta}{\sqrt{n}} \qquad (9\text{-}89)$$

where K and $\delta$ are from Tables 9-2 and 9-3, t is the standard normal deviate for the required confidence level and n is the sample size.  Values from Equation 9-89 approximate those given by Bobée and Morin.

## References

1.  Abramowitz, M. and I. A. Stegun, 1965, Handbook of Mathematical Functions, Dover Publications, New York.

2.  Bobée, B, 1973, Sample Error of T-year Events Computed by Fitting a Pearson Type 3 Distribution, Wat. Res. Res., Vol. 9, No. 5, pp. 1264-1270.

3.  Bobée, B. and G. Morin, 1973, Statistique D'ordre de la Loi Pearson III et de sa Forme Derivée à Asymétrie Négative, Revue de Statistique Appliquee, Vol. XXI, No. 4, pp. 69-80.

4.  Bobée, B. and R. Robitaille, 1976, The Use of the Pearson Type 3 and Log-Pearson Type 3 Distributions Revisited, Wat. Res. Res., Vol. 13, No. 2, April, pp. 427-443.

5.  Chin, W. O., 1967, Formulae and Tables for Computing and Plotting Drought Frequency Curves, Technical Bulletin No. 8, Inland Waters Branch, Ottawa.

6.  Condie, R. and G. Nix, 1975, Modelling of Low Flow Frequency Distributions and Parameter Estimation, Int. Wat. Res. Assoc., Proc. Symp. on Water for Arid Lands, Tehran, Dec. 8 and 9.

7.  Greenwood, J. A. and D. Durand, 1960, Aids for Fitting the Gamma Distribution, Technometrics, Vol. 2, No. 1, pp. 55-66.

8.  Kendall, M. G. and A. Stuart, 1963, The Advanced Theory of Statistics, Vol. I, Griffin, London.

9.  Matalas, N. C., 1963, Probability Distribution of Low Flows, USGS Professional Paper No. 434-A.

10. Matalas, N. C. and J. R. Wallis, 1973, Eureka! It Fits a Pearson Type 3 Distribution, Wat. Res. Res., Vol. 9, No. 2, pp. 281-289.

11. Moran, P. A. P., 1957, The Statistical Treatment of Flood Flows, Trans. American Geophysical Union, Vol. 38, No. 4, pp. 519-523.

12. Santos, A., 1970, The Statistical Treatment of Flood Flows, Water Power, Vol. 22, No. 2, pp. 63-67.

13. Soil Conservation Service, U. S. Dept. of Agriculture, 1968, New Tables of Percentage Points of the Pearson Type III Distribution, Technical Release No. 38, Central Technical Unit.

```
      PROGRAM PT3(INPUT,OUTPUT,TAPE5=INPUT,TAPE6=OUTPUT)      A   1
C                                                             A   2
C                                                             A   3
C     COMPUTES METHOD OF MOMENTS AND MAXIMUM LIKELIHOOD ESTIMATES FOR  A   4
C     T YEAR EVENTS AND STANDARD ERRORS FOR PEARSON TYPE 3  A   5
C     DISTRIBUTION                                           A   6
C     INPUT                                                  A   7
C     TITLE                                                  A   8
C     N NUMBER OF ANNUAL MAXIMUM EVENTS                      A   9
C     X SERIES OF EVENTS                                     A  10
C                                                            A  11
C                                                            A  12
      REAL M1,M2,M3,K                                        A  13
      DIMENSION SND(6), X(100)                               A  14
      DIMENSION XT(6), SX(6), TITLE(80)                      A  15
      SND(1)=0.                                              A  16
      SND(2)=0.84162                                         A  17
      SND(3)=1.28155                                         A  18
      SND(4)=1.64485                                         A  19
      SND(5)=2.05375                                         A  20
      SND(6)=2.32635                                         A  21
      READ (5,9) TITLE                                       A  22
      READ (5,10) N                                          A  23
      READ (5,11) (X(I),I=1,N)                               A  24
      XN=N                                                   A  25
      A=0.0                                                  A  26
      B=0.0                                                  A  27
      C=0.0                                                  A  28
      DO 1 I=1,N                                             A  29
      A=A+X(I)                                               A  30
      B=B+X(I)**2                                            A  31
      C=C+X(I)**3                                            A  32
    1 CONTINUE                                               A  33
      M1=A/XN                                                A  34
      M2=(B/XN)-(A/XN)**2                                    A  35
      M3=(C/XN)+2.0*M1**3-3.0*M1*(B/XN)                      A  36
      SKEW=M3/(M2**1.5)                                      A  37
      C1=(SQRT(XN*(XN-1.0)))/(XN-2.0)                        A  38
      C2=1.0+8.5/XN                                          A  39
      C3=XN/(XN-1.0)                                         A  40
      SKEW=SKEW*C1*C2                                        A  41
      M2=M2*C3                                               A  42
      BETA=(2.0/SKEW)**2                                     A  43
      ALPHA=(M2**0.5)/(BETA**0.5)                            A  44
      GAMMA=M1-(M2**0.5)*(BETA**0.5)                         A  45
      WRITE (6,12) TITLE                                     A  46
      WRITE (6,13)                                           A  47
      WRITE (6,22) ALPHA,M1                                  A  48
      WRITE (6,23) BETA,M2                                   A  49
      WRITE (6,24) GAMMA,SKEW                                A  50
      DO 2 J=1,6                                             A  51
      T=SND(J)                                               A  52
      T1=T                                                   A  53
      T2=(T**2-1.0)/6.0                                      A  54
      T3=2.0*(T**3-6.0*T)/6.0**3                             A  55
      T4=(T**2-1.0)/6.0**3                                   A  56
      T5=T/6.0**4                                            A  57
      T6=2.0/6.0**6                                          A  58
      K=T1+T2*SKEW+T3*SKEW**2-T4*SKEW**3+T5*SKEW**4-T6*SKEW**5   A  59
      SLOPE=T2+T3*2.0*SKEW-T4*3.0*SKEW**2+T5*4.0*SKEW**3-T6*5.0*SKEW**4   A  60
      T7=(1.0+0.75*SKEW**2)*(0.5*K**2)                       A  61
      T8=K*SKEW                                              A  62
      DELTA=1.0+T7+T8+T9*T10                                 A  63
      T9=1.0+(1.0+1.25*SKEW**2)*SLOPE                        A  63
      T10=SLOPE*(1.0+1.25*SKEW**2)+(SKEW*K/2.0)              A  64
      DELTA=T7+T8+T9*T10                                     A  65
      XT(J)=M1+K*SQRT(M2)                                    A  66
    2 SX(J)=SQRT(M2*DELTA/XN)                                A  67
      WRITE (6,14)                                           A  68
      WRITE (6,15) (XT(J),J=1,6)                             A  69
      WRITE (6,16) (SX(J),J=1,6)                             A  70
      WRITE (6,17)                                           A  71
      WRITE (6,18)                                           A  72
      ICOUNT=0                                               A  73
      XMIN=10000000.                                         A  74
      DO 3 I=1,N                                             A  75
    3 IF (X(I).LT.XMIN) XMIN=X(I)                            A  76
      GML=XMIN*0.99                                          A  77
    4 ICOUNT=ICOUNT+1                                        A  78
      A=0.0                                                  A  79
      H=0.0                                                  A  80
```

130

```
          C=0.0                                                          A  81
          R=0.0                                                          A  82
          DO 5 I=1,N                                                     A  83
          A=A+1.0/(X(I)-GML)                                             A  84
          B=B+(X(I)-GML)                                                 A  85
          C=C+ALOG(X(I)-GML)                                             A  86
          R=R+1.0/((X(I)-GML)**2)                                        A  87
    5     CONTINUE                                                       A  88
          BETA=A/TA-(XN**2)/B)                                           A  89
          ALPHA=B/(XN*BETA)                                              A  90
          D=BETA+2.0                                                     A  91
          PSI=ALOG(D)-(1.0/(2.0*D))-(1.0/(12.0*D**2))+(1.0/(120.0*D**4))-(1.  A  92
         10/(252.0*D**6))-(1.0/(BETA+1.0))-(1.0/BETA)                    A  93
          FCN=XN*PSI+C-XN*ALOG(ALPHA)                                    A  94
          TRI=(1.0/D)+(1.0/(2.0*D**2))+(1.0/(6.0*D**3))-(1.0/(30.0*D**5))+(1  A  95
         1.0/(42.0*D**7))-(1.0/(30.0*D**9))+(1.0/((BETA+1.0)**2))+(1.0/(BETA  A  96
         2**2))                                                          A  97
          V=A-(XN**2)/B                                                  A  98
          U=A                                                            A  99
          W=(B/XN)-(XN/A)                                                A 100
          DU=R                                                           A 101
          DV=R+(XN**3)/(B**2)                                            A 102
          DW=1.0+(XN*R)/(A**2)                                           A 103
          FPN=XN*TRI*((V*DU-U*DV)/(V**2))-A-XN*DW/W                      A 104
          AS=GML-(FCN/FPN)                                               A 105
          WRITE (6,19) ICOUNT,AS,FCN                                     A 106
          DELTA=ABS(0.00000001*AS)                                       A 107
          IF (ABS(AS-GML).LT.DELTA) GO TO 6                             A 108
          IF (ICOUNT.GT.25) GO TO 8                                      A 109
          GML=AS                                                         A 110
          GO TO 4                                                        A 111
    6     CONTINUE                                                       A 112
          GAMMA=AS                                                       A 113
          M1=GAMMA+ALPHA*BETA                                            A 114
          M2=BETA*ALPHA**2                                               A 115
          SKEW=2.0/SQRT(BETA)                                            A 116
          WRITE (6,20)                                                   A 117
          WRITE (6,22) ALPHA,M1                                          A 118
          WRITE (6,23) BETA,M2                                           A 119
          WRITE (6,24) GAMMA,SKEW                                        A 119
          WRITE (6,20)                                                   A 120
          D=BETA+2.0                                                     A 121
          TRI=(1.0/D)+(1.0/(2.0*D**2))+(1.0/(6.0*D**3))-(1.0/(30.0*D**5))+(1  A 122
         1.0/(42.0*D**7))-(1.0/(30.0*D**9))+(1.0/((BETA+1.0)**2))+(1.0/((BETA  A 123
         2**2))                                                          A 124
          H=(BETA-2.0)*ALPHA**4                                          A 125
          P=2.0*TRI-(2.0*BETA-3.0)/((BETA-1.0)**2)                       A 126
          DET=P/H                                                        A 127
          VARA=(1.0/(XN*(ALPHA**2)*DET))*((TRI/(BETA-2.0))-1.0/((BETA-1.0)**  A 128
         12))                                                            A 129
          VARB=2.0/(XN*DET*(BETA-2.0)*ALPHA**4)                          A 130
          VARG=(BETA*TRI-1.0)/(XN*DET*ALPHA**2)                          A 131
          COVAB=(-1.0/(XN*DET*ALPHA**3))*(1.0/(BETA-2.0))-(1.0/(BETA-1.0)))   A 132
          COVAG=(1.0/(XN*DET*ALPHA**2))*((1.0/(BETA-1.0))-TRI)           A 133
          COVBG=(-1.0/(XN*DET*ALPHA**3))*((BETA/(BETA-1.0))-1.0)         A 134
          DO 7 J=1,6                                                     A 135
          T=SND(J)                                                       A 136
          E=BETA**(1./3.)-1.0/(9.0*BETA**(2./3.))+T/(3.0*BETA**(1./6.))  A 137
          F=1.0/(3.0*BETA**(2./3.))+2.0/(27.0*BETA**(5./3.))-T/(18.0*BETA**(  A 138
         17./6.))                                                        A 139
          XT(J)=GAMMA+ALPHA*E**3                                         A 140
          DXDA=E**3                                                      A 141
          DXDB=3.0*ALPHA*E**2*F                                          A 142
          DXDG=1.0                                                       A 143
          SX(J)=SQRT(VARA*DXDA**2+VARB*DXDB**2+VARG*DXDG**2+2.0*DXDA*DXDB*CO  A 144
         1VAB+2.0*DXDA*DXDG*COVAG+2.0*DXDB*DXDG*COVBG)                   A 145
    7     CONTINUE                                                       A 146
          WRITE (6,14)                                                   A 147
          WRITE (6,15) (XT(J),J=1,6)                                     A 148
          WRITE (6,16) (SX(J),J=1,6)                                     A 149
          WRITE (6,21)                                                   A 150
    8     CONTINUE                                                       A 151
          STOP                                                           A 152
    C                                                                    A 153
    C                                                                    A 154
    9     FORMAT (80A1)                                                  A 155
   10     FORMAT (I5)                                                    A 156
   11     FORMAT (8F10.0)                                                A 157
   12     FORMAT (1H1,/,80A1//,26X,28HPEARSON TYPE 3 DISTRIBUTION ,/)    A 158
   13     FORMAT (31X,17HMETHOD OF MOMENTS,//)                           A 159
```

131

```
 14    FORMAT (3X,7HT,YEARS,4X,1H2,11X,1H5,10X,2H10,10X,2H20,10X,2H50,9X,    A 160
       13H100,/)                                                            A 161
 15    FORMAT (3X,1HX,3X,6E12.5,/,4X,1HT)                                    A 162
 16    FORMAT (3X,1HS,3X,6E12.5,/,4X,1HT,//)                                 A 163
 17    FORMAT (25X,28HMAXIMUM LIKELIHOOD PROCEDURE,//)                       A 164
 18    FORMAT (21X,5HTRIAL,11X,1HG,11X,4HF(G),/)                             A 165
 19    FORMAT (22X,I2,8X,E12.5,1X,E12.5)                                     A 166
 20    FORMAT (//)                                                          A 167
 21    FORMAT (/1H1)                                                        A 168
 22    FORMAT (9X,5HALPHA,5X,E12.5,14X,4HM1  ,6X,E12.5)                      A 169
 23    FORMAT (9X,5HBETA ,5X,E12.5,14X,4HM2  ,6X,E12.5)                      A 170
 24    FORMAT (9X,5HGAMMA,5X,E12.5,14X,4H5KEW,6X,E12.5,/)                    A 171
       END                                                                  A 172-

 10 '
 20 ' Program PT3
 30 ' Copyright G.W. Kite 1986
 40 '
 50 ' Compute method of moments and maximum likelihood estimates
 60 ' for T year events and standard errors
 70 ' for a Pearson Type III Distribution.
 80 '
 90 DIM T#(6),X#(250),XT#(6),SX#(6),SND#(6)
 91 FORM$="  ##.###^^^^"
 92 VV$=CHR$(179)
 100 DATA 0.#,0.84162#,1.28155#,1.64485#,2.05375#,2.32635#
 110 FOR I%=1 TO 6
 120 READ SND#(I%)
 130 NEXT I%
 140 T$="Pearson Type III Distribution"
 150 IERR%=0
 160 GOSUB 3000
 170 IF IERR% = 1 GOTO 2140
 230 FLAG%=0:IF YN$ = "Y" OR YN$ = "y" THEN FLAG%=1
 240 NREC%=0
 250 OPEN FILE$ FOR INPUT AS #1
 260 LINE INPUT#1,TITLE$
 270 IF EOF(1) GOTO 310
 280 NREC%=NREC%+1
 290 LINE INPUT#1,I$
 291 X#(NREC%)=VAL(MID$(I$,5,12))
 300 GOTO 270
 310 CLOSE#1
 311 NREC#=NREC%
 320 A%=LEN(TITLE$)
 330 B%=(80-A%)/2
 340 U$=STRING$(A%,205)
 350 M$="Method of Moments"
 360 E%=LEN(M$)
 370 F%=(80-E%)/2
 380 W$=STRING$(E%,205)
 390 CLS
 400 PRINT TAB(B%) TITLE$:PRINT TAB(B%) U$
 410 PRINT TAB(D%) T$:PRINT TAB(D%) V$
 420 PRINT TAB(F%) M$:PRINT TAB(F%) W$
 430 IF FLAG% <> 1 GOTO 480
 440 LPRINT CHR$(12):LPRINT:LPRINT
 450 LPRINT TAB(B%) TITLE$:LPRINT TAB(B%) U$
 460 LPRINT:LPRINT TAB(D%) T$:LPRINT TAB(D%) V$
 470 LPRINT:LPRINT TAB(F%) M$:LPRINT TAB(F%) W$
 480 A#=0#:B#=0#:C#=0#
 490 FOR I%=1 TO NREC%
 500 A#=A#+X#(I%)
 510 B#=B#+X#(I%)*X#(I%)
 520 C#=C#+X#(I%)*X#(I%)*X#(I%)
 530 NEXT I%
 540 M1#=A#/NREC#
 550 M2#=(B#/NREC#)-M1#*M1#
 570 M3#=(C#/NREC#)+2#*M1#*M1#*M1#-3#*M1#*(B#/NREC#)
 580 SKEW#=M3#/(M2#^1.5#)
 590 C1#=(SQR(NREC#*(NREC#-1#)))/(NREC#-2#)
 600 C2#=1#+8.5#/NREC#
 610 C3#=NREC#/(NREC#-1#)
 620 SKEW#=SKEW#*C1#*C2#
 630 M2#=M2#*C3#
 640 BETA#=(2#/SKEW#)*(2#/SKEW#)
 650 ALPHA#=SQR(M2#)/SQR(BETA#)
 660 GAMMA#=M1#-SQR(M2#)*SQR(BETA#)
 670 PRINT
```

132

```
680 PRINT TAB(20) "Mean is               ";:PRINT USING FORM$;M1#
690 PRINT TAB(20) "Variance is           ";:PRINT USING FORM$;M2#
700 PRINT TAB(20) "Coefficient of skew is ";:PRINT USING FORM$;SKEW#
710 PRINT TAB(20) "Parameter Alpha is    ";:PRINT USING FORM$;ALPHA#
720 PRINT TAB(20) "Parameter Beta is     ";:PRINT USING FORM$;BETA#
730 PRINT TAB(20) "Parameter Gamma is    ";:PRINT USING FORM$;GAMMA#
740 IF FLAG% <> 1 GOTO 820
750 LPRINT
760 LPRINT TAB(20) "Mean is               ";:LPRINT USING FORM$;M1#
770 LPRINT TAB(20) "Variance is           ";:LPRINT USING FORM$;M2#
780 LPRINT TAB(20) "Coefficient of skew is ";:LPRINT USING FORM$;SKEW#
790 LPRINT TAB(20) "Parameter Alpha is    ";:LPRINT USING FORM$;ALPHA#
800 LPRINT TAB(20) "Parameter Beta is     ";:LPRINT USING FORM$;BETA#
810 LPRINT TAB(20) "Parameter Gamma is    ";:LPRINT USING FORM$;GAMMA#
820 FOR J%=1 TO 6
830 T#=SND#(J%)
840 T1#=T#
850 T2#=(T#*T#-1)/6#
860 T3#=2#*(T#*T#*T#-6#*T#)/216#
870 T4#=(T#*T#-1#)/216#
880 T5#=T#/1296#
890 T6#=.0000428669#
900 K#=T1#+T2#*SKEW#+T3#*SKEW#*SKEW#-T4#*SKEW#^3#+T5#*SKEW#^4#-T6#*SKEW#^5#
910 SLOPE#=T2#+T3#*2#*SKEW#-T4#*3#*SKEW#*SKEW#+T5#*4#*SKEW#^3#-T6#*5#*SKEW#^4#
920 T7#=(1#+.75#*SKEW#*SKEW#)*(.5#*K#*K#)
930 T8#=K#*SKEW#
940 T9#=6#*(1#+.25#*SKEW#*SKEW#)*SLOPE#
941 T10#=SLOPE#*(1#+1.25#*SKEW#*SKEW#)+(SKEW#*K#/2#)
950  DELTA#=1.0#+T7#+T8#+T9#*T10#
960 XT#(J%)=M1#+K#*SQR(M2#)
970 SX#(J%)=SQR(M2#*DELTA#/NREC#)
980 NEXT J%
981 ROW$=CHR$(218)+STRING$(72,196)
990 PRINT:PRINT TAB(2) "T, years  2           5          10         20
     50          100":PRINT TAB(5) ROW$
1000 PRINT TAB(2) "X  " VV$;:FOR J%=1 TO 6:PRINT USING FORM$;XT#(J%);:NEXT J%
1010 PRINT TAB(2) " t " VV$
1020 PRINT TAB(2) "S  " VV$;:FOR J%=1 TO 6:PRINT USING FORM$;SX#(J%);:NEXT J%
1030 PRINT TAB(2) " t " VV$
1040 IF FLAG% <> 1 GOTO 1100
1050 LPRINT:LPRINT TAB(2) "T, years  2           5          10         20
     50          100":LPRINT TAB(5) ROW$
1060 LPRINT TAB(2) "X  " VV$;:FOR J%=1 TO 6:LPRINT USING FORM$;XT#(J%);:NEXT J%
1070 LPRINT TAB(2) " t " VV$
1080 LPRINT TAB(2) "S  " VV$;:FOR J%=1 TO 6:LPRINT USING FORM$;SX#(J%);:NEXT J%
1090 LPRINT TAB(2) " t " VV$
1100 M$="Method of Maximum Likelihood"
1110 E%=LEN(M$)
1120 F%=(80-E%)/2
1130 W$=STRING$(E%,205)
1140 PRINT:PRINT"Press any key to continue"
1150 Q$=INKEY$:IF Q$ ="" GOTO 1150
1160 CLS
1170 PRINT:PRINT TAB(F%) M$:PRINT TAB(F%) W$
1180 PRINT:PRINT TAB(19) "Iteration" TAB(37) "G" TAB(53) "f(G)"
1190 PRINT TAB(19) "---------" TAB(37) "-" TAB(53) "----"
1200 IF FLAG% <> 1 GOTO 1240
1210 LPRINT:LPRINT TAB(F%) M$:LPRINT TAB(F%) W$
1220 LPRINT:LPRINT TAB(19) "Iteration" TAB(37) "G" TAB(53) "f(G)"
1230 LPRINT TAB(19)"---------" TAB(37) "-" TAB(53) "----"
1240 COUNT%=0
1250 XMIN#=100000000#
1260 FOR I% = 1 TO NREC%
1270 IF X#(I%) < XMIN# THEN XMIN#=X#(I%)
1280 NEXT I%
1290 GML#=XMIN#*.99#
1300 COUNT%=COUNT%+1
1310 A#=0#:B#=0#:C#=0#:R#=0#
1320 FOR I%=1 TO NREC%
1330 TEMP#=X#(I%)-GML#
1340 A#=A#+1#/TEMP#
1350 B#=B#+TEMP#
1360 C#=C#+LOG(TEMP#)
1370 R#=R#+1#/(TEMP#*TEMP#)
1380 NEXT I%
1390 BETA#=A#/(A#-(NREC#*NREC#/B#))
1400 ALPHA#=B#/(NREC#*BETA#)
1410 D#=BETA#+2#
1420 PSI#=LOG(D#)-(1#/(2#*D#))-(1#/(12#*D#*D#))+(1#/(120#*D#^4#))-(1#/(252#*D#^6
```

133

```
#))-(1#/(BETA#+1#))-(1#/BETA#)
1430 FCN#=-NREC#*PSI#+C#-NREC%*LOG(ALPHA#)
1440 TRI#=(1#/D#)+(1#/(2#*D#*D#))+(1#/(6#*D#*D#*D#))-(1#/(30#*D#^5#))+(1#/(42#*D
#^7#))-(1#/(30#*D#^9#))+(1#/((BETA#+1#)*(BETA#+1#)))+(1#/(BETA#*BETA#)))
1450 V#=A#-(NREC#*NREC#)/B#
1460 U#=A#
1470 W#=(B#/NREC#)-(NREC#/A#)
1480 DU#=R#
1481 DV#=R#-(NREC#*NREC#*NREC#)/(B#*B#)
1490 DW#=-1#+(NREC#*R#)/(A#*A#)
1500 FPN#=-NREC#*TRI#*((V#*DU#-U#*DV#)/(V#*V#))-A#-NREC#*DW#/W#
1520 AAS#=GML#-(FCN#/FPN#)
1530 PRINT TAB(22);:PRINT USING "##";COUNT%;:PRINT USING "        ##.###^^^^";AA
S#,FCN#
1540 IF FLAG% = 1 THEN LPRINT TAB(22);:LPRINT USING "##";COUNT%;:LPRINT USING "
     ##.###^^^^";AAS#,FCN#
1550 DELTA#=ABS(.00000001#*AAS#)
1560 IF ABS(AAS#-GML#) < DELTA# GOTO 1600
1570 IF COUNT% > 24 THEN PRINT:PRINT "Procedure does not converge":PRINT:GOTO 21
20
1580 GML#=AAS#
1590 GOTO 1300
1600 GAMMA#=AAS#
1610 M1#=GAMMA#+ALPHA#*BETA#
1620 M2#=BETA#*ALPHA#*ALPHA#
1630 SKEW#=2#/SQR(BETA#)
1640 D#=BETA#+2#
1650 TRI#=(1#/D#)+(1#/(2#*D#*D#))+(1#/(6#*D#*D#*D#))-(1#/(30#*D#^5#))+(1#/(42#*D
#^7#))-(1#/(30#*D#^9#))+(1#/((BETA#+1#)*(BETA#+1#)))+(1#/(BETA#*BETA#)))
1660 H#=(BETA#-2#)*ALPHA#^4#
1670 P#=2#*TRI#-(2#*BETA#-3#)/((BETA#-1#)*(BETA#-1#))
1680 DET#=P#/H#
1690 VARA#=(1#/(NREC#*ALPHA#*ALPHA#*DET#))*((TRI#/(BETA#-2#))-1#/((BETA#-1#)*(BE
TA#-1#)))
1700 VARB#=2#/(NREC#*DET#*(BETA#-2#)*ALPHA#^4#)
1710 VARG#=(BETA#*TRI#-1#)/(NREC#*DET#*ALPHA#*ALPHA#)
1720 COVAB#=(-1#/(NREC#*DET#*ALPHA#*ALPHA#*ALPHA#))*((1#/(BETA#-2#))-(1#/(BETA#-
1#)))
1730 COVAG#=(1#/(NREC#*DET#*ALPHA#*ALPHA#))*((1#/(BETA#-1#))-TRI#)
1740 COVBG#=(-1#/(NREC#*DET#*ALPHA#*ALPHA#*ALPHA#))*((BETA#/(BETA#-1#))-1#)
1750 FOR J% = 1 TO 6
1760 T#=SND#(J%)
1770 E#=BETA#^(1#/3#)-1#/(9#*BETA#^(2#/3#))+T#/(3#*BETA#^(1#/6#))
1780 F#=1#/(3#*BETA#^(2#/3#))+2#/(27#*BETA#^(5#/3#))-T#/(18#*BETA#^(7#/6#))
1790 XT#(J%)=GAMMA#+ALPHA#*E#*E#*E#
1800 DXDA#=E#*E#*E#
1810 DXDB#=3#*ALPHA#*E#*E#*F#
1820 DXDG#=1#
1830 SX#(J%)=SQR(VARA#*DXDA#*DXDA#+VARB#*DXDB#*DXDB#+VARG#*DXDG#*DXDG#+2#*DXDA#*
DXDB#*COVAB#+2#*DXDA#*DXDG#*COVAG#+2#*DXDB#*DXDG#*COVBG#)
1840 NEXT J%
1870 CLS
1880 PRINT:PRINT TAB(B%) TITLE$:PRINT TAB(B%) U$
1890 PRINT TAB(D%) T$:PRINT TAB(D%) V$
1900 PRINT TAB(F%) M$:PRINT TAB(F%) W$
1910 PRINT TAB(19) "Mean is                   ";:PRINT USING FORM$;M1#
1920 PRINT TAB(19) "Variance is               ";:PRINT USING FORM$;M2#
1921 PRINT TAB(19) "Coefficient of skew is    ";:PRINT USING FORM$;SKEW#
1930 PRINT TAB(19) "Parameter Alpha is        ";:PRINT USING FORM$;ALPHA#
1940 PRINT TAB(19) "Parameter Beta is         ";:PRINT USING FORM$;BETA#
1941 PRINT TAB(19) "Parameter Gamma is        ";:PRINT USING FORM$;GAMMA#
1950 IF FLAG% <> 1 GOTO 2010
1960 LPRINT
1970 LPRINT TAB(19) "Mean is                   ";:LPRINT USING FORM$;M1#
1980 LPRINT TAB(19) "Variance is               ";:LPRINT USING FORM$;M2#
1981 LPRINT TAB(19) "Coefficient of skew is    ";:LPRINT USING FORM$;SKEW#
1990 LPRINT TAB(19) "Parameter Alpha is        ";:LPRINT USING FORM$;ALPHA#

2000 LPRINT TAB(19) "Parameter Beta is         ";:LPRINT USING FORM$;BETA#
2001 LPRINT TAB(19) "Parameter Gamma is        ";:LPRINT USING FORM$;GAMMA#

2010 PRINT:PRINT TAB(2) "T, years  2            5           10          20
  50         100":PRINT TAB(5) ROW$
2020 PRINT TAB(2) "X  " VV$;:FOR J%=1 TO 6:PRINT USING FORM$;XT#(J%);:NEXT J%
2030 PRINT TAB(2) " t " VV$
2040 PRINT TAB(2) "S " VV$;:FOR J%=1 TO 6:PRINT USING FORM$;SX#(J%);:NEXT J%
2050 PRINT TAB(2) " t " VV$
2060 IF FLAG% <> 1 GOTO 2120
2070 LPRINT:LPRINT TAB(2) "T, years  2            5           10          20
  50         100":LPRINT TAB(5) ROW$
```

```
2080 LPRINT TAB(2) "X  " VV$;:FOR J%=1 TO 6:LPRINT USING FORM$;XT#(J%);:NEXT J%
2090 LPRINT TAB(2) " t " VV$
2100 LPRINT TAB(2) "S  " VV$;:FOR J%=1 TO 6:LPRINT USING FORM$;SX#(J%);:NEXT J%
2110 LPRINT TAB(2) " t " VV$
2120 PRINT:PRINT"Press any key to continue"
2130 Q$=INKEY$:IF Q$ ="" GOTO 2130
2140 CLS
2150 SYSTEM
3000 '
3010 ' Subroutine for standard screen format
3020 '
3030 CLS:PRINT:PRINT
3040 ROWS=STRING$(78,205)
3050 BOXTOP$=CHR$(201)+ROW$+CHR$(187)
3060 PRINT BOXTOP$;
3070 PRINT CHR$(186) TAB(80) CHR$(186);
3080 PRINT CHR$(186) TAB(80) CHR$(186);
3090 C%=LEN(T$)
3100 D%=(80-C%)/2
3110 V$=STRING$(C%,205)
3140 PRINT CHR$(186) TAB(D%) T$ TAB(80) CHR$(186);
3150 PRINT CHR$(186) TAB(D%) V$ TAB(80) CHR$(186);
3160 PRINT CHR$(186) TAB(80) CHR$(186);
3170 PRINT CHR$(186) TAB(80) CHR$(186);
3180 PRINT CHR$(186) "   What is the name of your data file?" TAB(80) CHR$(186);
3190 PRINT CHR$(186) TAB(80) CHR$(186);
3200 PRINT CHR$(186) "   Do you want printer output (Y/N)?" TAB(80) CHR$(186);
3201 PRINT CHR$(186) "   (press Alt Q to quit at this stage)" TAB(80) CHR$(186+;
3210 PRINT CHR$(186) TAB(80) CHR$(186);
3220 PRINT CHR$(186) TAB(80) CHR$(186+;
3230 BOXBOT$=CHR$(200)+ROW$+CHR$(188)
3240 PRINT BOXBOT$;
3241 LOCATE 24,1,0,0,0
3242 PRINT "G Kite";
3250 LOCATE 10,41,1,0,13
3260 FILE$=""
3270 I$=INPUT$(1)
3280 IF I$ = CHR$(13) GOTO 3380                    ' ENTER key
3290 IF I$ = CHR$(8)  GOTO 3330                    ' BACKSPACE key
3300 PRINT I$;
3310 FILE$=FILE$+I$
3320 GOTO 3270
3330 H%=POS(0)
3340 LOCATE 10,H%-1,1,0,13
3350 L%=LEN(FILE$)
3360 FILE$=LEFT$(FILE$,L%-1)
3370 GOTO 3270
3380 LOCATE 12,39,1,0,13
3381 YN$=""
3390 I$=INKEY$:IF I$ = "" GOTO 3390
3400 IF LEN(I$) = 1 GOTO 3410
3401 I$=RIGHT$(I$,1)
3402 IF I$ = CHR$(16) THEN IERR%=1:GOTO 3490        ' ALT Q key
3403 GOTO 3380
3410 IF I$ = CHR$(13) GOTO 3440                    ' ENTER key
3420 PRINT I$;
3421 YN$=I$
3430 GOTO 3390
3440 IF YN$ = "Y" OR YN$ = "y" OR YN$ = "N" OR YN$ = "n" GOTO 3460
3450 GOTO 3380
3460 LOCATE ,,0,13,13
3490 RETURN
```

## Pearson Type III Distribution

### Method of Moments

| | |
|---|---|
| Mean is | 4.132D+02 |
| Variance is | 2.067D+04 |
| Coefficient of skew is | 1.394D+00 |
| Parameter Alpha is | 1.002D+02 |
| Parameter Beta is | 2.058D+00 |
| Parameter Gamma is | 2.069D+02 |

| T, years | 2 | 5 | 10 | 20 | 50 | 100 |
|---|---|---|---|---|---|---|
| $X_t$ | 3.815D+02 | 5.137D+02 | 6.038D+02 | 6.902D+02 | 8.015D+02 | 8.844D+02 |
| $s_t$ | 1.254D+01 | 2.150D+01 | 3.260D+01 | 5.071D+01 | 8.157D+01 | 1.085D+02 |

### Method of Maximum Likelihood

| Iteration | G | f(G) |
|---|---|---|
| 1 | 1.849D+02 | 9.025D+00 |
| 2 | 1.794D+02 | 4.240D+00 |
| 3 | 1.718D+02 | 1.863D+00 |
| 4 | 1.638D+02 | 7.245D-01 |
| 5 | 1.585D+02 | 2.255D-01 |
| 6 | 1.569D+02 | 4.309D-02 |
| 7 | 1.568D+02 | 2.495D-03 |
| 8 | 1.568D+02 | 9.642D-06 |
| 9 | 1.568D+02 | 1.326D-10 |

| | |
|---|---|
| Mean is | 4.132D+02 |
| Variance is | 2.030D+04 |
| Coefficient of skew is | 1.111D+00 |
| Parameter Alpha is | 7.918D+01 |
| Parameter Beta is | 3.238D+00 |
| Parameter Gamma is | 1.568D+02 |

| T, years | 2 | 5 | 10 | 20 | 50 | 100 |
|---|---|---|---|---|---|---|
| $X_t$ | 3.877D+02 | 5.185D+02 | 6.032D+02 | 6.824D+02 | 7.822D+02 | 8.552D+02 |
| $s_t$ | 1.609D+01 | 2.463D+01 | 3.330D+01 | 4.309D+01 | 5.714D+01 | 6.838D+01 |

# CHAPTER 10
## LOG-PEARSON TYPE III DISTRIBUTION

Introduction

The U. S. Federal Water Resources Council recommended in
1967 that the log-Pearson type III distribution be adopted as
the standard flood frequency distribution by all U. S. government agencies. In describing the investigations behind this
recommendation, Benson (3) explained that no rigorous statistical criteria exist on which a comparison of distributions can
be based and therefore the choice of the log-Pearson type III
was, to some extent, subjective.

If the logarithms, ln x, of a variable x are distributed
as a Pearson type III variate then the variable x will be distributed as a log-Pearson type III with probability density
function

$$p(x) = \frac{1}{\alpha x \Gamma(\beta)} \left\{ \frac{\ln x - \gamma}{\alpha} \right\}^{\beta-1} e^{-\left\{ \frac{\ln x - \gamma}{\alpha} \right\}} \qquad (10-1)$$

where $\alpha$, $\beta$ and $\gamma$ are the scale, shape and location parameters
respectively.

Being a 3-parameter distribution operating upon the logarithms of the variable the log-Pearson type III would appear to
be an extremely versatile distribution. However, its applicability in hydrology is strictly limited. As shown by Bobée
(1) the pdf of the log-Pearson type III may be J-shaped, reverse
J-shaped, U-shaped, inverted U-shaped, inverted U with inflex-
ions, bell shaped with an upper bound, bell shaped with a lower
bound, etc, etc.. For flood frequency analysis the only shape
of interest is that which is unimodal, continuous from 0 to $+\infty$,
has either an infinitely high order ($\partial p/\partial x = \infty$) or smooth con-
tact ($\partial p/\partial x = 0$) with the lower limit and is unbounded at the
upper limit. The log-Pearson type III falls within these criteria only when $\beta > 1$ and when $1/\alpha > 0$. If the coefficient of
skew, $\gamma$, is negative this corresponds to a negative value of $\alpha$
and is not suitable. Reich (5) has given examples of the application of the log-Pearson type III distribution to samples with
negative skew in which the computed upper bounds have been lower
than the maximum observed events.

Estimation of Parameters

As with the lognormal distribution, two procedures are
available to estimate the parameters of a log-Pearson type III
distribution. Either the distribution can be fitted directly
to the data or else the Pearson type III distribution can be

fitted to the logarithms of the events. In the first case the method of moments or maximum likelihood can be applied. In the second case the maximum likelihood method is the same as in the first case and only the method of moments gives new results. As with the lognormal distribution care must be taken with events of zero magnitude. Jennings and Benson (4) describe a procedure of coping with these events.

*Method of Moments (Direct Application of Log-Pearson type III)* - Applying the moment generating function to the pdf of the log-Pearson type III distribution yields, following Bobée (1)

$$\mu_r' = \frac{e^{\gamma r}}{(1 - r\alpha)^\beta} \tag{10-2}$$

Evaluating this expression for the first, second and third moments about the origin gives

$$\ln \mu_1' = \gamma - \beta \ln (1 - \alpha) \tag{10-3}$$

$$\ln \mu_2' = 2\gamma - \beta \ln (1 - 2\alpha) \tag{10-4}$$

$$\ln \mu_3' = 3\gamma - \beta \ln (1 - 3\alpha) \tag{10-5}$$

Manipulating these three equations gives

$$\frac{\ln \mu_3' - 3 \ln \mu_1'}{\ln \mu_2' - 2 \ln \mu_1'} = \frac{\ln [(1 - \alpha)^3/(1 - 3\alpha)]}{\ln [(1 - \alpha)^2/(1 - 2\alpha)]} \tag{10-6}$$

The left hand side of Equation 10-6, call it B,

$$B = \frac{\ln \mu_3' - 3 \ln \mu_1'}{\ln \mu_2' - 2 \ln \mu_1'} \tag{10-7}$$

can be calculated directly from the sample moments about the origin. It is then necessary to compute $\alpha$ from B. This is difficult analytically in Equation 10-6 but by substituting

$$A = 1/\alpha - 3 \tag{10-8}$$

and

$$C = 1/(B - 3) \tag{10-9}$$

a series of polynomials can be derived.

Figure 10-1 shows the shape of the function relating $\alpha$ to B. From this figure and noting the restrictions on the useful range of $\alpha$, it is only necessary to relate A and C for $3 < B \leq 6$. Two regressions have been developed to cover this range:

138

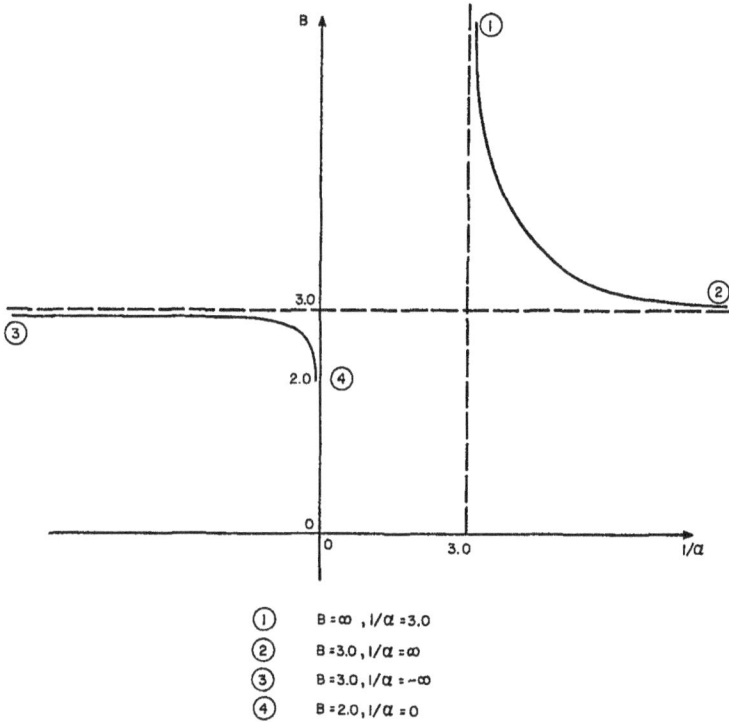

|     |                         |
|-----|-------------------------|
| ①   | $B = \infty$ , $1/\alpha = 3.0$ |
| ②   | $B = 3.0$ , $1/\alpha = \infty$ |
| ③   | $B = 3.0$ , $1/\alpha = -\infty$ |
| ④   | $B = 2.0$ , $1/\alpha = 0$ |

Figure 10-1.   Relationship between $\alpha$ and B in equation 10-8. (not to scale)

(a)   for $3.5 < B \leq 6.0$

$$A = -0.23019 + 1.65262C + 0.20911C^2 - 0.04557C^3 \quad (10\text{-}10)$$

(b)   for $3.0 < B \leq 3.5$

$$A = -0.47157 + 1.99955C \quad (10\text{-}11)$$

Having computed B and $\alpha$ the remaining parameters, $\beta$ and $\gamma$, can be obtained from reorganizations of Equations 10-3 and 10-4 as

$$\beta = \frac{\ln \mu'_2 - 2 \ln \mu'_1}{\ln (1 - \alpha)^2 - \ln (1 - 2\alpha)} \quad (10\text{-}12)$$

$$\gamma = \ln \mu'_1 + \beta \ln (1 - \alpha) \quad (10\text{-}13)$$

The parameters $\alpha$, $\beta$ and $\gamma$ are then used to compute the mean, $\mu_y$, standard deviation, $\sigma_y$, and coefficient of skew, $\gamma_y$, of the logarithms of x as:

$$\mu_y = \gamma + \alpha\beta \qquad (10\text{-}14)$$

$$\sigma_y = \alpha\sqrt{\beta} \qquad (10\text{-}15)$$

and

$$\gamma_y = 2.0/\sqrt{\beta} \qquad (10\text{-}16)$$

The coefficient of skew, $\gamma_y$, should then be corrected for bias as in Equation 9.22 and the standard deviation should be multiplied by $\sqrt{N/(N-1)}$.

Bobée (2) has compared the application of the direct method of moments, indirect method of moments and method of maximum likelihood to 18 long term records of annual flood peaks and has concluded that the direct method of moments gives the best results.

*Method of Moments (Application of Pearson type III to the Logarithm of x)* - Calculating the mean, $\mu_y$, the standard deviation, $\sigma_y$, and the coefficient of skew, $\gamma_y$, of the logarithms of x, the parameters $\alpha$, $\beta$ and $\gamma$ are found following the procedure outlined in Chapter 9 using the relationships given in Equations 10-14 - 10-16. The standard deviation and coefficient of skew should be corrected for bias, as before.

*Maximum Likelihood* - For a data sample $x_1, \ldots, x_n$ the logarithm of the likelihood function is given by

$$\ln L = -\sum_{i=1}^{n} \ln x - n \ln \Gamma(\beta) - \frac{\sum_{i=1}^{n} (\ln x_i - \gamma)}{\alpha}$$

$$+ (\beta - 1) \sum_{i=1}^{n} \ln (\ln x_i - \gamma) - n\beta \ln \alpha \qquad (10\text{-}17)$$

Differentiating with respect to $\alpha$, $\beta$ and $\gamma$ and equating to zero yields:

$$\sum_{i=1}^{n} (\ln x_i - \alpha) = n\alpha\beta \qquad (10\text{-}18)$$

$$n\psi(\beta) = \sum_{i=1}^{n} \ln [(\ln x - \gamma)/\alpha] \qquad (10\text{-}19)$$

$$n = \alpha(\beta - 1) \sum_{i=1}^{n} 1/(\ln x_i - \gamma) \qquad (10\text{-}20)$$

where, as before, $\psi(\beta)$ is the psi or digamma function which can be calculated by an asymptotic expansion.

Eliminating $\alpha$ in 10-20 by means by 10-18 yields an expression in x, $\beta$ and $\gamma$. Substitution in 10-19 leaves an expression in $\gamma$ only which can be solved numerically.

### Frequency Factor

Applying the Pearson type III distribution to the logarithms of the sample events the T-year event can be computed from

$$y_T = \ln x_T = \mu_y + K\sigma_y \qquad (10\text{-}21)$$

where $\mu_y$ and $\sigma_y$ are the mean and standard deviation of the logarithms of x and K is obtained from Table 9-2 or Equation 9-52 using $\gamma_y$, the coefficient of skew of the logarithms, in place of $\gamma_1$, the coefficient of skew of the events.

There is no need to derive a frequency factor without logarithmic components, as was done with the lognormal distribution, since the logarithmic moments can be derived directly from the moments of the observations using equations 10-14 - 10-16.

### Standard Error

*Method of Moments* - The standard error of the T-year event for the log-Pearson type III distribution may be computed using first Equation 9-56 to obtain $S_{T,y}$ in log units from the standard normal deviate and the coefficient of skew of the logarithms of the observed events.

The standard error is then converted back to linear units as

$$S_{T,x} = \frac{x_T (e^{S_{T,y}} - e^{-S_{T,y}})}{2.0} \qquad (10\text{-}22)$$

where $S_{T,x}$ is now the average of the positive and negative standard errors in linear units and $x_T$ is the T-year event.

*Maximum Likelihood* - The standard error in log units is obtained from Equation 9-57 and subsequent development using parameters $\alpha$, $\beta$ and $\gamma$ derived from section on Maximum Likelihood (p. 126). The logarithmic standard error is then converted to linear units using equation 10-22 above.

## References

1.  Bobée, B., 1975, The Log-Pearson Type III Distribution and its Application in Hydrology, Wat. Res. Res., Vol. 11, No. 5, pp. 681-689.

2.  Bobée, B. and R. Robitaille, 1976, The Use of the Pearson Type 3 and log-Pearson Type 3 Distributions Revisited, Wat. Res. Res., Vol. 13, No. 2, April, pp. 427-443.

3.  Benson, M. A., 1968, Uniform Flood Frequency Estimating Methods for Federal Agencies, Wat. Res. Res., Vol. 4, No. 5, pp. 891-908.

4.  Jennings, M. E. and M. A. Benson, 1969, Frequency Curves for Annual Flood Series With Some Zero Events or Incomplete Data, Wat. Res. Res., Vol. 5, No. 1, pp. 276-280.

5.  Reich, B. M., 1972, Log-Pearson Type 3 and Gumbel Analysis of Floods, Second International Symposium in Hydrology, Fort Collins, Colorado, pp. 290-303.

```
      PROGRAM LP3(INPUT,OUTPUT,TAPE5=INPUT,TAPE6=OUTPUT)          A    1
C                                                                 A    2
C                                                                 A    3
C     COMPUTES METHOD OF MOMENTS AND MAXIMUM LIKELIHOOD ESTIMATES FOR   A    4
C     T YEAR EVENTS AND STANDARD ERRORS FOR LOG-PEARSON TYPE 3   A    5
C     DISTRIBUTION                                                A    6
C     INPUT                                                       A    7
C     TITLE                                                       A    8
C     N NUMBER OF ANNUAL MAXIMUM EVENTS                           A    9
C     X SERIES OF EVENTS                                          A   10
C                                                                 A   11
C                                                                 A   12
      REAL M1,M2,M3,K,NSX,L1,L2,L3                                A   13
      DIMENSION SND(6), X(100)                                    A   14
      DIMENSION XT(6), TITLE(80), ST(6)                           A   15
      SND(1)=0.                                                   A   16
      SND(2)=0.84162                                              A   17
      SND(3)=1.28155                                              A   18
      SND(4)=1.64485                                              A   19
      SND(5)=2.05375                                              A   20
      SND(6)=2.32635                                              A   21
      READ (5,16) TITLE                                           A   22
      READ (5,17) N                                               A   23
      XN=N                                                        A   24
      READ (5,18) (X(I),I=1,N)                                    A   25
      C1=(SQRT(XN*(XN-1.0)))/(XN-2.0)                             A   26
      C2=1.0+8.5/XN                                               A   27
      C3=XN/(XN-1.0)                                              A   28
      WRITE (6,19) TITLE                                          A   29
      WRITE (6,20)                                                A   30
      A=0.0                                                       A   31
      B=0.0                                                       A   32
      C=0.0                                                       A   33
      DO 1 I=1,N                                                  A   34
      A=A+X(I)                                                    A   35
      B=B+X(I)**2                                                 A   36
      C=C+X(I)**3                                                 A   37
1     CONTINUE                                                    A   38
      L1=A/XN                                                     A   39
      L2=B/XN                                                     A   40
      L3=C/XN                                                     A   41
      M1=A/XN                                                     A   42
      M2=(B/XN)-(A/XN)**2                                         A   43
      M3=(C/XN)-2.0*M1**3-3.0*M1*(B/XN)                           A   44
      SKEW=M3/(M2**1.5)                                           A   45
      B=(ALOG(L3)-3.0*ALOG(L1))/(ALOG(L2)-2.0*ALOG(L1))          A   46
      WRITE (6,34) L1,M1                                          A   47
      WRITE (6,35) L2,M2                                          A   48
      WRITE (6,36) L3,SKEW                                        A   49
      C=1.0/(B-3.0)                                               A   50
      IF (B.GT.6.0) GO TO 3                                       A   51
      IF (B.LE.3.0) GO TO 3                                       A   52
      IF (B.LE.3.5) GO TO 2                                       A   53
      A=-0.23019+1.65262*C+0.20911*C**2-0.04557*C**3             A   54
      GO TO 4                                                     A   55
2     A=-0.47157+1.99955*C                                       A   56
      GO TO 4                                                     A   57
3     WRITE (6,21) B                                             A   58
      GO TO 6                                                     A   59
4     ALPHA=1.0/(A+3.0)                                          A   60
      A1=ALOG(1.0-ALPHA)                                          A   61
      A2=ALOG(1.0-2.0*ALPHA)                                      A   62
      BETA=(ALOG(L2)-2.0*ALOG(L1))/(2.0*A1-A2)                   A   63
      GAMMA=ALOG(L1)+BETA*A1                                      A   64
      M1=GAMMA+ALPHA*BETA                                         A   65
      M2=BETA*ALPHA**2                                            A   66
      SKEW=2.0/SQRT(BETA)                                         A   67
      WRITE (6,23) ALPHA,M1                                       A   68
      WRITE (6,24) BETA,M2                                        A   69
      WRITE (6,25) GAMMA,SKEW                                     A   70
      IF (SKEW.LT.0.0) WRITE (6,33)                               A   71
      DO 5 J=1,6                                                  A   72
      T=SND(J)                                                    A   73
      T1=T                                                        A   74
      T2=(T**2-1.0)/6.0                                           A   75
      T3=2.0*(T**3-6.0*T)/6.0**3                                  A   76
      T4=(T**2-1.0)/6.0**3                                        A   77
      T5=T/6.0**4                                                 A   78
      T6=2.0/6.0**6                                               A   79
      K=T1+T2*SKEW+T3*SKEW**2-T4*SKEW**3+T5*SKEW**4-T6*SKEW**5    A   80
```

143

```
          SLOPE=T2+T3+2.0*SKEW-T4+3.0*SKEW**2+T5*4.0*SKEW**3-T6*5.0*SKEW**4   A  81
          T7=1.0                                                               A  82
          T8=SKEW*K                                                            A  83
          T9=(1.0+0.75*SKEW**2)*((K**2)/2.0)                                   A  84
          T10=3.0*SLOPE*K*(SKEW+0.25*SKEW**3)                                  A  85
          T11=3.0*(SLOPE**2)*(2.0+3.0*SKEW**2+(5.0/8.0)*SKEW**4)               A  86
          DELTA=T7+T8+T9+T10+T11                                              A  87
          XT(J)=EXP(M1+K*SQRT(M2))                                            A  88
          SX=SQRT(M2*DELTA/XN)                                                A  89
          PSX=XT(J)*(EXP(SX)-1.0)                                             A  90
          NSX=-XT(J)*(EXP(-SX)-1.0)                                           A  91
          ST(J)=(PSX+NSX)/2.0                                                 A  92
    5     CONTINUE                                                            A  93
          WRITE (6,26)                                                        A  94
          WRITE (6,27) (XT(J),J=1,6)                                          A  95
          WRITE (6,28) (ST(J),J=1,6)                                          A  96
    6     DO 7 I=1,N                                                          A  97
    7     X(I)=ALOG(X(I))                                                     A  98
          A=0.0                                                               A  99
          B=0.0                                                               A 100
          C=0.0                                                               A 101
          DO 8 I=1,N                                                          A 102
          A=A+X(I)                                                            A 103
          B=B+X(I)**2                                                         A 104
          C=C+X(I)**3                                                         A 105
    8     CONTINUE                                                            A 106
          M1=A/XN                                                             A 107
          M2=(B/XN)-(A/XN)**2                                                 A 108
          M3=(C/XN)+2.0*M1**3-3.0*M1*(B/XN)                                   A 109
          SKEW=M3/(M2**1.5)                                                   A 110
          SKEW=SKEW*C1*C2                                                     A 111
          M2=M2*C3                                                            A 112
          BETA=(2.0/SKEW)**2                                                  A 113
          ALPHA=(M2**0.5)/(BETA**0.5)                                         A 114
          GAMMA=M1-(M2**0.5)*(BETA**0.5)                                      A 115
          WRITE (6,22)                                                        A 116
          WRITE (6,23) ALPHA,M1                                              A 117
          WRITE (6,24) BETA,M2                                               A 118
          WRITE (6,25) GAMMA,SKEW                                            A 119
          IF (SKEW.LT.0.0) WRITE (6,33)                                       A 120
          DO 9 J=1,6                                                          A 121
          T=SND(J)                                                            A 122
          T1=T                                                                A 123
          T2=(T**2-1.0)/6.0                                                   A 124
          T3=2.0*(T**3-6.0*T)/6.0**3                                          A 125
          T4=(T**2-1.0)/6.0**3                                                A 126
          T5=T/6.0**4                                                         A 127
          T6=2.0/6.0**6                                                       A 128
          K=T1+T2*SKEW+T3*SKEW**2-T4*SKEW**3+T5*SKEW**4-T6*SKEW**5            A 129
          SLOPE=T2+T3+2.0*SKEW-T4+3.0*SKEW**2+T5*4.0*SKEW**3-T6*5.0*SKEW**4   A 130
          T7=1.0                                                              A 131
          T8=SKEW*K                                                           A 132
          T9=(1.0+0.75*SKEW**2)*((K**2)/2.0)                                  A 133
          T10=3.0*SLOPE*K*(SKEW+0.25*SKEW**3)                                 A 134
          T11=3.0*(SLOPE**2)*(2.0+3.0*SKEW**2+(5.0/8.0)*SKEW**4)              A 135
          DELTA=T7+T8+T9+T10+T11                                             A 136
          XT(J)=EXP(M1+K*SQRT(M2))                                           A 137
          SX=SQRT(M2*DELTA/XN)                                               A 138
          PSX=XT(J)*(EXP(SX)-1.0)                                            A 139
          NSX=-XT(J)*(EXP(-SX)-1.0)                                          A 140
          ST(J)=(PSX+NSX)/2.0                                                A 141
    9     CONTINUE                                                           A 142
          WRITE (6,26)                                                       A 143
          WRITE (6,27) (XT(J),J=1,6)                                         A 144
          WRITE (6,28) (ST(J),J=1,6)                                         A 145
          WRITE (6,29)                                                       A 146
          WRITE (6,30)                                                       A 147
          ICOUNT=0                                                           A 148
          XMIN=10000000.                                                     A 149
          DO 10 I=1,N                                                        A 150
   10     IF (X(I).LT.XMIN) XMIN=X(I)                                        A 151
          GML=XMIN*0.99                                                      A 152
   11     ICOUNT=ICOUNT+1                                                    A 153
          A=0.0                                                              A 154
          B=0.0                                                              A 155
          C=0.0                                                              A 156
          R=0.0                                                              A 157
          DO 12 I=1,N                                                        A 158
          A=A+1.0/(X(I)-GML)                                                 A 159
          B=B+(X(I)-GML)                                                     A 160
```

144

```
          C=C+ALOG(X(I)-GML)                                              A 161
          R=R+1.0/((X(I)-GML)**2)                                         A 162
12        CONTINUE                                                        A 163
          BETA=1.0/(1.0-(XN**2)/(B*A))                                    A 164
          ALPHA=(B/XN)-(XN/A)                                             A 165
          D=BETA+2.0                                                      A 166
          PSI=ALOG(D)-(1.0/(2.0*D))-(1.0/(12.0*D**2))+(1.0/(120.0*D**4))-(1. A 167
         10/(252.0*D**6))-(1.0/(BETA+1.0))-(1.0/BETA)                     A 168
          FCN=XN*PSI+C-XN*ALOG(ALPHA)                                     A 169
          TRI=(1.0/D)+(1.0/(2.0*D**2))+(1.0/(6.0*D**3))-(1.0/(30.0*D**5))+(1 A 170
         1.0/(42.0*D**7))-(1.0/(30.0*D**9))+(1.0/((BETA+1.0)**2))+(1.0/(BETA A 171
         2**2))                                                           A 172
          V=A-(XN**2)/B                                                   A 173
          U=A                                                            A 174
          W=(B/XN)-(XN/A)                                                 A 175
          DU=R                                                            A 176
          DV=R-(XN**3)/(B**2)                                             A 177
          DW=-1.0+(XN*R)/(A**2)                                           A 178
          FPN=XN*TRI*((V*DU-U*DV)/(V**2))-A-XN*DW/W                       A 179
          AS=GML-(FCN/FPN)                                                A 180
          WRITE (6,31) ICOUNT,AS,FCN                                      A 181
          DELTA=ABS(0.00000001*AS)                                        A 182
          IF (ABS(AS-GML).LT.DELTA) GO TO 13                             A 183
          IF (ICOUNT.GE.25) GO TO 15                                     A 184
          GML=AS                                                         A 185
          GO TO 11                                                       A 186
13        CONTINUE                                                        A 187
          GAMMA=AS                                                       A 188
          M1=GAMMA+ALPHA*BETA                                            A 189
          M2=BETA*ALPHA**2                                               A 190
          SKEW=2.0/SQRT(BETA)                                            A 191
          WRITE (6,32)                                                    A 192
          WRITE (6,23) ALPHA,M1                                          A 193
          WRITE (6,24) BETA,M2                                           A 194
          WRITE (6,25) GAMMA,SKEW                                        A 195
          IF (SKEW.LT.0.0) WRITE (6,33)                                  A 196
          D=BETA+2.0                                                     A 197
          TRI=(1.0/D)+(1.0/(2.0*D**2))+(1.0/(6.0*D**3))-(1.0/(30.0*D**5))+(1 A 198
         1.0/(42.0*D**7))-(1.0/(30.0*D**9))+(1.0/((BETA+1.0)**2))+(1.0/(BETA A 199
         2**2))                                                           A 200
          H=(BETA-2.0)*ALPHA**4                                          A 201
          P=2.0*TRI-(2.0*BETA-3.0)/((BETA-1.0)**2)                       A 202
          DET=P/H                                                        A 203
          VARA=(1.0/(XN*(ALPHA**2)*DET))*((TRI/(BETA-2.0))-1.0/((BETA-1.0)** A 204
         12))                                                             A 205
          VARB=2.0/(XN*DET*(BETA-2.0)*ALPHA**4)                          A 206
          VARG=(BETA*TRI-1.0)/(XN*DET*ALPHA**2)                          A 207
          COVAB=(-1.0/(XN*DET*ALPHA**3))*((1.0/(BETA-2.0))-(1.0/(BETA-1.0))) A 208
          COVAG=(1.0/(XN*DET*ALPHA**2))*((1.0/(BETA-1.0))-TRI)           A 209
          COVBG=(-1.0/(XN*DET*ALPHA**3))*((BETA/(BETA-1.0))-1.0)         A 210
          DO 14 J=1,6                                                    A 211
          T=SND(J)                                                       A 212
          E=BETA**(1./3.)-1.0/(9.0*BETA**(2./3.))+T/(3.0*BETA**(1./6.))  A 213
          F=1.0/(3.0*BETA**(2./3.))+2.0/(27.0*BETA**(5./3.))-T/(18.0*BETA**( A 214
         17./6.))                                                        A 215
          XT(J)=EXP(ALPHA*E**3+GAMMA)                                    A 216
          DXDA=E**3                                                      A 217
          DXDB=3.0*ALPHA*E**2*F                                          A 218
          DXDG=1.0                                                       A 219
          SX=SQRT(VARA*DXDA**2+VARB*DXDB**2+VARG*DXDG**2+2.0*DXDA*DXDB*COVAB A 220
         1+2.0*DXDA*DXDG*COVAG+2.0*DXDB*DXDG*COVBG)                      A 221
          PSX=XT(J)*(EXP(SX)-1.0)                                         A 222
          NSX=XT(J)*(EXP(-SX)-1.0)                                        A 223
          ST(J)=(PSX+NSX)/2.0                                            A 224
14        CONTINUE                                                        A 225
          WRITE (6,26)                                                    A 226
          WRITE (6,27) (XT(J),J=1,6)                                      A 227
          WRITE (6,28) (ST(J),J=1,6)                                      A 228
15        CONTINUE                                                        A 229
          STOP                                                           A 230
C                                                                         A 231
C                                                                         A 232
16        FORMAT (80A1)                                                   A 233
17        FORMAT (I5)                                                     A 234
18        FORMAT (8F10.0)                                                 A 235
19        FORMAT (1H1,80A1,/,24X,31HLOG-PEARSON TYPE 3 DISTRIBUTION,/)    A 236
20        FORMAT (28X,26HMETHOD OF MOMENTS (DIRECT),/)                    A 237
21        FORMAT (/,3X,43HMETHOD NOT APPLICABLE BECAUSE OF B VALUE OF,9X,E12 A 238
         1.5,/)                                                          A 239
```

145

```
22      FORMAT (27X,28HMETHOD OF MOMENTS (INDIRECT),/)                    A 240
23      FORMAT (9X,5HALPHA,5X,E12.5,14X,4HM1   ,6X,E12.5)                 A 241
24      FORMAT (9X,5HBETA ,5X,E12.5,14X,4HM2   ,6X,E12.5)                 A 242
25      FORMAT (9X,5HGAMMA,5X,E12.5,14X,4HSKEW,6X,E12.5,/)                A 243
26      FORMAT (3X,7HT,YEARS,4X,1H2,11X,1H5,10X,2H10,10X,2H20,10X,2H50,9X, A 244
        13H100,/)                                                         A 245
27      FORMAT (3X,1HX,3X,6E12.5,/,4X,1HT)                                A 246
28      FORMAT (3X,1HS,3X,6E12.5,/,4X,1HT,/)                              A 247
29      FORMAT (27X,28HMAXIMUM LIKELIHOOD PROCEDURE,/)                    A 248
30      FORMAT (21X,5HTRIAL,11X,1HG,11X,4HF(G),/)                         A 249
31      FORMAT (22X,I2,8X,E12.5,1X,E12.5)                                 A 250
32      FORMAT (/)                                                        A 251
33      FORMAT (/,3X,50HSKEW IS NEGATIVE - DISTRIBUTION HAS AN UPPER BOUND A 252
        1),/)                                                             A 253
34      FORMAT (9X,5HL1   ,5X,E12.5,14X,4HM1   ,6X,E12.5)                 A 254
35      FORMAT (9X,5HL2   ,5X,E12.5,14X,4HM2   ,6X,E12.5)                 A 255
36      FORMAT (9X,5HL3   ,5X,E12.5,14X,4HSKEW,6X,E12.5,/)                A 256
        END                                                              A 257

10 '
20 ' Program LP3
30 ' Copyright G.W. Kite 1986
40 '
50 ' Compute method of moments and maximum likelihood estimates
60 ' for T year events and standard errors
70 ' for a Log Pearson Type III Distribution.
80 '
90 DIM T#(6),X#(250),XT#(6),SX#(6),SND#(6)
100 FORM$="  ##.###^^^^"
110 VV$=CHR$(179)
120 DATA 0.#,0.84162#,1.28155#,1.64485#,2.05375#,2.32635#
130 FOR I%=1 TO 6
140 READ SND#(I%)
150 NEXT I%
160 T$="Log Pearson Type III Distribution"
170 IERR%=0
180 GOSUB 3230
190 IF IERR% = 1 GOTO 3210
200 FLAG%=0:IF YN$ = "Y" OR YN$ = "y" THEN FLAG%=1
210 NREC%=0
220 OPEN FILE$ FOR INPUT AS #1
230 LINE INPUT#1,TITLE$
240 IF EOF(1) GOTO 290
250 NREC%=NREC%+1
260 LINE INPUT#1,I$
270 X#(NREC%)=VAL(MID$(I$,5,12))
280 GOTO 240
290 CLOSE#1
300 NREC#=NREC%
310 A%=LEN(TITLE$)
320 C1#=(SQR(NREC#*(NREC#-1#)))/(NREC#-2#)
330 C2#=1#+8.5#/NREC#
340 C3#=NREC#/(NREC#-1#)
350 B%=(80-A%)/2
360 U$=STRING$(A%,205)
370 M$="Method of Moments (Direct)"
380 E%=LEN(M$)
390 F%=(80-E%)/2
400 W$=STRING$(E%,205)
410 CLS
420 PRINT TAB(B%) TITLE$:PRINT TAB(B%) U$
430 PRINT TAB(D%) T$:PRINT TAB(D%) V$
440 PRINT TAB(F%) M$:PRINT TAB(F%) W$
450 IF FLAG% <> 1 GOTO 500
460 LPRINT CHR$(12):LPRINT:LPRINT
470 LPRINT TAB(B%) TITLE$:LPRINT TAB(B%) U$
480 LPRINT:LPRINT TAB(D%) T$:LPRINT TAB(D%) V$
490 LPRINT:LPRINT TAB(F%) M$:LPRINT TAB(F%) W$
500 A#=0#:B#=0#:C#=0#
510 FOR I%=1 TO NREC%
520 A#=A#+X#(I%)
530 B#=B#+X#(I%)*X#(I%)
540 C#=C#+X#(I%)*X#(I%)*X#(I%)
550 NEXT I%
560 L1#=A#/NREC#
570 L2#=B#/NREC#
580 L3#=C#/NREC#
590 B#=(LOG(L3#)-3#*LOG(L1#))/(LOG(L2#)-2#*LOG(L1#))
600 C#=1#/(B#-3#)
```

146

```
610 IF B# > 6# GOTO 680
620 IF B# <= 3# GOTO 680
630 IF B# <= 3.5# GOTO 660
640 A#=-.23019#+1.65262#*C#+.20911#*C#*C#-.04557#*C#*C#*C#
650 GOTO 700
660 A#=-.47157#+1.99955#*C#
670 GOTO 700
680 PRINT "Method is not applicable because of a B value of ";:PRINT USING FORM$
;B#
690 GOTO 3210
700 ALPHA#=1#/(A#+3#)
710 A1#=LOG(1#-ALPHA#)
720 A2#=LOG(1#-2#*ALPHA#)
730 BETA#=(LOG(L2#)-2#*LOG(L1#))/(2#*A1#-A2#)
740 GAMMA#=LOG(L1#)+BETA#*A1#
750 M1#=GAMMA#+ALPHA#*BETA#
760 M2#=BETA#*ALPHA#*ALPHA#
770 SKEW#=2#/SQR(BETA#)
780 PRINT
790 PRINT TAB(20) "Mean of the logs is        ";:PRINT USING FORM$;M1#
800 PRINT TAB(20) "Variance of the logs is    ";:PRINT USING FORM$;M2#
810 PRINT TAB(20) "Coeff. of skew of logs is ";:PRINT USING FORM$;SKEW#
820 PRINT TAB(20) "Parameter Alpha is         ";:PRINT USING FORM$;ALPHA#
830 PRINT TAB(20) "Parameter Beta is          ";:PRINT USING FORM$;BETA#
840 PRINT TAB(20) "Parameter Gamma is         ";:PRINT USING FORM$;GAMMA#
850 IF FLAG% <> 1 GOTO 960
860 LPRINT
870 LPRINT TAB(20) "Mean of the logs is        ";:LPRINT USING FORM$;M1#
880 LPRINT TAB(20) "Variance of the logs is    ";:LPRINT USING FORM$;M2#
890 LPRINT TAB(20) "Coeff. of skew of logs is ";:LPRINT USING FORM$;SKEW#
900 LPRINT TAB(20) "Parameter Alpha is         ";:LPRINT USING FORM$;ALPHA#
910 LPRINT TAB(20) "Parameter Beta is          ";:LPRINT USING FORM$;BETA#
920 LPRINT TAB(20) "Parameter Gamma is         ";:LPRINT USING FORM$;GAMMA#
930 IF SKEW# > 0# GOTO 960
940 PRINT TAB(10) "Coeff. of skew is -ve; distribution has an upper bound"
950 IF FLAG% = 1 THEN LPRINT TAB(10) "Coeff. of skew is -ve; distribution has an
upper bound"
960 FOR J%=1 TO 6
970 T#=SND#(J%)
980 T1#=T#
990 T2#=(T#*T#-1)/6#
1000 T3#=2#*(T#*T#*T#-6#*T#)/216#
1010 T4#=(T#*T#*T#-1#)/216#
1020 T5#=T#/1296#
1030 T6#=.0000428669#
1040 K#=T1#+T2#*SKEW#+T3#*SKEW#*SKEW#-T4#*SKEW#^3#+T5#*SKEW#^4#-T6#*SKEW#^5#
1050 SLOPE#=T2#+T3#*2#*SKEW#-T4#*3#*SKEW#*SKEW#+T5#*4#*SKEW#^3#-T6#*5#*SKEW#^4#
1060 T7#=1#
1070 T8#=K#*SKEW#
1080 T9#=(1#+.75#*SKEW#*SKEW#)*((K#*K#)/2#)
1090 T10#=3#*SLOPE#*K#*(SKEW#+.25#*SKEW#*SKEW#*SKEW#)
1100 T11#=3#*(SLOPE#*SLOPE#)*(2#+3#*SKEW#*SKEW#+.625#*SKEW#*SKEW#*SKEW#*SKEW#)
1110 DELTA#=T7#+T8#+T9#+T10#+T11#
1120 XT#(J%)=EXP(M1#+K#*SQR(M2#))
1130 ST#=SQR(M2#*DELTA#/NREC#)
1140 PSX#=XT#(J%)*(EXP(ST#)-1#)
1150 NSX#=-XT#(J%)*(EXP(-ST#)-1#)
1160 SX#(J%)=(PSX#+NSX#)/2#
1170 NEXT J%
1180 ROW$=CHR$(218)+STRING$(72,196)
1190 PRINT:PRINT TAB(2) "T, years  2            5           10          20
   50          100":PRINT TAB(5) ROW$
1200 PRINT TAB(2) "X  " VV$;:FOR J%=1 TO 6:PRINT USING FORM$;XT#(J%);:NEXT J%
1210 PRINT TAB(2) " t " VV$
1220 PRINT TAB(2) "S  " VV$;:FOR J%=1 TO 6:PRINT USING FORM$;SX#(J%);:NEXT J%
1230 PRINT TAB(2) " t " VV$
1240 IF FLAG% <> 1 GOTO 1300
1250 LPRINT:LPRINT TAB(2) "T, years  2            5           10          20
   50          100":LPRINT TAB(5) ROW$
1260 LPRINT TAB(2) "X  " VV$;:FOR J%=1 TO 6:LPRINT USING FORM$;XT#(J%);:NEXT J%
1270 LPRINT TAB(2) " t " VV$
1280 LPRINT TAB(2) "S  " VV$;:FOR J%=1 TO 6:LPRINT USING FORM$;SX#(J%);:NEXT J%
1290 LPRINT TAB(2) " t " VV$
1300 M$="Method of Moments (Indirect)"
1310 E%=LEN(M$)
1320 F%=(80-E%)/2
1330 W$=STRING$(E%,205)
1340 PRINT:PRINT "Press any key to continue"
1350 Q$=INKEY$:IF Q$="" GOTO 1350
```

147

```
1360 CLS
1370 PRINT:PRINT
1380 PRINT TAB(F%) M$:PRINT TAB(F%) W$
1390 IF FLAG% <> 1 GOTO 1410
1400 LPRINT:LPRINT TAB(F%) M$:LPRINT TAB(F%) W$
1410 A#=0#:B#=0#:C#=0#
1420 FOR I%=1 TO NREC%
1430 X#(I%)=LOG(X#(I%))
1440 A#=A#+X#(I%)
1450 B#=B#+X#(I%)*X#(I%)
1460 C#=C#+X#(I%)*X#(I%)*X#(I%)
1470 NEXT I%
1480 M1#=A#/NREC#
1490 M2#=(B#/NREC#)-M1#*M1#
1500 M3#=(C#/NREC#)+2#*M1#*M1#*M1#-3#*M1#*(B#/NREC#)
1510 SKEW#=M3#/(M2#^1.5#)
1520 SKEW#=SKEW#*C1#*C2#
1530 M2#=M2#*C3#
1540 BETA#=(2#/SKEW#)*(2#/SKEW#)
1550 ALPHA#=SQR(M2#)/SQR(BETA#)
1560 GAMMA#=M1#-SQR(M2#)*SQR(BETA#)
1570 PRINT
1580 PRINT TAB(20) "Mean of the logs is        ";:PRINT USING FORM$;M1#
1590 PRINT TAB(20) "Variance of the logs is    ";:PRINT USING FORM$;M2#
1600 PRINT TAB(20) "Coeff. of skew of logs is ";:PRINT USING FORM$;SKEW#
1610 PRINT TAB(20) "Parameter Alpha is         ";:PRINT USING FORM$;ALPHA#
1620 PRINT TAB(20) "Parameter Beta is          ";:PRINT USING FORM$;BETA#
1630 PRINT TAB(20) "Parameter Gamma is         ";:PRINT USING FORM$;GAMMA#
1640 IF FLAG% <> 1 GOTO 1720
1650 LPRINT
1660 LPRINT TAB(20) "Mean of the logs is        ";:LPRINT USING FORM$;M1#
1670 LPRINT TAB(20) "Variance of the logs is    ";:LPRINT USING FORM$;M2#
1680 LPRINT TAB(20) "Coeff. of skew of logs is ";:LPRINT USING FORM$;SKEW#
1690 LPRINT TAB(20) "Parameter Alpha is         ";:LPRINT USING FORM$;ALPHA#
1700 LPRINT TAB(20) "Parameter Beta is          ";:LPRINT USING FORM$;BETA#
1710 LPRINT TAB(20) "Parameter Gamma is         ";:LPRINT USING FORM$;GAMMA#
1720 IF SKEW# > 0# GOTO 1750
1730 PRINT TAB(10) "Coeff. of skew is -ve; distribution has an upper bound"
1740 IF FLAG% = 1 THEN LPRINT TAB(10) "Coeff. of skew is -ve; distribution has a
n upper bound"
1750 FOR J%=1 TO 6
1760 T#=SND#(J%)
1770 T1#=T#
1780 T2#=(T#*T#-1)/6#
1790 T3#=2#*(T#*T#*T#-6#*T#)/216#
1800 T4#=(T#*T#-1#)/216#
1810 T5#=T#/1296#
1820 T6#=.0000428669#
1830 K#=T1#+T2#*SKEW#+T3#*SKEW#*SKEW#-T4#*SKEW#^3#+T5#*SKEW#^4#-T6#*SKEW#^5#
1840 SLOPE#=T2#+T3#*2#*SKEW#-T4#*3#*SKEW#*SKEW#+T5#*4#*SKEW#^3#-T6#*5#*SKEW#^4#
1850 T7#=1#
1860 T8#=K#*SKEW#
1870 T9#=(1#+.75#*SKEW#*SKEW#)*((K#*K#)/2#)
1880 T10#=3#*SLOPE#*K#*(SKEW#+.25#*SKEW#*SKEW#*SKEW#)
1890 T11#=3#*(SLOPE#*SLOPE#)*(2#+3#*SKEW#*SKEW#+.625#*SKEW#^4#)
1900 DELTA#=T7#+T8#+T9#+T10#+T11#
1910 XT#(J%)=EXP(M1#+K#*SQR(M2#))
1920 ST#=SQR(M2#*DELTA#/NREC#)
1930 PSX#=XT#(J%)*(EXP(ST#)-1#)
1940 NSX#=-XT#(J%)*(EXP(-ST#)-1#)
1950 SX#(J%)=(PSX#+NSX#)/2#
1960 NEXT J%
1970 ROW$=CHR$(218)+STRING$(72,196)
1980 PRINT:PRINT TAB(2) "T, years  2          5          10         20
   50        100":PRINT TAB(5) ROW$
1990 PRINT TAB(2) "X  " VV$;:FOR J%=1 TO 6:PRINT USING FORM$;XT#(J%);:NEXT J%
2000 PRINT TAB(2) " t " VV$
2010 PRINT TAB(2) "S  " VV$;:FOR J%=1 TO 6:PRINT USING FORM$;SX#(J%);:NEXT J%
2020 PRINT TAB(2) " t " VV$
2030 IF FLAG% <> 1 GOTO 2090
2040 LPRINT:LPRINT TAB(2) "T, years  2          5          10         20
   50        100":LPRINT TAB(5) ROW$
2050 LPRINT TAB(2) "X  " VV$;:FOR J%=1 TO 6:LPRINT USING FORM$;XT#(J%);:NEXT J%
2060 LPRINT TAB(2) " t " VV$
2070 LPRINT TAB(2) "S  " VV$;:FOR J%=1 TO 6:LPRINT USING FORM$;SX#(J%);:NEXT J%
2080 LPRINT TAB(2) " t " VV$
2090 M$="Method of Maximum Likelihood"
2100 E%=LEN(M$)
2110 F%=(80-E%)/2
2120 W$=STRING$(E%,205)
```

```
2130 PRINT:PRINT"Press any key to continue"
2140 Q$=INKEY$:IF Q$ ="" GOTO 2140
2150 CLS
2160 PRINT:PRINT TAB(F%) M$:PRINT TAB(F%) W$
2170 PRINT:PRINT TAB(19) "Iteration" TAB(37) "G" TAB(53) "f(G)"
2180 PRINT TAB(19) "---------" TAB(37) "-" TAB(53) "----"
2190 IF FLAG% <> 1 GOTO 2230
2200 LPRINT CHR$(12):LPRINT:LPRINT TAB(F%) M$:LPRINT TAB(F%) W$
2210 LPRINT:LPRINT TAB(19) "Iteration" TAB(37) "G" TAB(53) "f(G)"
2220 LPRINT TAB(19)"---------" TAB(37) "-" TAB(53) "----"
2230 COUNT%=0
2240 XMIN#=100000000#
2250 FOR I% = 1 TO NREC%
2260 IF X#(I%) < XMIN# THEN XMIN#=X#(I%)
2270 NEXT I%
2280 GML#=XMIN#*.99#
2290 COUNT%=COUNT%+1
2300 A#=0#:B#=0#:C#=0#:R#=0#
2310 FOR I%=1 TO NREC%
2320 TEMP#=X#(I%)-GML#
2330 A#=A#+1#/TEMP#
2340 B#=B#+TEMP#
2350 C#=C#+LOG(TEMP#)
2360 R#=R#+1#/(TEMP#*TEMP#)
2370 NEXT I%
2380 BETA#=1#/(1#-(NREC#*NREC#)/(B#*A#))
2390 ALPHA#=(B#/NREC#)-(NREC#/A#)
2400 D#=BETA#+2#
2410 PSI#=LOG(D#)-(1#/(2#*D#))-(1#/(12#*D#*D#))+(1#/(120#*D#^4#))-(1#/(252#*D#^6
#))-(1#/(BETA#+1#))-(1#/BETA#)
2420 FCN#=-NREC#*PSI#+C#-NREC#*LOG(ALPHA#)
2430 TRI#=(1#/D#)+(1#/(2#*D#*D#))+(1#/(6#*D#*D#*D#))-(1#/(30#*D#^5#))+(1#/(42#*D
#^7#))-(1#/(30#*D#^9#))+(1#/((BETA#+1#)*(BETA#+1#)))+(1#/(BETA#*BETA#))
2440 V#=A#-(NREC#*NREC#)/B#
2450 U#=A#
2460 W#=(B#/NREC#)-(NREC#/A#)
2470 DU#=R#
2480 DV#=R#-(NREC#*NREC#*NREC#)/(B#*B#)
2490 DW#=-1#+(NREC#*R#)/(A#*A#)
2500 FPN#=-NREC#*TRI#*((V#*DU#-U#*DV#)/(V#*V#))-A#-NREC#*DW#/W#
2510 AAS#=GML#-(FCN#/FPN#)
2520 PRINT TAB(22);:PRINT USING "##";COUNT%;:PRINT USING "       ##.###^^^^";AA
S#,FCN#
2530 IF FLAG% = 1 THEN LPRINT TAB(22);:LPRINT USING "##";COUNT%;:LPRINT USING "
       ##.###^^^^";AAS#,FCN#
2540 DELTA#=ABS(.00000001#*AAS#)
2550 IF ABS(AAS#-GML#) < DELTA# GOTO 2590
2560 IF COUNT% > 24 THEN PRINT:PRINT "Procedure does not converge":PRINT:GOTO 31
90
2570 GML#=AAS#
2580 GOTO 2290
2590 GAMMA#=AAS#
2600 M1#=GAMMA#+ALPHA#*BETA#
2610 M2#=BETA#*ALPHA#*ALPHA#
2620 SKEW#=2#/SQR(BETA#)
2630 D#=BETA#+2#
2640 TRI#=(1#/D#)+(1#/(2#*D#*D#))+(1#/(6#*D#*D#*D#))-(1#/(30#*D#^5#))+(1#/(42#*D
#^7#))-(1#/(30#*D#^9#))+(1#/((BETA#+1#)*(BETA#+1#)))+(1#/(BETA#*BETA#))
2650 H#=(BETA#-2#)*ALPHA#^4#
2660 P#=2#*TRI#-(2#*BETA#-3#)/((BETA#-1#)*(BETA#-1#))
2670 DET#=P#/H#
2680 VARA#=(1#/(NREC#*ALPHA#*ALPHA#*DET#))*((TRI#/(BETA#-2#))-1#/((BETA#-1#)*(BE
TA#-1#)))
2690 VARB#=2#/(NREC#*DET#*(BETA#-2#)*ALPHA#^4#)
2700 VARG#=(BETA#*TRI#-1#)/(NREC#*DET#*ALPHA#*ALPHA#)
2710 COVAB#=(-1#/(NREC#*DET#*ALPHA#*ALPHA#*ALPHA#))*((1#/(BETA#-2#))-(1#/(BETA#-
1#)))
2720 COVAG#=(1#/(NREC#*DET#*ALPHA#*ALPHA#))*((1#/(BETA#-1#))-TRI#)
2730 COVBG#=(-1#/(NREC#*DET#*ALPHA#*ALPHA#*ALPHA#))*((BETA#/(BETA#-1#))-1#)
2740 FOR J% = 1 TO 6
2750 T#=SND(J%)
2760 E#=BETA#^(1#/3#)-1#/(9#*BETA#^(2#/3#))+T#/(3#*BETA#^(1#/6#))
2770 F#=1#/(3#*BETA#^(2#/3#))+2#/(27#*BETA#^(5#/3#))-T#/(18#*BETA#^(7#/6#))
2780 XT#(J%)=EXP(GAMMA#+ALPHA#*E#*E#*E#)
2790 DXDA#=E#*E#*E#
2800 DXDB#=3#*ALPHA#*E#*E#*F#
2810 DXDG#=1#
2820 ST#=SQR(VARA#*DXDA#*DXDA#+VARB#*DXDB#*DXDB#+VARG#*DXDG#*DXDG#+2#*DXDA#*DXDB
#*COVAB#+2#*DXDA#*DXDG#*COVAG#+2#*DXDB#*DXDG#*COVBG#)
2830 PSX#=XT#(J%)*(EXP(ST#)-1#)
```

149

```
2840 NSX#=-XT#(J%)*(EXP(-ST#)-1#)
2850 SX#(J%)=(PSX#+NSX#)/2#
2860 NEXT J%
2870 CLS
2880 PRINT:PRINT TAB(B%) TITLE$:PRINT TAB(B%) U$
2890 PRINT TAB(D%) T$:PRINT TAB(D%) V$
2900 PRINT TAB(F%) M$:PRINT TAB(F%) W$
2910 PRINT TAB(19) "Mean of the logs is          ";:PRINT USING FORM$;M1#
2920 PRINT TAB(19) "Variance of the logs is      ";:PRINT USING FORM$;M2#
2930 PRINT TAB(19) "Coeff. of skew of logs is    ";:PRINT USING FORM$;SKEW#
2940 PRINT TAB(19) "Parameter Alpha is           ";:PRINT USING FORM$;ALPHA#
2950 PRINT TAB(19) "Parameter Beta is            ";:PRINT USING FORM$;BETA#
2960 PRINT TAB(19) "Parameter Gamma is           ";:PRINT USING FORM$;GAMMA#
2970 IF FLAG% <> 1 GOTO 3080
2980 LPRINT
2990 LPRINT TAB(19) "Mean of the logs is             ";:LPRINT USING FORM$;M1#
3000 LPRINT TAB(19) "Variance of the logs is         ";:LPRINT USING FORM$;M2#
3010 LPRINT TAB(19) "Coeff. of skew of logs is       ";:LPRINT USING FORM$;SKEW#

3020 LPRINT TAB(19) "Parameter Alpha is              ";:LPRINT USING FORM$;ALPHA
#
3030 LPRINT TAB(19) "Parameter Beta is               ";:LPRINT USING FORM$;BETA#

3040 LPRINT TAB(19) "Parameter Gamma is              ";:LPRINT USING FORM$;GAMMA
#
3050 IF SKEW# > 0# GOTO 3080
3060 PRINT TAB(10) "Coeff. of skew is -ve; distribution has an upper bound"
3070 IF FLAG% = 1 THEN LPRINT TAB(10) "Coeff. of skew is -ve; distribution has a
n upper bound"
3080 PRINT:PRINT TAB(2) "T, years  2             5            10           20
  50        100":PRINT TAB(5) ROW$
3090 PRINT TAB(2) "X  " VV$;:FOR J%=1 TO 6:PRINT USING FORM$;XT#(J%);:NEXT J%
3100 PRINT TAB(2) " t " VV$
3110 PRINT TAB(2) "S  " VV$;:FOR J%=1 TO 6:PRINT USING FORM$;SX#(J%);:NEXT J%
3120 PRINT TAB(2) " t " VV$
3130 IF FLAG% <> 1 GOTO 3190
3140 LPRINT:LPRINT TAB(2) "T, years  2             5            10           20
  50        100":LPRINT TAB(5) ROW$
3150 LPRINT TAB(2) "X  " VV$;:FOR J%=1 TO 6:LPRINT USING FORM$;XT#(J%);:NEXT J%
3160 LPRINT TAB(2) " t " VV$
3170 LPRINT TAB(2) "S  " VV$;:FOR J%=1 TO 6:LPRINT USING FORM$;SX#(J%);:NEXT J%
3180 LPRINT TAB(2) " t " VV$
3190 PRINT:PRINT"Press any key to continue"
3200 Q$=INKEY$:IF Q$ ="" GOTO 3200
3210 CLS
3220 SYSTEM
3230 '
3240 ' Subroutine for standard screen format
3250 '
3260 CLS:PRINT:PRINT
3270 ROW$=STRING$(78,205)
3280 BOXTOP$=CHR$(201)+ROW$+CHR$(187)
3290 PRINT BOXTOP$;
3300 PRINT CHR$(186) TAB(80) CHR$(186);
3310 PRINT CHR$(186) TAB(80) CHR$(186);
3320 C%=LEN(T$)
3330 D%=(80-C%)/2
3340 V$=STRING$(C%,205)
3350 PRINT CHR$(186) TAB(D%) T$ TAB(80) CHR$(186);
3360 PRINT CHR$(186) TAB(D%) V$ TAB(80) CHR$(186);
3370 PRINT CHR$(186) TAB(80) CHR$(186);
3380 PRINT CHR$(186) TAB(80) CHR$(186);
3390 PRINT CHR$(186) "   What is the name of your data file?" TAB(80) CHR$(186);

3400 PRINT CHR$(186) TAB(80) CHR$(186);
3410 PRINT CHR$(186) "   Do you want printer output (Y/N)?" TAB(80) CHR$(186);
3420 PRINT CHR$(186) "   (press Alt Q to quit at this stage)" TAB(80) CHR$(186);

3430 PRINT CHR$(186) TAB(80) CHR$(186);
3440 PRINT CHR$(186) TAB(80) CHR$(186);
3450 BOXBOT$=CHR$(200)+ROW$+CHR$(188)
3460 PRINT BOXBOT$;
3470 LOCATE 24,1,0,0,0
3480 PRINT "G Kite";
3490 LOCATE 10,41,1,0,13
3500 FILE$=""
3510 I$=INPUT$(1)
3520 IF I$ = CHR$(13) GOTO 3620                    ' ENTER key
3530 IF I$ = CHR$(8)  GOTO 3570                    ' BACKSPACE key
```

```
3540 PRINT I$;
3550 FILE$=FILE$+I$
3560 GOTO 3510
3570 H%=POS(0)
3580 LOCATE 10,H%-1,1,0,13
3590 L%=LEN(FILE$)
3600 FILE$=LEFT$(FILE$,L%-1)
3610 GOTO 3510
3620 LOCATE 12,39,1,0,13
3630 YN$=""
3640 I$=INKEY$:IF I$ = "" GOTO 3640
3650 IF LEN(I$) = 1 GOTO 3690
3660 I$=RIGHT$(I$,1)
3670 IF I$ = CHR$(16) THEN IERR%=1:GOTO 3760        ' ALT Q key
3680 GOTO 3620
3690 IF I$ = CHR$(13) GOTO 3730                     ' ENTER key
3700 PRINT I$;
3710 YN$=I$
3720 GOTO 3640
3730 IF YN$ = "Y" OR YN$ = "y" OR YN$ = "N" OR YN$ = "n" GOTO 3750
3740 GOTO 3620
3750 LOCATE ,,0,13,13
3760 RETURN
```

St. Mary's River at Stillwater, Nova Scotia. Station No. 01E0001. m**3/s.

### Log Pearson Type III Distribution

#### Method of Moments (Direct)

| | |
|---|---|
| Mean of the logs is | 5.969D+00 |
| Variance of the logs is | 1.088D-01 |
| Coeff. of skew of logs is | 1.078D-01 |
| Parameter Alpha is | 1.778D-02 |
| Parameter Beta is | 3.443D+02 |
| Parameter Gamma is | -1.512D-01 |

| T, years | 2 | 5 | 10 | 20 | 50 | 100 |
|---|---|---|---|---|---|---|
| $X_t$ | 3.887D+02 | 5.152D+02 | 5.989D+02 | 6.794D+02 | 7.846D+02 | 8.645D+02 |
| $S_t$ | 1.634D+01 | 2.410D+01 | 3.358D+01 | 4.728D+01 | 7.189D+01 | 9.535D+01 |

#### Method of Moments (Indirect)

| | |
|---|---|
| Mean of the logs is | 5.969D+00 |
| Variance of the logs is | 1.088D-01 |
| Coeff. of skew of logs is | 1.938D-01 |
| Parameter Alpha is | 3.197D-02 |
| Parameter Beta is | 1.065D+02 |
| Parameter Gamma is | 2.565D+00 |

| T, years | 2 | 5 | 10 | 20 | 50 | 100 |
|---|---|---|---|---|---|---|
| $X_t$ | 3.871D+02 | 5.145D+02 | 6.008D+02 | 6.850D+02 | 7.968D+02 | 8.832D+02 |
| $S_t$ | 1.630D+01 | 2.460D+01 | 3.494D+01 | 4.999D+01 | 7.722D+01 | 1.034D+02 |

## Method of Maximum Likelihood

| Iteration | G | f(G) |
|-----------|-----------|-----------|
| 1 | 5.143D+00 | 2.817D+00 |
| 2 | 5.058D+00 | 1.307D+00 |
| 3 | 4.935D+00 | 5.798D-01 |
| 4 | 4.766D+00 | 2.479D-01 |
| 5 | 4.551D+00 | 1.032D-01 |
| 6 | 4.286D+00 | 4.206D-02 |
| 7 | 3.974D+00 | 1.682D-02 |
| 8 | 3.624D+00 | 6.578D-03 |
| 9 | 3.261D+00 | 2.485D-03 |
| 10 | 2.931D+00 | 8.832D-04 |
| 11 | 2.703D+00 | 2.758D-04 |
| 12 | 2.614D+00 | 6.287D-05 |
| 13 | 2.603D+00 | 6.260D-06 |
| 14 | 2.603D+00 | 8.226D-08 |
| 15 | 2.603D+00 | 1.489D-11 |
| 16 | 2.603D+00 | -1.457D-13 |

| | |
|---|---|
| Mean of the logs is | 5.969D+00 |
| Variance of the logs is | 1.074D-01 |
| Coeff. of skew of logs is | 1.947D-01 |
| Parameter Alpha is | 3.191D-02 |
| Parameter Beta is | 1.055D+02 |
| Parameter Gamma is | 2.603D+00 |

| T, years | 2 | 5 | 10 | 20 | 50 | 100 |
|----------|---|---|----|----|----|-----|
| $x_t$ | 3.871D+02 | 5.136D+02 | 5.991D+02 | 6.826D+02 | 7.933D+02 | 8.787D+02 |
| $s_t$ | 1.610D+01 | 2.439D+01 | 3.459D+01 | 4.920D+01 | 7.543D+01 | 1.005D+02 |

# CHAPTER 11
## TYPE III EXTREMAL DISTRIBUTION

Introduction

   The distributions discussed so far in Chapters 5-10 have been
used primarily for frequency analysis of maximum events such as
flood peaks.  Two of these distributions (the 3-parameter lognormal and Pearson type III) have also been considered (Matalas
(10)) suitable for analysis of minimum events such as low flows.
The descriptions given for flood events remain valid for low
flows except that the notation of probabilities must be changed.

   The notation used so far for distributions of flood events
has been

$$P(x) = 1 - \frac{m}{n+1} = 1 - 1/T \qquad (11-1)$$

where $P(x)$ is the cumulative probability of an event being less
than or equal to x and m is the order number of the recorded
event, m being 1 for the maximum event and m being n for the
minimum event.  Following this notation, the larger the return
period, T, the larger is the magnitude of the expected event.
In the analysis of low flows, however, it is required that smaller
events be associated with larger return periods and so a different
notation must be used.  If the recorded events are arranged in
order of increasing magnitude with m being 1 for the minimum
event and m being n for the maximum event then the cumulative
probability of an event being less than or equal to x is given
by

$$P(x) = \frac{m}{n+1} = \frac{1}{T} \qquad (11-2)$$

   Another of the distributions considered by Matalas as suitable for low flow analysis is the type III extremal or Weibull
distribution.  Becuase this distribution has not been covered
in any of the earlier chapters, a brief description follows.  It
has been noted by Condie (4) that if $\gamma = 0$ in the type III extremal distribution the lower limit is zero and the distribution
might be suitable for flood analysis also.

   The cumulative probability distribution is, (3):

$$P(x) = e^{-\{\frac{x-\gamma}{\beta-\gamma}\}^{\alpha}} \qquad (11-3)$$

and the probability density function is:

$$p(x) = \frac{\alpha}{\beta-\gamma} \{\frac{x-\gamma}{\beta-\gamma}\}^{\alpha-1} e^{-\{\frac{x-\gamma}{\beta-\gamma}\}^{\alpha}} \qquad (11-4)$$

where $\alpha$ is a scale parameter equal to the order of the lowest derivative of the probability function that is not zero at $x=\gamma$, $\beta$ is the characteristic drought (a location or central value parameter) and $\gamma$ is the lower limit to x.

Commonly, the transformation

$$y = \{\frac{x-\gamma}{\beta-\gamma}\}^{\alpha} \qquad (11-5)$$

is made (8) reducing the cumulative probability and probability density equations to

$$P(x) = e^{-y} \qquad (11-6)$$

and

$$p(x) = \frac{\alpha}{\beta-\gamma} y^{(\alpha-1)/\alpha} e^{-y} \qquad (11-7)$$

## Estimation of Parameters

*Method of Moments* - The general expression for the rth moment, $\mu_r^\gamma$, about the lower bound, $\gamma$, of the type III extremal distribution is obtained from Equations 3-1 and 11-4 using $\gamma$ in place of the origin

$$\mu_r^\gamma = \int_0^\infty (x-\gamma)^r \frac{\alpha}{(\beta-\gamma)} \left\{\frac{x-\gamma}{\beta-\gamma}\right\}^{\alpha-1} \exp\left\{\frac{x-\gamma}{\beta-\gamma}\right\}^\alpha dx \qquad (11-8)$$

substituting y for $\{(x-\gamma)/(\beta-\gamma)\}^\alpha$ and simplifying gives

$$\mu_r^\gamma = (\beta-\gamma)^r \int_0^\infty y^{r/\alpha} e^{-y} dy \qquad (11-9)$$

$$\mu_r^\gamma = (\beta-\gamma)^r \Gamma(1+r/\alpha) \qquad (11-10)$$

where $\Gamma(x)$ is the gamma function equal to $(x-1)!$ for integer x. For non-integer values $\Gamma(x)$ can be computed from

$$\Gamma(x) = (x-1)(x-2)...(x-r)\Gamma(y) \qquad (11-11)$$

for x>1, and for x<1 by

$$\Gamma(x) = \frac{\Gamma(y)}{x(x+1)(x+2)...(x+r-1)} \qquad (11-12)$$

where $1 \le y \le 2$ and $\Gamma(y)$ can be approximated by

$$\Gamma(y) = (y + 4.5)^{(y-0.5)} e^{-(y+4.5)} \sqrt{2\pi} [1 + \sum_{i+0}^{5} (\frac{c_i}{(y+i)})] \qquad (11\text{-}13)$$

where

$$c_0 = 76.18009173 \qquad c_3 = -1.231739516$$

$$c_1 = -86.50532033 \qquad c_4 = 0.1208580 \times 10^{-2}$$

$$c_2 = 24.01409822 \qquad c_5 = -0.536382 \times 10^{-5}$$

(after Lanczos, 1964)

A computer subroutine is available (9) which calculates $\Gamma(x)$ using similar relationships.

The moments about $\gamma$ may then be converted to origin moments or central moments as needed. As examples, for r=1

$$\mu_1^\gamma = (\beta-\gamma)\Gamma(1+1/\alpha) \qquad (11\text{-}14)$$

and so the first moment about the origin, the mean, is

$$\mu_1' = \gamma + (\beta-\gamma)\Gamma(1+1/\alpha) \qquad (11\text{-}15)$$

for r=2

$$\mu_2' = (\beta-\gamma)^2 \Gamma(1+2/\alpha) \qquad (11\text{-}16)$$

and from Equation 3-13

$$\mu_2 = \mu_2^\gamma - (\mu_1^\gamma)^2 \qquad (11\text{-}17)$$

$$\mu_2 = \sigma^2 = (\beta-\gamma)^2 \{\Gamma(1+2/\alpha) - \Gamma^2(1+1/\alpha)\} \qquad (11\text{-}18)$$

similarly

$$\mu_3 = (\beta-\gamma)^3 \{\Gamma(1+3/\alpha) - 3\Gamma(1+2/\alpha)\Gamma(1+1/\alpha) + 2\Gamma^3(1+1/\alpha)\}$$

$$(11\text{-}19)$$

$$\mu_4 = (\beta-\gamma)^4 \{\Gamma(1+4/\alpha) - 4\Gamma(1+3/\alpha)\Gamma(1+1/\alpha)$$

$$+ 6\Gamma(1+2/\alpha)\Gamma^2(1+1/\alpha) - 3\Gamma^4(1+1/\alpha)\} \qquad (11\text{-}20)$$

$$\mu_5 = (\beta-\gamma)^5 \{\Gamma(1+5/\alpha) - 5\Gamma(1+4/\alpha)\Gamma(1+1/\alpha) + 10\Gamma(1+3/\alpha)\Gamma^2(1+1/\alpha)$$

$$- 10\Gamma(1+2/\alpha)\Gamma^3(1+1/\alpha) + 4\Gamma^5(1+1/\alpha)\} \qquad (11\text{-}21)$$

$$\mu_6 = (\beta-\gamma)^6 \{\Gamma(1+6/\alpha) - 6\Gamma(1+5/\alpha)\Gamma(1+1/\alpha) + 15\Gamma(1+4/\alpha)\Gamma^2(1+1/\alpha)$$

$$- 20\Gamma(1+3/\alpha)\Gamma^3(1+1/\alpha) + 15\Gamma(1+2/\alpha)\Gamma^4(1+1/\alpha) - 5\Gamma^6(1+1/\alpha)\}$$

$$(11\text{-}22)$$

If two new variables are defined, $A_\alpha$ and $B_\alpha$, such that $A_\alpha$ is the standardized difference between the characteristic value and the mean and $B_\alpha$ is the standardized difference between the lower limit and the characteristic value,

$$A_\alpha = \frac{\beta-\mu}{\sigma} \qquad (11\text{-}23)$$

and

$$B_\alpha = \frac{\beta-\gamma}{\sigma} \qquad (11\text{-}24)$$

then, by substituting $\mu$ and $\sigma$ from Equations 11-15 and 11-18

$$B_\alpha = \{\Gamma(1+2/\alpha) - \Gamma^2(1+1/\alpha)\}^{-1/2} \qquad (11\text{-}25)$$

and

$$A_\alpha = \{1-\Gamma(1+1/\alpha)\} \, B_\alpha \qquad (11\text{-}26)$$

If the coefficient of skew, $\gamma_1$, is defined as usual

$$\gamma_1 = \frac{\mu_3}{\mu_2^{3/2}} \qquad (11\text{-}27)$$

then from Equations 11-18, 11-19 and 11-25

$$\gamma_1 = \{\Gamma(1+3/\alpha) - 3\Gamma(1+2/\alpha)\Gamma(1+1/\alpha) + 2\Gamma^3(1+1/\alpha)\} \, B_\alpha^3 \qquad (11\text{-}28)$$

an expression involving only functions of $\alpha$.

Thus if the sample coefficient of skew is computed as:

$$\hat{\gamma}_1 = \frac{n \, \Sigma \, (x-\bar{x})^3}{(n-2)[\Sigma(x-\bar{x})^2]^{3/2}} \qquad (11\text{-}29)$$

then $\alpha$ can be found by the solution of Equation 11-28. Knowing $\alpha$, the parameter $\beta$ can be obtained from Equation 11-23 as

$$\beta = \mu_1' + A_\alpha \sqrt{\mu_2} \qquad (11\text{-}30)$$

and $\gamma$ can be found from Equation 11-24 as

$$\gamma = \beta - B_\alpha \sqrt{\mu_2} \qquad (11\text{-}31)$$

To solve Equation 11-28 tables are available, (7), (2), relating $\alpha$ (generally as $1/\alpha$) to $\gamma_1$, $A_\alpha$ and $B_\alpha$. These tables are usually arranged in incremental steps of $1/\alpha$ so that for a computed sample skew a great deal of interpolation is needed to determine the corresponding values of $1/\alpha$, $A_\alpha$ and $B_\alpha$. In order to avoid this interpolation the following regression equation has been developed to enable $\alpha$ to be calculated directly from $\gamma_1$:

$$\alpha = 1/[a_1 + a_2\gamma_1 + a_3\gamma_1^2 + a_4\gamma_1^3 + a_5\gamma_1^4] \qquad (11\text{-}32)$$

$$a_1 = 0.2777757913 \quad a_4 = -0.0013038566$$
$$a_2 = 0.3132617714 \quad a_5 = -0.0081523408$$
$$a_3 = 0.0575670910$$

This polynomial is valid for a range of $\gamma_1$ from -1.02 to +2.00, has a multiple correlation coefficient of 0.9999 and a standard error of 0.0006575. Table 11-1 has been derived from this equation. For convenience Table 11-1 also gives values of the parameters, $A_\alpha$ and $B_\alpha$ as computed from Equations 11-26 and 11-25.

If the parameter $\gamma$ can be assumed to be zero, then a much simpler, if less accurate, method of estimating the remaining parameters, $\alpha$ and $\beta$, exists. If $x = \beta$ then, in Equation 11-3

$$P(\beta) = 1-e^{-1} = 0.632 \qquad (11-33)$$

The median of the type III extremal distribution, M, is obtained by substituting $P = 0.50$ in Equation 11-3,

$$M = \beta(\ln 2)^{1/\alpha} \qquad (11-34)$$

Table 11-1

Parameter $\alpha$ for Type III Extremal Distribution Tabulated as a Function of the Sample Coefficient of Skewness, $\gamma_1$

| $\gamma_1$ | $\alpha$ | $A_\alpha$ | $B_\alpha$ |
|---|---|---|---|
| -1.00 | 65.63043 | 0.44760 | 52.24465 |
| -0.90 | 26.26360 | 0.44229 | 21.47978 |
| -0.80 | 16.30207 | 0.43629 | 13.68443 |
| -0.70 | 11.73785 | 0.42952 | 10.10381 |
| -0.60 | 9.10978 | 0.42193 | 8.03409 |
| -0.50 | 7.39676 | 0.41343 | 6.67757 |
| -0.40 | 6.18962 | 0.40397 | 5.71462 |
| -0.30 | 5.29236 | 0.39350 | 4.99218 |
| -0.20 | 4.59923 | 0.38198 | 4.42770 |
| -0.10 | 4.04809 | 0.36938 | 3.97273 |
| 0.00 | 3.59997 | 0.35571 | 3.59692 |
| 0.10 | 3.22914 | 0.34098 | 3.28029 |
| 0.20 | 2.91791 | 0.32523 | 3.00911 |
| 0.30 | 2.65366 | 0.30851 | 2.77366 |
| 0.40 | 2.42717 | 0.29089 | 2.56682 |
| 0.50 | 2.23149 | 0.27246 | 2.38329 |
| 0.60 | 2.06133 | 0.25334 | 2.21910 |
| 0.70 | 1.91253 | 0.23367 | 2.07116 |
| 0.80 | 1.78181 | 0.21360 | 1.93718 |
| 0.90 | 1.66654 | 0.19329 | 1.81524 |
| 1.00 | 1.56457 | 0.17291 | 1.70391 |
| 1.10 | 1.47416 | 0.15265 | 1.60204 |
| 1.20 | 1.39386 | 0.13268 | 1.50873 |
| 1.30 | 1.32247 | 0.11318 | 1.42324 |
| 1.40 | 1.25900 | 0.09432 | 1.34501 |
| 1.50 | 1.20261 | 0.07626 | 1.27360 |
| 1.60 | 1.15260 | 0.05914 | 1.20866 |
| 1.70 | 1.10840 | 0.04311 | 1.14991 |
| 1.80 | 1.06954 | 0.02828 | 1.09714 |
| 1.90 | 1.03562 | 0.01477 | 1.05020 |
| 2.00 | 1.00634 | 0.00268 | 1.00900 |

If, therefore, a graph of original variable, x, versus cumulative probability is drawn, where cumulative probability is estimated from the sample using Equation 11-2, then values of $\beta$ and M can be extracted from the graph at P = 0.632 and P = 0.50. Substitution of $\beta$ and M in Equation 11-34 will then yield the sample estimate of $\alpha$.

*Maximum Likelihood* - Taking the logarithm of the product of the pdf (Equation 11-4) for $x_1 \rightarrow x_n$ the likelihood function is derived as:

$$\ln L = n \ln \alpha - n \ln (\beta-\gamma) + (\alpha-1) \sum_{i=1}^{n} \ln (x_i-\gamma)$$

$$- n (\alpha-1) \ln (\beta-\gamma) - (\beta-\gamma)^{-\alpha} \sum_{i=1}^{n} (x_i-\gamma)^\alpha \quad (11-35)$$

Taking partial derivatives with respect to $\alpha$, $\beta$ and $\gamma$ yields:

$$\frac{\partial \ln L}{\partial \alpha} = \frac{n}{\alpha} + \sum_{i=1}^{n} \ln (x_i-\gamma) - n \ln (\beta-\gamma) - (\beta-\gamma)^{-\alpha} [ \sum_{i=1}^{n} (x_i-\gamma)^\alpha \cdot$$

$$\ln (x_i-\gamma) + \ln (\beta-\gamma) \sum (x_i-\gamma)^\alpha] \quad (11-36)$$

$$\frac{\partial \ln L}{\partial \beta} = - \frac{n}{\beta-\gamma} - \frac{n(\alpha-1)}{\beta-\gamma} + \alpha(\beta-\gamma)^{-(\alpha+1)} \sum_{i=1}^{n} (x_i-\gamma)^\alpha \quad (11-37)$$

$$\frac{\partial \ln L}{\partial \gamma} = \frac{n}{\beta-\gamma} - (\alpha-1) \sum_{i=1}^{n} (x_i-\gamma)^{-1} + \frac{n(\alpha-1)}{\beta-\gamma}$$

$$- \alpha(\beta-\gamma)^{-(\alpha+1)} \sum_{i=1}^{n} (x_i-\gamma)^\alpha + \alpha(\beta-\gamma)^{-\alpha} \sum_{i=1}^{n} (x_i-\gamma)^{\alpha-1} \quad (11-38)$$

From 11-37 the expression

$$n(\beta-\gamma)^\alpha - \sum_{i=1}^{n} (x_i-\gamma)^\alpha = 0 \quad (11-39)$$

is obtained. Using this as a substitution to eliminate $\beta$, Equations 11-36 and 11-38 reduce to

$$(\alpha - 1) \sum_{i=1}^{n} (x_i-\gamma)^{-1} - \frac{n\alpha \sum_{i=1}^{n} (x_i-\gamma)^{\alpha-1}}{\sum_{i=1}^{n} (x_i-\gamma)^\alpha} = 0 \quad (11-40)$$

and

$$n + \alpha \sum_{i=1}^{n} \ln (x_i - \gamma) - \frac{n\alpha \sum_{i=1}^{n} (x_i - \gamma)^{\alpha} \ln (x_i - \gamma)}{\sum_{i=1}^{n} (x_i - \gamma)^{\alpha}} = 0 \quad (11\text{-}41)$$

No further simplification is possible and Equations 11-40 and 11-41 must be solved as simultaneous equations. The procedure used in the computer program at the end of this chapter broadly follows the method of Condie and Nix (5). The interval between zero and the lowest recorded event is searched for the interval within which functions 11-40 and 11-41 cross. This is found by assuming successive values of $\gamma$ and calculating $\alpha_1$ from 11-40 and $\alpha_2$ from 11-41 both by Newton's method. Successive iterations then refine the interval containing the root until the required accuracy is found. Knowing $\alpha$ and $\gamma$, Equation 11-39 can be solved to give

$$\beta = \gamma + \left[ \frac{\sum_{i=1}^{n} (x_i - \gamma)^{\alpha}}{n} \right]^{1/\alpha} \quad (11\text{-}42)$$

*Other Methods* Various other methods of estimating parameters of the type III extremal distribution are available such as order statistics, smallest observed drought and least squares. Deininger and Westfield (6) compared the performance of several of these methods.

The method of smallest observed drought as described by Gumbel (7) leads to the three equations

$$\frac{\mu_1' - x_o}{\sqrt{\mu_2}} - \frac{(1 - n^{-1/\alpha}) \Gamma(1 + 1/\alpha)}{[\Gamma(1 + 2/\alpha) - \Gamma^2(1 + 1/\alpha)]^{1/2}} = 0 \quad (11\text{-}43)$$

$$\gamma = x_o - \frac{\mu_1' - x_o}{(n^{1/\alpha} - 1)} \quad (11\text{-}44)$$

$$\beta = \frac{\mu_1' - \gamma}{\Gamma(1 + 1/\alpha)} + \gamma \quad (11\text{-}45)$$

where $x_o$ is the smallest observed flow. Equation 11-43 can be solved for $\alpha$ by iteration using Newton's method and then $\gamma$ and $\beta$ can be found from 11-44 and 11-45. Gumbel (7) has also provided tables for the solution of 11-43.

159

## Frequency Factor

From Equation 11-5, 11-6, and 11-1 the following expressions are derived for the type III extremal distribution:

$$x = \gamma + y^{1/\alpha} \, (\beta - \gamma) \qquad (11\text{-}46)$$

and

$$y = - \ln (1 - 1/T) \qquad (11\text{-}47)$$

Values of $y$, the reduced variable, are given in Table 11-2 for some commonly used return periods.

Table 11-2

Values of the Reduced Variable, $y$, of the Type III Extremal Distribution for some Commonly Used Return Periods, $T$

| Return Period T | Reduced Variable y |
|---|---|
| 2 | 0.69315 |
| 5 | 0.22314 |
| 10 | 0.10536 |
| 20 | 0.05129 |
| 50 | 0.02020 |
| 100 | 0.01005 |

Knowing $\alpha$, $\beta$ and $\gamma$, and obtaining the reduced variable, $y$, from Table 11-2, the event magnitude, $x$, corresponding to return period, $T$, can be computed using Equation 11-46.

If Equations 11-46 and 11-47 are combined, then the following expression results:

$$x = \gamma + [- \ln (1 - 1/T)]^{1/\alpha} \, (\beta - \gamma)$$
$$(11\text{-}48)$$

But from previous developments,

$$\beta - \gamma = B_\alpha \sigma \qquad (11\text{-}49)$$

and

$$\gamma = \mu - \sigma \, (B_\alpha - A_\alpha) \qquad (11\text{-}50)$$

where $\mu$ and $\sigma$ are the mean and standard deviation of the population event magnitudes as estimated from the sample, and $A_\alpha$ and $B_\alpha$ are as defined in Equations 11-26 and 11-25.

Substituting Equations 11-49 and 11-50 into Equation 11-48

$$x = \mu + \sigma \, \{(A_\alpha - B_\alpha) + B_\alpha [- \ln (1 - 1/T)]^{1/\alpha}\} \qquad (11\text{-}51)$$

Comparing Equation 11-51 with the standard frequency equation it is apparent that K, the frequency factor, is given by

$$K = A_\alpha + B_\alpha \{[- \ln (1 - 1/T)]^{1/\alpha} - 1\} \qquad (11\text{-}52)$$

This expression is dependent only upon the return period, $T$, and the coefficient of skew, $\gamma_1$, of the recorded events. Table 11-3 provides values of K for some typical values of T and $\gamma_1$.

Table 11-3

Frequency Factor for Use in Type III Extremal Distribution

| Coefficient of skew $\gamma_1$ | Cumulative Probability, P, % | | | | | |
|---|---|---|---|---|---|---|
| | 50 | 80 | 90 | 95 | 98 | 99 |
| | Corresponding Return Period, T, Years | | | | | |
| | 2 | 5 | 10 | 20 | 50 | 100 |
| -1.00 | 0.1567 | -0.7329 | -1.3134 | -1.8641 | -2.5680 | -3.0889 |
| -0.90 | 0.1446 | -0.7501 | -1.3215 | -1.8546 | -2.5232 | -3.0089 |
| -0.80 | 0.1321 | -0.7666 | -1.3282 | -1.8430 | -2.4766 | -2.9282 |
| -0.70 | 0.1189 | -0.7825 | -1.3332 | -1.8294 | -2.4280 | -2.8465 |
| -0.60 | 0.1051 | -0.7977 | -1.3366 | -1.8134 | -2.3771 | -2.7634 |
| -0.50 | 0.0906 | -0.8122 | -1.3382 | -1.7950 | -2.3239 | -2.6788 |
| -0.40 | 0.0754 | -0.8258 | -1.3379 | -1.7741 | -2.2683 | -2.5928 |
| -0.30 | 0.0595 | -0.8385 | -1.3356 | -1.7506 | -2.2103 | -2.5055 |
| -0.20 | 0.0428 | -0.8502 | -1.3313 | -1.7245 | -2.1502 | -2.4172 |
| -0.10 | 0.0255 | -0.8607 | -1.3248 | -1.6960 | -2.0881 | -2.3282 |
| 0.00 | 0.0075 | -0.8699 | -1.3161 | -1.6650 | -2.0244 | -2.2390 |
| 0.10 | -0.0110 | -0.8778 | -1.3053 | -1.6318 | -1.9595 | -2.1500 |
| 0.20 | -0.0300 | -0.8842 | -1.2923 | -1.5966 | -1.8938 | -2.0619 |
| 0.30 | -0.0493 | -0.8891 | -1.2773 | -1.5595 | -1.8277 | -1.9752 |
| 0.40 | -0.0689 | -0.8923 | -1.2603 | -1.5210 | -1.7616 | -1.8902 |
| 0.50 | -0.0885 | -0.8939 | -1.2415 | -1.4812 | -1.6961 | -1.8075 |
| 0.60 | -0.1081 | -0.8938 | -1.2209 | -1.4405 | -1.6315 | -1.7275 |
| 0.70 | -0.1275 | -0.8921 | -1.1989 | -1.3992 | -1.5682 | -1.6506 |
| 0.80 | -0.1466 | -0.8888 | -1.1757 | -1.3578 | -1.5068 | -1.5770 |
| 0.90 | -0.1651 | -0.8840 | -1.1515 | -1.3165 | -1.4473 | -1.5071 |
| 1.00 | -0.1829 | -0.8777 | -1.1266 | -1.2757 | -1.3903 | -1.4409 |
| 1.10 | -0.2000 | -0.8703 | -1.1013 | -1.2358 | -1.3359 | -1.3787 |
| 1.20 | -0.2162 | -0.8617 | -1.0758 | -1.1969 | -1.2842 | -1.3204 |
| 1.30 | -0.2313 | -0.8522 | -1.0505 | -1.1594 | -1.2356 | -1.2661 |
| 1.40 | -0.2454 | -0.8421 | -1.0255 | -1.1236 | -1.1901 | -1.2159 |
| 1.50 | -0.2583 | -0.8314 | -1.0013 | -1.0896 | -1.1477 | -1.1696 |
| 1.60 | -0.2701 | -0.8206 | -0.9780 | -1.0577 | -1.1086 | -1.1272 |
| 1.70 | -0.2807 | -0.8097 | -0.9558 | -1.0279 | -1.0728 | -1.0887 |
| 1.80 | -0.2900 | -0.7990 | -0.9351 | -1.0006 | -1.0403 | -1.0540 |
| 1.90 | -0.2983 | -0.7887 | -0.9159 | -0.9758 | -1.0112 | -1.0231 |
| 2.00 | -0.3053 | -0.7790 | -0.8985 | -0.9536 | -0.9854 | -0.9959 |

## Standard Error of Estimate

*Method of Moments* - The general equation for the standard error of estimate of a 3-parameter distribution has been given as

$$S_T^2 = \frac{\mu_2}{n}\left\{ 1 + K\gamma_1 + \frac{K^2}{4}\left[\gamma_2-1\right] + \frac{\partial K}{\partial \gamma_1}\left[2\gamma_2 - 3\gamma_1^2 - 6\right.\right.$$

$$\left. + K\left(\gamma_3 - 6\gamma_1\gamma_2/4 - 10\gamma_1/4\right)\right] + \left(\frac{\partial K}{\partial \gamma_1}\right)^2\left[\gamma_4\right.$$

$$\left.\left. - 3\gamma_3\gamma_1 - 6\gamma_2 + 9\gamma_1^2\gamma_2/4 + 35\gamma_1^2/4 + 9\right]\right\} \qquad (11\text{-}53)$$

161

where $\gamma_1$ is the coefficient of skew $\mu_3/\mu_2^{3/2}$; $\gamma_2$, the coefficient of kurtosis, $\mu_4/\mu_2^2$. $\gamma_3$ is $\mu_5/\mu_2^{5/2}$ and $\gamma_4$ is $\mu_6/\mu_2^3$ all obtainable from Equations 11-18 - 11-22. The frequency factor, K, can be substituted from Equation 11-52 leaving only $\partial K/\partial\gamma_1$ to be found. This may be done in two ways:

(a) analytically as

$$\frac{\partial K}{\partial \gamma_1} = \frac{\partial K}{\partial (1/\alpha)} \frac{\partial (1/\alpha)}{\partial \gamma_1} \tag{11-54}$$

where $\partial K/\partial(1/\alpha)$ can be derived from Equation 11-52 and $\partial\gamma_1/\partial(1/\alpha)$ can be derived from Equation 11-28 noting that

$$\frac{\partial\ \Gamma(1+r/\alpha)}{\partial(1/\alpha)} = r\Gamma(1+r/\alpha).\psi(1+r/\alpha) \tag{11-55}$$

where $\psi$ is the psi or digamma function. This function can be evaluated by an asymptotic equation (1). Condie and Nix (5) found that to preserve accuracy at low values it was necessary to include a recurrence equation so that

$$\psi(z) \approx \ln\ (z+2) - \frac{1}{2(z+2)} - \frac{1}{12(z+2)^2} + \frac{1}{120(z+2)^4}$$

$$- \frac{1}{252(z+2)^6} - \frac{1}{(z+1)} - \frac{1}{z} \tag{11-56}$$

If $y$ is $-\ln\ (1-1/T)$, $G_r$ is $\Gamma(1+r/\alpha)$ and $P_r$ is $\psi(1+r/\alpha)$ then the two partial derivatives are given by

$$\frac{\partial K}{\partial (1/\alpha)} = \frac{[\ln y.y^{1/\alpha} - G_1P_1] - [y^{1/\alpha} - G_1].[G_2 - G_1^2]^{-1}.[G_2P_2 - G_1^2P_1]}{[G_2 - G_1^2]^{1/2}} \tag{11-57}$$

and

$$\frac{\partial\gamma_1}{\partial(1/\alpha)} = 3\{[G_2 - G_1^2].[G_3P_3 - G_1G_2(P_1+2P_2) + 2G_1^3P_1]$$

$$- [G_3 - 3G_2G_1 + 2G_1^3].[G_2P_2 - G_1^2P_1]\}/[G_2 - G_1^2]^{5/2} \tag{11-58}$$

so that

$$\frac{\partial K}{\partial\gamma_1} = \{[\ln y.y^{1/\alpha} - G_1P_1] - [y^{1/\alpha} - G_1].[G_2 - G_1^2]^{-1}.[G_2P_2 - G_1^2P_1]\}.$$

$$[G_2 - G_1^2]^2/\{3([G_2 - G_1^2].[G_3P_3 - G_1G_2(P_1+2P_2)$$

$$+ 2G_1^3P_1] - [G_3 - 3G_2G_1 + 2G_1^3].[G_2P_2 - G_1^2P_1])\} \tag{11-59}$$

162

(b)  numerically by computing $K_+$ and $K_-$ for $\gamma_{+,-} = \gamma_1 \pm \Delta\gamma$, from Equation 11-52.  Then

$$\frac{\partial K}{\partial \gamma_1} \approx \frac{K_+ - K_-}{2\Delta\gamma_1} \qquad (11\text{-}60)$$

Using $\Delta\gamma_1 = 0.05$ this method gives good accuracy very simply.

Table 11-4 lists values of $\delta$ in

$$S_T = \delta\sqrt{\frac{\mu_2}{n}} \qquad (11\text{-}61)$$

for several values of return period, T, and coefficient of skew, $\gamma_1$, computed analytically.

Table 11-4

Values of Parameter $\delta$ for Use in Standard Error of
Type III Extremal Distribution

| Coeff. of skew | Cumulative Probability, P, % | | | | | |
|---|---|---|---|---|---|---|
| | 50 | 80 | 90 | 95 | 98 | 99 |
| | Corresponding Return Period, T, Years | | | | | |
| | 2 | 5 | 10 | 20 | 50 | 100 |
| -0.80 | 1.04273 | 1.40472 | 1.79759 | 2.41475 | 3.60124 | 4.75123 |
| -0.70 | 1.05090 | 1.38989 | 1.74781 | 2.29490 | 3.32718 | 4.31273 |
| -0.60 | 1.05997 | 1.37113 | 1.69262 | 2.17590 | 3.07597 | 3.92416 |
| -0.50 | 1.06944 | 1.34863 | 1.63326 | 2.05871 | 2.84530 | 3.57834 |
| -0.40 | 1.07900 | 1.32271 | 1.57072 | 1.94409 | 2.63360 | 3.27021 |
| -0.30 | 1.08850 | 1.29375 | 1.50586 | 1.83275 | 2.43992 | 2.99623 |
| -0.20 | 1.09783 | 1.26206 | 1.43941 | 1.72541 | 2.26381 | 2.75389 |
| -0.10 | 1.10691 | 1.22792 | 1.37202 | 1.62289 | 2.10517 | 2.54140 |
| -0.00 | 1.11570 | 1.19156 | 1.30438 | 1.52607 | 1.96411 | 2.35730 |
| 0.10 | 1.12409 | 1.15320 | 1.23720 | 1.43598 | 1.84085 | 2.20019 |
| 0.20 | 1.13201 | 1.11302 | 1.17126 | 1.35375 | 1.73552 | 2.06859 |
| 0.30 | 1.13934 | 1.07123 | 1.10748 | 1.28057 | 1.64806 | 1.96077 |
| 0.40 | 1.14595 | 1.02809 | 1.04694 | 1.21767 | 1.57815 | 1.87475 |
| 0.50 | 1.15174 | 0.98392 | 0.99089 | 1.16623 | 1.52518 | 1.80844 |
| 0.60 | 1.15663 | 0.93914 | 0.94075 | 1.12728 | 1.48829 | 1.75972 |
| 0.70 | 1.16057 | 0.89429 | 0.89813 | 1.10160 | 1.46640 | 1.72664 |
| 0.80 | 1.16357 | 0.85004 | 0.86474 | 1.08963 | 1.45835 | 1.70742 |
| 0.90 | 1.16570 | 0.80724 | 0.84220 | 1.09137 | 1.46290 | 1.70055 |
| 1.00 | 1.16707 | 0.76693 | 0.83191 | 1.10636 | 1.47881 | 1.70473 |
| 1.10 | 1.16782 | 0.73034 | 0.83472 | 1.13367 | 1.50479 | 1.71873 |
| 1.20 | 1.16809 | 0.69886 | 0.85072 | 1.17195 | 1.53949 | 1.74139 |
| 1.30 | 1.16799 | 0.67395 | 0.87920 | 1.21957 | 1.58151 | 1.77148 |
| 1.40 | 1.16757 | 0.65692 | 0.91867 | 1.27464 | 1.62932 | 1.80767 |
| 1.50 | 1.16685 | 0.64869 | 0.96708 | 1.33512 | 1.68130 | 1.84849 |
| 1.60 | 1.16578 | 0.64944 | 1.02197 | 1.39889 | 1.73572 | 1.89236 |
| 1.70 | 1.16426 | 0.65849 | 1.08071 | 1.46374 | 1.79078 | 1.93757 |
| 1.80 | 1.16221 | 0.67416 | 1.14055 | 1.52742 | 1.84458 | 1.98232 |
| 1.90 | 1.15957 | 0.69397 | 1.19869 | 1.58765 | 1.89516 | 2.02472 |
| 2.00 | 1.15637 | 0.71482 | 1.25235 | 1.64214 | 1.94054 | 2.06283 |

$S_T^2$, **Maximum Likelihood** - The variance of the T-year event, $S_T^2$, by maximum likelihood is

$$S_T^2 = \left(\frac{\partial x}{\partial \alpha}\right)^2 \text{var } \alpha + \left(\frac{\partial x}{\partial \beta}\right)^2 \text{var } \beta + \left(\frac{\partial x}{\partial \gamma}\right)^2 \text{var } \gamma$$

$$+ 2 \frac{\partial x}{\partial \alpha} \frac{\partial x}{\partial \beta} \text{ cov } (\alpha,\beta) + 2 \frac{\partial x}{\partial \alpha} \frac{\partial x}{\partial \gamma} \text{ cov } (\alpha,\gamma)$$

$$+ 2 \frac{\partial x}{\partial \beta} \frac{\partial x}{\partial \gamma} \text{ cov } (\beta,\gamma) \tag{11-62}$$

From the logarithm of the likelihood equation

$$\ln L = n \ln \alpha - n \ln (\beta-\gamma) + (\alpha-1) \sum_{i=1}^{n} \ln (x_i-\gamma)$$

$$- n(\alpha-1) \ln (\beta-\gamma) - (\beta-\gamma)^{-\alpha} \sum_{i=1}^{n} (x_i-\gamma)^{\alpha} \tag{11-63}$$

the following derivatives are obtained:

$$\frac{\partial^2 \ln L}{\partial \alpha^2} = - \frac{n}{\alpha^2} - (\beta-\gamma)^{-\alpha} \left\{ \sum_{i=1}^{n} (x_i-\gamma)^{\alpha} \ln^2 (x_i-\gamma) \right.$$

$$\left. - \ln^2 (\beta-\gamma) \sum_{i=1}^{n} (x_i-\gamma)^{\alpha} \right\} \tag{11-64}$$

$$\frac{\partial^2 \ln L}{\partial \alpha \partial \beta} = - \frac{n}{(\beta-\gamma)} + (\beta-\gamma)^{-(\alpha+1)} \{[\ln (\beta-\gamma) - 1] \sum_{i=1}^{n} (x_i-\gamma)^{\alpha}$$

$$+ \sum_{i=1}^{n} (x_i-\gamma)^{\alpha} \ln (x_i-\gamma)\} \tag{11-65}$$

$$\frac{\partial^2 \ln L}{\partial \alpha \partial \gamma} = \frac{n}{(\beta-\gamma)} - \sum_{i=1}^{n} (x_i-\gamma)^{-1} + (\beta-\gamma)^{-\alpha} \left\{ \right.$$

$$\sum_{i=1}^{n} (x_i-\gamma)^{\alpha-1} [1 + \alpha \ln (x_i-\gamma)] + \alpha \ln (\beta-\gamma).$$

$$\left. \sum_{i=1}^{n} (x_i-\gamma)^{\alpha-1} \right\} + (\beta-\gamma)^{-(\alpha+1)} [1 - \alpha \ln (\beta-\gamma)].$$

$$\sum_{i=1}^{n} (x_i-\gamma)^{\alpha} - \alpha(\beta-\gamma)^{\alpha-1} \sum_{i=1}^{n} (x_i-\gamma)^{\alpha} \ln (x_i-\gamma) \tag{11-66}$$

$$\frac{\partial^2 \ln L}{\partial \beta^2} = \frac{n\alpha}{(\beta-\gamma)^2} - \alpha(\alpha+1)(\beta-\gamma)^{-(\alpha+2)} \sum_{i=1}^{n} (x_i-\gamma)^{\alpha} \quad (11\text{-}67)$$

$$\frac{\partial^2 \ln L}{\partial \beta \partial \gamma} = - \frac{n\alpha}{(\beta-\gamma)^2} - \alpha^2(\beta-\gamma)^{-(\alpha+1)} \sum_{i=1}^{n} (x_i-\gamma)^{\alpha-1}$$

$$+ \alpha(\alpha+1)(\beta-\gamma)^{-(\alpha+2)} \sum_{i=1}^{n} (x_i-\gamma)^{\alpha} \quad (11\text{-}68)$$

$$\frac{\partial^2 \ln L}{\partial \gamma^2} = \frac{n\alpha}{(\beta-\gamma)^2} - (\alpha-1) \sum_{i=1}^{n} (x_i-\gamma)^{-2} - \alpha(\alpha+1)(\beta-\gamma)^{-(\alpha+2)} \cdot$$

$$\sum_{i=1}^{n} (x_i-\gamma)^{\alpha} + \alpha^2(\beta-\gamma)^{-(\alpha+1)} \sum_{i=1}^{n} (x_i-\gamma)^{\alpha-1}$$

$$- \alpha(\alpha-1)(\beta-\gamma)^{-\alpha} \sum_{i=1}^{n} (x_i-\gamma)^{\alpha-2} \quad (11\text{-}69)$$

now from Equation 11-46

$$\frac{\partial X}{\partial \alpha} = - \frac{1}{\alpha^2} (\beta-\gamma) \, y^{1/\alpha} \ln y \quad (11\text{-}70)$$

$$\frac{\partial X}{\partial \beta} = y^{1/\alpha} \quad (11\text{-}71)$$

and

$$\frac{\partial X}{\partial \gamma} = 1 - y^{1/\alpha} \quad (11\text{-}72)$$

where y is - ln (1-1/T) and T is the return period.

The information matrix is made up of the negative expectations of each of the above double derivatives. These expressions could be simplified, as has been done for other distributions, by finding analytical expressions for the expectations of

$$\sum_{i=1}^{n} (x_i-\gamma)^{\alpha} \quad (11\text{-}73)$$

$$\sum_{i=1}^{n} (x_i-\gamma)^{\alpha-1} \quad (11\text{-}74)$$

$$\sum_{i=1}^{n} [(x_i-\gamma)^{\alpha} \cdot \ln (x_i-\gamma)] \quad (11\text{-}75)$$

and

$$\sum_{i=1}^{n} [(x_i-\gamma)^{\alpha} \cdot \ln^2 (x_i-\gamma)] \quad (11\text{-}76)$$

so that, by inverting the matrix, analytic expressions for the variances and covariances needed in 11-62 could be found. For the type III extremal distribution this procedure has not been followed because of the complexity of expressions 11-64 - 11-69. Instead, the program at the end of this chapter evaluates each summation directly, inverts the numerical matrix and computes the variances and covariances in terms of the matrix values.

Substitution of the numerical variances and covariances together with Equations 11-70 - 11-72 in Equation 11-62 yields the variance of the T-year event.


References

1.  Abramowitz, M. and I. A. Stegun, 1965, Handbook of Mathematical Functions, Dover Publications, New York, 1046 p.

2.  Chin, W. O., 1967, Formulae and Tables for Computing and Plotting Drought Frequency Curves, Technical Bulletin No. 8, Inland Waters Branch, Ottawa.

3.  Chow, V. T., 1964, Editor-in-Chief, Handbook of Applied Hydrology, McGraw-Hill.

4.  Condie, R., 1976, personal communication.

5.  Condie, R. and G. Nix, 1975, Modelling of Low Flow Frequency Distributions and Parameter Estimation, Int. Wat. Res. Assoc., Proc. Symp. on Water for Arid Lands, Tehran, Dec. 8-9.

6.  Deininger, R. A. and J. D. Westfield, 1969, Estimation of the Parameters of Gumbel's Third Asymptotic Distribution by Different Methods, Wat. Res. Res., Vol. 5, No. 6, pp. 1238-1243.

7.  Gumbel, E. J., 1958, Statistics of Extremes, Columbia University Press, 375 p.

8.  Gumbel, E. J., 1966, Extreme Value Analysis of Hydrologic Data, Proc. Hydrology Symp. No. 5, NRC, Ottawa, pp. 149-169.

9.  IBM, 1968, System/360 Scientific Subroutine Package (360A-CM-03X) Version III, Programmer's Manual, White Plains, New York, 454 p.

10. Lanczos, C., 1964. A precision approximation of the gamma function. Society for Industrial and Applied Mathematics, Numerical Analysis, Series B., 1:86.

11. Matalas, N. C., 1963, Probability Distribution of Low Flows, USGS Professional Paper No. 434-A, 26 p.

```
      PROGRAM T3E(INPUT,OUTPUT,TAPE5=INPUT,TAPE6=OUTPUT)              A   1
C                                                                     A   2
C                                                                     A   3
C     COMPUTES METHOD OF MOMENTS AND MAXIMUM LIKELIHOOD ESTIMATES FOR A   4
C     T YEAR EVENTS AND STANDARD ERRORS FOR TYPE 3 EXTREMAL           A   5
C     DISTRIBUTION                                                    A   6
C     INPUT                                                           A   7
C     TITLE                                                           A   8
C     N NUMBER OF ANNUAL MAXIMUM EVENTS                               A   9
C     X SERIES OF EVENTS                                              A  10
C                                                                     A  11
C                                                                     A  12
      REAL M1,M2,M3,K,IM,ML1,ML2,MLSKEW                               A  13
      DIMENSION X(100), T(6), DUM(6), PS(3)                           A  14
      DIMENSION XT(6), TITLE(80), ST(6), IM(9)                        A  15
      DIMENSION AL1(25), AL2(25), DIF(25)                             A  16
      T(1)=2                                                          A  17
      T(2)=5                                                          A  18
      T(3)=10                                                         A  19
      T(4)=20                                                         A  20
      T(5)=50                                                         A  21
      T(6)=100                                                        A  22
      A1=0.2777757913                                                 A  23
      A2=0.3132617714                                                 A  24
      A3=0.0575670910                                                 A  25
      A4=-0.0013038566                                                A  26
      A5=-0.0081523408                                                A  27
      READ (5,30) TITLE                                               A  28
      READ (5,31) N                                                   A  29
      XN=N                                                            A  30
      READ (5,32) (X(I),I=1,N)                                        A  31
      WRITE (6,33) TITLE                                              A  32
      WRITE (6,34)                                                    A  33
      A=0.0                                                           A  34
      B=0.0                                                           A  35
      C=0.0                                                           A  36
      DO 1 I=1,N                                                      A  37
      A=A+X(I)                                                        A  38
      B=B+X(I)**2                                                     A  39
      C=C+X(I)**3                                                     A  40
1     CONTINUE                                                        A  41
      M1=A/XN                                                         A  42
      M2=(B/XN)-(A/XN)**2                                             A  43
      M3=(C/XN)+2.0*M1**3-3.0*M1*(B/XN)                               A  44
      SKEW=M3/(M2**1.5)                                               A  45
      IF (SKEW.GT.2.00) WRITE (6,38)                                  A  46
      IF (SKEW.LT.-1.02) WRITE (6,38)                                 A  47
      AINV=A1+A2*SKEW+A3*SKEW**2+A4*SKEW**3+A5*SKEW**4                A  48
      ALPHA=1.0/AINV                                                  A  49
      DO 2 JJ=1,6                                                     A  50
      Z=1.0+JJ*AINV                                                   A  51
      DUM(JJ)=0.0                                                     A  52
2     CALL GMMMA (Z,DUM(JJ),IER)                                      A  53
      BA=1.0/((DUM(2)-DUM(1)**2)**0.5)                                A  54
      AA=(1.0-DUM(1))*BA                                              A  55
      TEMP=DUM(2)-DUM(1)**2                                           A  56
      GA=SKEW                                                         A  57
      GB=(DUM(4)-4.0*DUM(3)*DUM(1)+6.0*DUM(2)*DUM(1)**2-3.0*DUM(1)**4)/T A  58
     1EMP**2                                                          A  59
      GC=(DUM(5)-5.0*DUM(4)*DUM(1)+10.0*DUM(3)*DUM(1)**2-10.0*DUM(2)*DUM A  60
     1(1)**3+4.0*DUM(1)**5)/TEMP**2.5                                 A  61
      GD=(DUM(6)-6.0*DUM(5)*DUM(1)+15.0*DUM(4)*DUM(1)**2-20.0*DUM(3)*DUM A  62
     1(1)**3+15.0*DUM(2)*DUM(1)**4-5.0*DUM(1)**6)/TEMP**3             A  63
      DO 4 J=1,6                                                      A  64
      Y=(-ALOG(1.0-1.0/T(J)))                                         A  65
      K=AA+BA*(Y**AINV-1.0)                                           A  66
      T1=1.0+K*GA+(GB-1.0)*(K**2)/4.0                                 A  67
      T2=2.0*GB-3.0*GA**2-6.0                                         A  68
      T3=K*(GC-1.5*GA*GB-2.5*GA)                                      A  69
      T4=GD-3.0*GC*GA-6.0*GB+2.25*GB*GA**2+8.75*GA**4+9.0             A  70
      DO 3 JJ=1,3                                                     A  71
      Z=1.0+JJ*AINV                                                   A  72
      DUM(JJ)=0.0                                                     A  73
      CALL GMMMA (Z,DUM(JJ),IER)                                      A  74
      PS(JJ)=0.0                                                      A  75
3     CALL PSI (Z,PS(JJ))                                            A  76
      G1=DUM(1)                                                       A  77
      G2=DUM(2)                                                       A  78
      G3=DUM(3)                                                       A  79
      P1=PS(1)                                                        A  80
```

167

```
        P2=PS(2)                                                          A  81
        P3=PS(3)                                                          A  82
        T5=(ALOG(Y)*Y**AINV)-G1*PI                                        A  83
        T6=(Y**AINV)-G1                                                   A  84
        T7=G2-G1**2                                                       A  85
        T8=G2*P2-P1*G1**2                                                 A  86
        T9=G3*P3-G1*G2*(P1+2.0*P2)+2.0*P1*G1**3                           A  87
        T10=G3-3.0*G2*G1+2.0*G1**3                                        A  88
        SLOPE=((T5-(T6/T7)*T8)*T7**2)/(3.0*(T7*T9-T10*T8))                A  89
        DELTA=T1*SLOPE*(T2*T3)+T4*SLOPE**2                                A  90
        XT(J)=M1+K*SQRT(M2)                                               A  91
        ST(J)=SQRT(M2*DELTA/XN)                                           A  92
4       CONTINUE                                                          A  93
        BETA=AA*SQRT(M2)+M1                                               A  94
        GAMMA=BETA-BA*SQRT(M2)                                            A  95
        WRITE (6,35) ALPHA,M1                                             A  96
        WRITE (6,36) BETA,M2                                              A  97
        WRITE (6,37) GAMMA,SKEW                                           A  98
        WRITE (6,39)                                                      A  99
        WRITE (6,40) (XT(J),J=1,6)                                        A 100
        WRITE (6,41) (ST(J),J=1,6)                                        A 101
        WRITE (6,42)                                                      A 102
        WRITE (6,47)                                                      A 103
        AL=ALPHA                                                          A 104
        GA=GAMMA                                                          A 105
        XMIN=1000000.                                                     A 106
        DO 5 I=1,N                                                        A 107
5       IF (X(I).LT.XMIN) XMIN=X(I)                                       A 108
        DELTAX=XMIN/10.0                                                  A 109
        GA=XMIN+DELTAX/2.0                                                A 110
        KCOUNT=0                                                          A 111
6       KCOUNT=KCOUNT+1                                                   A 112
        DO 13 IJK=1,20                                                    A 113
        GA=GA-DELTAX                                                      A 114
        ICOUNT=0                                                          A 115
7       ICOUNT=ICOUNT+1                                                   A 116
        A=0.0                                                             A 117
        B=0.0                                                             A 118
        C=0.0                                                             A 119
        D=0.0                                                             A 120
        F=0.0                                                             A 121
        DO 8 I=1,N                                                        A 122
        A=A+1.0/(X(I)-GA)                                                 A 123
        B=B+(X(I)-GA)**AL                                                 A 124
        C=C+((X(I)-GA)**(AL-1.0))                                         A 125
        D=D+((X(I)-GA)**(AL-1.0))*ALOG(X(I)-GA)                           A 126
        F=F+((X(I)-GA)**AL)*ALOG(X(I)-GA)                                 A 127
8       CONTINUE                                                          A 128
        FCN=((AL-1.0)/(XN*AL))*A-C/B                                      A 129
        FPN=(1.0/(XN*AL**2))*A-(B*D-C*F)/(B**2)                           A 130
        BL=AL-FCN/FPN                                                     A 131
        DELTA=ABS(0.0000001*BL)                                           A 132
        IF (ABS(BL-AL).LT.DELTA) GO TO 9                                  A 133
        IF (ICOUNT.GE.25) GO TO 21                                        A 134
        AL=BL                                                             A 135
        GO TO 7                                                           A 136
9       AL1(IJK)=AL                                                       A 137
        JCOUNT=0                                                          A 138
10      JCOUNT=JCOUNT+1                                                   A 139
        A=0.0                                                             A 140
        B=0.0                                                             A 141
        C=0.0                                                             A 142
        D=0.0                                                             A 143
        DO 11 I=1,N                                                       A 144
        A=A+((X(I)-GA)**AL)                                               A 145
        B=B+((X(I)-GA)**AL)*ALOG(X(I)-GA)                                 A 146
        C=C+((X(I)-GA)**AL)*(ALOG(X(I)-GA))**2                            A 147
        D=D+ALOG(X(I)-GA)                                                 A 148
11      CONTINUE                                                          A 149
        FTN=XN*AL*D-XN*AL*B/A                                             A 150
        FUN=D*XN*AL*((A*C-B**2)/A**2)-XN*B/A                              A 151
        BL=AL-FTN/FUN                                                     A 152
        DELTA=ABS(0.0000001*BL)                                           A 153
        IF (ABS(BL-AL).LT.DELTA) GO TO 12                                 A 154
        IF (JCOUNT.GE.25) GO TO 21                                        A 155
        AL=BL                                                             A 156
        GO TO 10                                                          A 157
12      AL2(IJK)=AL                                                       A 158
        DIF(IJK)=AL1(IJK)-AL2(IJK)                                        A 159
        IF (IJK.EQ.1) GO TO 13                                            A 160
```

```
      DELTA=ABS(0.0000001*AL1(IJK))                                    A 161
      IF (ABS(DIF(IJK)).LT.DELTA) GO TO 15                             A 162
      IF (DIF(IJK-1).GE.0.0.AND.DIF(IJK).LT.0.0) GO TO 14             A 163
      IF (DIF(IJK-1).LT.0.0.AND.DIF(IJK).GT.0.0) GO TO 14             A 164
13    CONTINUE                                                         A 165
      GO TO 21                                                         A 167
14    GA=GA+DELTAX                                                     A 168
      AL=(AL1(IJK)+AL1(IJK-1)+AL2(IJK)+AL2(IJK-1))/4.0               A 169
      WRITE (6,43) KCOUNT,AL,FTN                                       A 170
      IF (KCOUNT.GE.25) GO TO 21                                       A 171
      DELTAX=DELTAX/20.0                                               A 172
      GO TO 6                                                          A 173
15    A=0.0                                                            A 174
      B=0.0                                                            A 175
      DO 16 I=1,N                                                      A 176
      A=A+ALOG(X(I)-GA)                                                A 177
      B=B+(X(I)-GA)**AL                                                A 178
16    CONTINUE                                                         A 179
      BE=GA+(B/XN)**(1.0/AL)                                          A 180
      ALPHA=AL                                                         A 181
      GAMMA=GA                                                         A 182
      BETA=BE                                                          A 183
      AINV=1.0/ALPHA                                                   A 184
      DO 17 JJ=1,3                                                     A 185
      DUM(JJ)=0.0                                                      A 186
      Z=1.0+JJ*AINV                                                    A 187
17    CALL GMMMA (Z,DUM(JJ),IER)                                       A 188
      BA=1.0/((DUM(2)-DUM(1)**2)**0.5)                               A 189
      AA=(1.0-DUM(1))*BA                                               A 190
      ML2=(BETA-GAMMA)/BA                                              A 191
      ML1=BETA-AA*ML2                                                  A 192
      ML2=ML2**2                                                       A 193
      MLSKEW=(DUM(3)-3.0*DUM(2)*DUM(1)+2.0*DUM(1)**3)*BA**3          A 194
      WRITE (6,44)                                                     A 195
      WRITE (6,35) ALPHA,ML1                                           A 196
      WRITE (6,36) BETA,ML2                                            A 197
      WRITE (6,37) GAMMA,MLSKEW                                        A 198
      A=0.0                                                            A 199
      B=0.0                                                            A 200
      C=0.0                                                            A 201
      D=0.0                                                            A 202
      E=0.0                                                            A 203
      F=0.0                                                            A 204
      G=0.0                                                            A 205
      H=0.0                                                            A 206
      P=0.0                                                            A 207
      Q=0.0                                                            A 208
      R=0.0                                                            A 209
      S=0.0                                                            A 210
      U=0.0                                                            A 211
      V=0.0                                                            A 212
      W=0.0                                                            A 213
      Z=0.0                                                            A 214
      DO 18 I=1,N                                                      A 215
      Z=X(I)-GAMMA                                                     A 216
      A=A+(Z**ALPHA)*ALOG(Z)                                          A 217
      B=B+(Z**ALPHA)*ALOG(Z)*ALOG(Z)                                 A 218
      C=C+Z**ALPHA                                                     A 219
      D=D+Z**(ALPHA-1.0)                                              A 220
      E=E+Z**(ALPHA-2.0)                                              A 221
      F=F+1.0/Z                                                        A 222
      G=G+1.0/(Z**2)                                                  A 223
      W=W+(Z**(ALPHA-1.0))*(1.0+ALPHA*ALOG(Z))                       A 224
18    CONTINUE                                                         A 225
      Z=BETA-GAMMA                                                     A 226
      H=Z**(-ALPHA)                                                    A 227
      P=Z**(-ALPHA-1.0)                                               A 228
      Q=Z**(-ALPHA-2.0)                                               A 229
      R=ALOG(Z)                                                        A 230
      S=Z**(ALPHA-1.0)                                                A 231
      U=XN/Z                                                           A 232
      V=XN/(Z**2)                                                     A 233
      IM(1)=-XN/(ALPHA**2)-H*(B-(R**2)*C)                            A 234
      IM(2)=-U+P*((R-1.0)*C+A*T)                                     A 235
      IM(3)=U-F+H*(W+ALPHA*R*D)*P*(1.0-ALPHA*R)*C-ALPHA*S*A         A 236
      TEMP1=ALPHA*(ALPHA+1.0)*Q*C                                    A 237
      TEMP3=ALPHA*(ALPHA-1.0)*H*E                                    A 238
      TEMP2=(ALPHA**2)*P*D                                           A 239
      IM(5)=ALPHA*V-TEMP1                                             A 240
      IM(6)=-ALPHA*V-TEMP2+TEMP1                                     A 241
```

```
      IM(9)=ALPHA*V-(ALPHA-1.0)*G-TEMP1+TEMP2-TEMP3           A 242
      IM(4)=IM(2)                                             A 243
      IM(7)=IM(3)                                             A 244
      IM(8)=IM(6)                                             A 245
      DO 19 KJ=1,9                                            A 246
 19   IM(KJ)=-IM(KJ)                                          A 247
      DET=IM(1)*(IM(5)*IM(9)-IM(6)*IM(8))-IM(2)*(IM(4)*IM(9)-IM(6)*IM(7)   A 248
     1)+IM(3)*(IM(4)*IM(8)-IM(5)*IM(7))                       A 249
      VARA=(IM(5)*IM(9)-IM(6)*IM(8))/DET                      A 250
      VARB=(IM(1)*IM(9)-IM(3)*IM(7))/DET                      A 251
      VARG=(IM(1)*IM(5)-IM(2)*IM(4))/DET                      A 252
      COVAB=-(IM(4)*IM(9)-IM(6)*IM(7))/DET                    A 253
      COVAG=(IM(4)*IM(8)-IM(5)*IM(7))/DET                     A 254
      COVBG=-(IM(1)*IM(8)-IM(2)*IM(7))/DET                    A 255
      DO 20 J=1,6                                             A 256
      Y=-ALOG(1.0/T(J))                                       A 257
      YA=Y**AINV                                              A 258
      YB=(-ALOG(1.0-1.0/T(J)))**AINV                          A 259
      DXDA=-(BETA-GAMMA)*YA*ALOG(Y)/(ALPHA**2)                A 260
      DXDB=YA                                                 A 261
      DXDG=1.0-YA                                             A 262
      XT(J)=GAMMA+(BETA-GAMMA)*YB                             A 263
      ST(J)=SQRT(VARA*DXDA**2+VARB*DXDB**2+VARG*DXDG**2+2.0*DXDA*DXDB*CO   A 264
     1VAB+2.0*DXDA*DXDG*COVAG+2.0*DXDB*DXDG*COVBG)            A 265
 20   CONTINUE                                                A 266
      WRITE (6,44)                                            A 267
      WRITE (6,39)                                            A 268
      WRITE (6,40)  (XT(J),J=1,6)                             A 269
      WRITE (6,41)  (ST(J),J=1,6)                             A 270
      GO TO 22                                                A 271
 21   WRITE (6,45)                                            A 272
 22   CONTINUE                                                A 273
      WRITE (6,46)                                            A 274
      WRITE (6,47)                                            A 275
      ASOD=ALPHA                                              A 276
      CONST=(M1-XMIN)/(SQRT(M2))                              A 277
      ICOUNT=0                                                A 278
 23   ICOUNT=ICOUNT+1                                         A 279
      F1=0.0                                                  A 280
      F2=0.0                                                  A 281
      P1=0.0                                                  A 282
      P2=0.0                                                  A 283
      T1=1.0+1.0/ASOD                                         A 284
      T2=1.0+2.0/ASOD                                         A 285
      CALL GMMMA (T1,F1,IER)                                  A 286
      CALL GMMMA (T2,F2,IER)                                  A 287
      CALL PSI (T1,P1)                                        A 288
      CALL PSI (T2,P2)                                        A 289
      F3=XN**(-1.0/ASOD)                                      A 290
      F4=ALOG(XN)                                             A 291
      F5=1.0/(ASOD**2)                                        A 292
      U=F1                                                    A 293
      V=(F2-F1**2)**0.5                                       A 294
      W=F1*F3                                                 A 295
      FCN=CONST-U/V+W/V                                       A 296
      DU=F5*P1*F1                                             A 297
      DV=(1.0/V)*F5*(F1*P1-F2*P2)                             A 298
      DW=F3*F1*F5*(F4-F1)                                     A 299
      FPN=(V*DW-W*DV)/(V**2)-(V*DU-U*DV)/(V**2)               A 300
      BSOD=ASOD-FCN/FPN                                       A 301
      WRITE (6,48) ICOUNT,BSOD,FCN                            A 302
      DELTA=0.000001*BSOD                                     A 303
      IF (ABS(BSOD-ASOD).LT.DELTA) GO TO 24                   A 304
      IF (ICOUNT.GE.25) GO TO 29                              A 305
      ASOD=BSOD                                               A 306
      GO TO 23                                                A 307
 24   ALPHA=BSOD                                              A 308
      AINV=1.0/ALPHA                                          A 309
      GAMMA=XMIN-(M1-XMIN)/(XN**AINV-1.0)                     A 310
      BETA=((M1-GAMMA)/F1)+GAMMA                              A 311
      DO 25 JJ=1,3                                            A 312
      DUM(JJ)=0.0                                             A 313
      Z=1.0+JJ*AINV                                           A 314
 25   CALL GMMMA (Z,DUM(JJ),IER)                              A 315
      BA=1.0/((DUM(2)-DUM(1)**2)**0.5)                        A 316
      AA=(1.0-DUM(1))*BA                                      A 317
      M2=(BETA-GAMMA)/BA                                      A 318
      M1=BETA-AA*M2                                           A 319
      M2=M2**2                                                A 320
      SKEW=(DUM(3)-3.0*DUM(2)*DUM(1)+2.0*DUM(1)**3)*BA**3     A 321
```

170

```
      WRITE (6,44)                                                      A 322
      WRITE (6,35) ALPHA,M1                                             A 323
      WRITE (6,36) BETA,M2                                              A 324
      WRITE (6,37) GAMMA,SKEW                                           A 325
      A=0.0                                                             A 326
      B=0.0                                                             A 327
      C=0.0                                                             A 328
      D=0.0                                                             A 329
      E=0.0                                                             A 330
      F=0.0                                                             A 331
      G=0.0                                                             A 332
      H=0.0                                                             A 333
      P=0.0                                                             A 334
      Q=0.0                                                             A 335
      R=0.0                                                             A 336
      S=0.0                                                             A 337
      U=0.0                                                             A 338
      V=0.0                                                             A 339
      W=0.0                                                             A 340
      Z=0.0                                                             A 341
      DO 26 I=1,N                                                       A 342
      Z=X(I)-GAMMA                                                      A 343
      A=A+(Z**ALPHA)*ALOG(Z)                                            A 344
      B=B+(Z**ALPHA)*ALOG(Z)*ALOG(Z)                                    A 345
      C=C+Z**ALPHA                                                      A 346
      D=D+Z**(ALPHA-1.0)                                                A 347
      E=E+Z**(ALPHA-2.0)                                                A 348
      F=F+1.0/Z                                                         A 349
      G=G+1.0/(Z**2)                                                    A 350
      W=W+(Z**(ALPHA-1.0))*(1.0+ALPHA*ALOG(Z))                          A 351
26    CONTINUE                                                          A 352
      Z=BETA-GAMMA                                                      A 353
      H=Z**(-ALPHA)                                                     A 354
      P=Z**(-ALPHA-1.0)                                                 A 355
      Q=Z**(-ALPHA-2.0)                                                 A 356
      R=ALOG(Z)                                                         A 357
      S=Z**(ALPHA-1.0)                                                  A 358
      U=XN/Z                                                            A 359
      V=XN/(Z**2)                                                       A 360
      IM(1)=-XN/(ALPHA**2)-H*(B-(R**2)*C)                               A 361
      IM(2)=-U+P*((R-1.0)*C*A)                                          A 362
      IM(3)=U-F+H*(W*ALPHA*R*D)+P*(1.0-ALPHA*R)*C-ALPHA*S*A             A 363
      TEMP1=ALPHA*(ALPHA+1.0)*Q*C                                       A 364
      TEMP3=ALPHA*(ALPHA-1.0)*H*E                                       A 365
      TEMP2=(ALPHA**2)*P*D                                              A 366
      IM(5)=ALPHA*V-TEMP1                                               A 367
      IM(6)=-ALPHA*V-TEMP2+TEMP1                                        A 368
      IM(9)=ALPHA*V-(ALPHA-1.0)*G-TEMP1+TEMP2-TEMP3                     A 369
      IM(4)=IM(2)                                                       A 370
      IM(7)=IM(3)                                                       A 371
      IM(8)=IM(6)                                                       A 372
      DO 27 KJ=1,9                                                      A 373
27    IM(KJ)=-IM(KJ)                                                    A 374
      DET=IM(1)*(IM(5)*IM(9)-IM(6)*IM(8))-IM(2)*(IM(4)*IM(9)-IM(6)*IM(7) A 375
     1)+IM(3)*(IM(4)*IM(8)-IM(5)*IM(7))                                 A 376
      VARA=(IM(5)*IM(9)-IM(6)*IM(8))/DET                                A 377
      VARB=(IM(1)*IM(9)-IM(3)*IM(7))/DET                                A 378
      VARG=(IM(1)*IM(5)-IM(2)*IM(4))/DET                                A 379
      COVAB=-(IM(4)*IM(9)-IM(6)*IM(7))/DET                              A 380
      COVAG=(IM(4)*IM(8)-IM(5)*IM(7))/DET                               A 381
      COVBG=(IM(1)*IM(8)-IM(2)*IM(7))/DET                               A 382
      DO 28 J=1,6                                                       A 383
      Y=-ALOG(1.0/T(J))                                                 A 384
      YA=Y**AINV                                                        A 385
      YB=(-ALOG(1.0-1.0/T(J)))**AINV                                    A 386
      DXDA=-(BETA-GAMMA)*YA*ALOG(Y)/(ALPHA**2)                          A 387
      DXDB=YA                                                           A 388
      DXDG=1.0-YA                                                       A 389
      XT(J)=GAMMA+(BETA-GAMMA)*YB                                       A 390
      ST(J)=SQRT(VARA*DXDA**2+VARB*DXDB**2+VARG*DXDG**2+2.0*DXDA*DXDB*CO A 391
     1VAB+2.0*DXDA*DXDG*COVAG+2.0*DXDB*DXDG*COVBG)                      A 392
28    CONTINUE                                                          A 393
      WRITE (6,39)                                                      A 394
      WRITE (6,40) (XT(J),J=1,6)                                        A 395
      WRITE (6,41) (ST(J),J=1,6)                                        A 396
29    CONTINUE                                                          A 397
      STOP                                                              A 398
C                                                                       A 399
30    FORMAT (80A1)                                                     A 400
```

171

```
31      FORMAT (I5)                                                       A 401
32      FORMAT (8F10.1)                                                   A 402
33      FORMAT (1H1,80A1,/,26X,28HTYPE 3 EXTREMAL DISTRIBUTION,//)        A 403
34      FORMAT (33X,17HMETHOD OF MOMENTS,/)                              A 404
35      FORMAT (9X,5HALPHA,5X,E12.5,14X,4HM1   ,6X,E12.5)                A 405
36      FORMAT (9X,5HBETA ,5X,E12.5,14X,4HM2  ,6X,E12.5)                 A 406
37      FORMAT (9X,5HGAMMA,5X,E12.5,14X,4HSKEW,6X,E12.5,/)               A 407
38      FORMAT (/,42H WARNING- SKEW IS BEYOND PERMISSIBLE LIMIT,/)       A 408
39      FORMAT (3X,7HT YEARS,4X,1H2,11X,1H5,10X,2H10,10X,2H20,10X,2H50,9X, A 409
       13H100,/)                                                          A 410
40      FORMAT (3X,1HX,3X,6E12.5,/,4X,1HT)                               A 411
41      FORMAT (3X,1HS,3X,6E12.5,/,4X,1HT/)                              A 412
42      FORMAT (27X,28HMAXIMUM LIKELIHOOD PROCEDURE,/)                   A 413
43      FORMAT (22X,I2,8X,E12.5,1X,E12.5)                                A 414
44      FORMAT (/)                                                       A 415
45      FORMAT (/,3X,45HMAXIMUM LIKELIHOOD DOES NOT CONVERGE IN ALPHA,/)  A 416
46      FORMAT (22X,35HMETHOD OF SMALLEST OBSERVED DROUGHT,/)            A 417
47      FORMAT (21X,5HTRIAL,12X,1HA,11X,4HF(A),/)                        A 418
48      FORMAT (22X,I2,8X,E12.5,1X,E12.5)                                A 419
        END                                                              A 420-

        SUBROUTINE GMMMA (XX,GX,IER)                                      B  1
        IF (XX-57.) 2,2,1                                                 B  2
1       GX=0.39909-0.43429*(XX-1.0)+(XX-0.5)*ALOG10(XX-1.0)             B  3
        GX=10.0**GX                                                       B  4
        RETURN                                                           B  5
2       X=XX                                                             B  6
        ERR=1.0E-6                                                       B  7
        IER=0                                                            B  8
        GX=1.0                                                           B  9
        IF (X-2.0) 5,5,4                                                 B 10
3       IF (X-2.0) 11,11,4                                               B 11
4       X=X-1.0                                                          B 12
        GX=GX*X                                                          B 13
        GO TO 3                                                          B 14
5       IF (X-1.0) 6,12,11                                               B 15
6       IF (X-ERR) 7,7,10                                                B 16
7       Y=FLOAT(INT(X))-X                                                B 17
        IF (ABS(Y)-ERR) 13,13,8                                          B 18
8       IF (1.0-Y-ERR) 13,13,9                                           B 19
9       IF (X-1.0) 10,10,11                                              B 20
10      GX=GX/X                                                          B 21
        X=X+1.0                                                          B 22
        GO TO 9                                                          B 23
11      Y=X-1.0                                                          B 24
        GY=1.0+Y*(-0.5771017+Y*(+0.9858540+Y*(-0.8764218+Y*(+0.8328212+Y*( B 25
       1-0.5684729+Y*(+0.2548205+Y*(-0.05149930)))))))                   B 26
        GX=GX*GY                                                         B 27
12      RETURN                                                          B 28
13      IER=1                                                           B 29
        RETURN                                                          B 30
        END                                                             B 31

        SUBROUTINE PSI (XX,YY)                                            C  1
        D=XX+2.0                                                         C  2
        YY=ALOG(D)-(1.0/(2.0*D))-(1.0/(12.0*D**2))+(1.0/(120.0*D**4))-(1.0 C  3
       1/(252.0*D**6))-(1.0/(XX+1.0))-(1.0/XX)                           C  4
        RETURN                                                          C  5
        END                                                             C  6
```

172

```
10 '
20 ' Program T3E
30 ' Copyright G.W. Kite 1986
40 '
50 ' Compute method of moments and maximum likelihood estimates
60 ' for T year events and standard errors
70 ' for a Type III Extremal Distribution.
80 '
90 DIM T#(6),X#(250),XT#(6),SX#(6),DUM#(6),PS#(3),IM#(9),AL1#(50),AL2#(50),DIF#(
50)
100 FORM$="  ##.###^^^^"
110 VV$=CHR$(179)
120 DATA 2#,5#,10#,20#,50#,100#
130 DATA 0.2777757913#,0.3132617714#,0.0575670910#,-0.0013038566#,-0.0081523408#

140 FOR I%=1 TO 6
150 READ T#(I%)
160 NEXT I%
170 READ A1#,A2#,A3#,A4#,A5#
180 T$="Type III Extremal Distribution"
190 IERR%=0
200 GOSUB 4710
210 IF IERR% = 1 GOTO 4690
220 FLAG%=0:IF YN$ = "Y" OR YN$ = "y" THEN FLAG%=1
230 NREC%=0
240 OPEN FILE$ FOR INPUT AS #1
250 LINE INPUT#1,TITLE$
260 IF EOF(1) GOTO 310
270 NREC%=NREC%+1
280 LINE INPUT#1,I$
290 X#(NREC%)=VAL(MID$(I$,26,12))
300 GOTO 260
310 CLOSE#1
320 NREC#=NREC%
330 A%=LEN(TITLE$)
340 B%=(80-A%)/2
350 U$=STRING$(A%,205)
360 M$="Method of Moments"
370 E%=LEN(M$)
380 F%=(80-E%)/2
390 W$=STRING$(E%,205)
400 CLS
410 PRINT TAB(B%) TITLE$:PRINT TAB(B%) U$
420 PRINT TAB(D%) T$:PRINT TAB(D%) V$
430 PRINT TAB(F%) M$:PRINT TAB(F%) W$
440 IF FLAG% <> 1 GOTO 490
450 LPRINT CHR$(12):LPRINT:LPRINT
460 LPRINT TAB(B%) TITLE$:LPRINT TAB(B%) U$
470 LPRINT:LPRINT TAB(D%) T$:LPRINT TAB(D%) V$
480 LPRINT:LPRINT TAB(F%) M$:LPRINT TAB(F%) W$
490 A#=0#:B#=0#:C#=0#
500 FOR I%=1 TO NREC%
510 A#=A#+X#(I%)
520 B#=B#+X#(I%)*X#(I%)
530 C#=C#+X#(I%)*X#(I%)*X#(I%)
540 NEXT I%
550 M1#=A#/NREC#
560 M2#=(B#/NREC#)-M1#*M1#
570 M3#=(C#/NREC#)+2#*M1#*M1#*M1#-3#*M1#*(B#/NREC#)
580 SKEW#=M3#/(M2#^1.5#)
590 IF SKEW# > 2# OR SKEW# < -1.02# THEN PRINT :PRINT "WARNING: Coeff. of skew i
s beyond permissible limit"
600 AINV#=A1#+A2#*SKEW#+A3#*SKEW#*SKEW#+A4#*SKEW#*SKEW#*SKEW#+A5#*SKEW#*SKEW#*SK
EW#*SKEW#
610 ALPHA#=1#/AINV#
620 FOR J%=1 TO 6
630 Z#=1#+J%*AINV#
640 DUM#(J%)=0#:SUBO#=0#
650 SUBI#=Z#:GOSUB 5250:DUM#(J%)=SUBO#
660 NEXT J%
670 TEMP#=DUM#(2)-DUM#(1)*DUM#(1)
680 BA#=1#/SQR(TEMP#)
690 AA#=BA#*(1#-DUM#(1))
700 BETA#=AA#*SQR(M2#)+M1#
710 GAMMA#=BETA#-BA#*SQR(M2#)
720 GA#=SKEW#
730 GB#=(DUM#(4)-4#*DUM#(3)*DUM#(1)+6#*DUM#(2)*DUM#(1)*DUM#(1)-3#*DUM#(1)^4)
740 GB#=GB#/(TEMP#*TEMP#)
750 GC#=DUM#(5)-5#*DUM#(4)*DUM#(1)+10#*DUM#(3)*DUM#(1)*DUM#(1)
```

173

```
760 GC#=(GC#-10#*DUM#(2)*DUM#(1)*DUM#(1)*DUM#(1)+4#*DUM#(1)^5)/TEMP#^2.5
770 GD#=DUM#(6)-6#*DUM#(5)*DUM#(1)+15#*DUM#(4)*DUM#(1)*DUM#(1)
780 GD#=GD#-20#*DUM#(3)*DUM#(1)*DUM#(1)*DUM#(1)+15#*DUM#(2)*DUM#(1)^4
790 GD#=(GD#-5#*DUM#(1)^6)/TEMP#^3
800 FOR J%=1 TO 6
810 Y#=(-LOG(1#-1#/T#(J%)))
820 K#=AA#+BA#*(Y#^AINV#-1#)
830 T1#=1#+K#*GA#+(GB#-1#)*(K#*K#/4#)
840 T2#=2#*GB#-3#*GA#*GA#-6#
850 T3#=K#*(GC#-1.5#*GA#*GB#-2.5#*GA#)
860 T4#=GD#-3#*GC#*GA#-6#*GB#+2.25#*GB#*GA#*GA#+8.75#*GA#*GA#+9#
870 FOR K%=1 TO 3
880 Z#=1#+K%*AINV#
890 DUM#(K%)=0#
900 SUBI#=Z#:GOSUB 5250:DUM#(K%)=SUBO#
910 PS#(K%)=0#
920 SUBI#=Z#:GOSUB 5570:PS#(K%)=SUBO#
930 NEXT K%
940 G1#=DUM#(1):G2#=DUM#(2):G3#=DUM#(3)
950 P1#=PS#(1):P2#=PS#(2):P3#=PS#(3)
960 T5#=(LOG(Y#)*Y#^AINV#)-G1#*P1#
970 T6#=(Y#^AINV#)-G1#
980 T7#=G2#-G1#*G1#
990 T8#=G2#*P2#-P1#*G1#*G1#
1000 T9#=G3#*P3#-G1#*G2#*(P1#+2#*P2#)+2#*P1#*G1#*G1#*G1#
1010 T10#=G3#-3#*G2#*G1#+2#*G1#*G1#*G1#
1020 SLOPE#=((T5#-(T6#/T7#)*T8#)*T7#*T7#)/(3#*(T7#*T9#-T10#*T8#))
1030 DELTA#=T1#+SLOPE#*(T2#+T3#)+T4#*SLOPE#*SLOPE#
1040 XT#(J%)=M1#+K#*SQR(M2#)
1050 SX#(J%)=SQR(M2#*DELTA#/NREC#)
1060 NEXT J%
1070 PRINT
1080 PRINT TAB(20) "Mean is                    ";:PRINT USING FORM$;M1#
1090 PRINT TAB(20) "Variance is                ";:PRINT USING FORM$;M2#
1100 PRINT TAB(20) "Coefficient of skew is ";:PRINT USING FORM$;SKEW#
1110 PRINT TAB(20) "Parameter Alpha is     ";:PRINT USING FORM$;ALPHA#
1120 PRINT TAB(20) "Parameter Beta is      ";:PRINT USING FORM$;BETA#
1130 PRINT TAB(20) "Parameter Gamma is     ";:PRINT USING FORM$;GAMMA#
1140 IF FLAG% <> 1 GOTO 1220
1150 LPRINT
1160 LPRINT TAB(20) "Mean is                    ";:LPRINT USING FORM$;M1#
1170 LPRINT TAB(20) "Variance is                ";:LPRINT USING FORM$;M2#
1180 LPRINT TAB(20) "Coefficient of skew is ";:LPRINT USING FORM$;SKEW#
1190 LPRINT TAB(20) "Parameter Alpha is     ";:LPRINT USING FORM$;ALPHA#
1200 LPRINT TAB(20) "Parameter Beta is      ";:LPRINT USING FORM$;BETA#
1210 LPRINT TAB(20) "Parameter Gamma is     ";:LPRINT USING FORM$;GAMMA#
1220 ROW$=CHR$(218)+STRING$(72,196)
1230 PRINT:PRINT TAB(2) "T, years  2            5           10          20
      50       100":PRINT TAB(5) ROW$
1240 PRINT TAB(2) "X  " VV$;:FOR J%=1 TO 6:PRINT USING FORM$;XT#(J%);:NEXT J%
1250 PRINT TAB(2) " t " VV$
1260 PRINT TAB(2) "S  " VV$;:FOR J%=1 TO 6:PRINT USING FORM$;SX#(J%);:NEXT J%
1270 PRINT TAB(2) " t " VV$
1280 IF FLAG% <> 1 GOTO 1340
1290 LPRINT:LPRINT TAB(2) "T, years  2            5           10          20
      50       100":LPRINT TAB(5) ROW$
1300 LPRINT TAB(2) "X  " VV$;:FOR J%=1 TO 6:LPRINT USING FORM$;XT#(J%);:NEXT J%
1310 LPRINT TAB(2) " t " VV$
1320 LPRINT TAB(2) "S  " VV$;:FOR J%=1 TO 6:LPRINT USING FORM$;SX#(J%);:NEXT J%
1330 LPRINT TAB(2) " t " VV$
1340 M$="Method of Smallest Observed Drought"
1350 E%=LEN(M$)
1360 F%=(80-E%)/2
1370 W$=STRING$(E%,205)
1380 PRINT:PRINT"Press any key to continue"
1390 Q$=INKEY$:IF Q$ ="" GOTO 1390
1400 CLS
1410 XMIN#=1000000000#
1420 FOR I% = 1 TO NREC%
1430 IF X#(I%) < XMIN# THEN XMIN#=X#(I%)
1440 NEXT I%
1450 PRINT:PRINT TAB(F%) M$:PRINT TAB(F%) W$
1460 PRINT:PRINT TAB(19) "Iteration" TAB(37) "A" TAB(53) "f(A)"
1470 PRINT TAB(19) "---------" TAB(37) "-" TAB(53) "----"
1480 IF FLAG% <> 1 GOTO 1520
1490 LPRINT:LPRINT TAB(F%) M$:LPRINT TAB(F%) W$
1500 LPRINT:LPRINT TAB(19) "Iteration" TAB(37) "A" TAB(53) "f(A)"
1510 LPRINT TAB(19)"---------" TAB(37) "-" TAB(53) "----"
1520 ASOD#=ALPHA#*.75
1530 CONST#=(M1#-XMIN#)/SQR(M2#)
```

174

```
1540 ICOUNT%=0
1550 ICOUNT%=ICOUNT%+1
1560 T1#=1#+1#/ASOD#
1570 T2#=1#+2#/ASOD#
1580 F1#=0#:F2#=0#:P1#=0#:P2#=0#
1590 SUBI#=T1#:GOSUB 5250:F1#=SUBO#
1600 SUBI#=T2#:GOSUB 5250:F2#=SUBO#
1610 SUBI#=T1#:GOSUB 5570:P1#=SUBO#
1620 SUBI#=T2#:GOSUB 5570:P2#=SUBO#
1630 F3#=NREC#^(-1#/ASOD#)
1640 F4#=LOG(NREC#)
1650 F5#=1#/(ASOD#*ASOD#)
1660 U#=F1#
1670 V#=SQR(F2#-F1#*F1#)
1680 W#=F1#*F3#
1690 FCN#=CONST#-U#/V#+W#/V#
1700 DU#=-F5#*P1#*F1#
1710 DV#=(1#/V#)*F5#*(F1#*P1#-F2#*P2#)
1720 DW#=F3#*F1#*F5#*(F4#-F1#)
1730 FPN#=(V#*DW#-W#*DV#)/(V#*V#)-(V#*DU#-U#*DV#)/(V#*V#)
1740 BSOD#=ASOD#-FCN#/FPN#
1750 PRINT TAB(22);:PRINT USING "##";ICOUNT%;:PRINT USING "        ##.###^^^^";B
SOD#,FCN#
1760 IF FLAG% = 1 THEN LPRINT TAB(22);:LPRINT USING "##";ICOUNT%;:LPRINT USING "
          ##.###^^^^";BSOD#,FCN#
1770 DELTA#=.000001#*BSOD#
1780 IF ABS(BSOD#-ASOD#) < DELTA# GOTO 1850
1790 IF ICOUNT% < 24 GOTO 1830
1800 PRINT:PRINT "Procedure does not converge":PRINT
1810 LPRINT:LPRINT "Procedure does not converge":LPRINT
1820 GOTO 2820
1830 ASOD#=BSOD#
1840 GOTO 1550
1850 ALPHA#=BSOD#
1860 AINV#=1#/ALPHA#
1870 GAMMA#=XMIN#-(M1#-XMIN#)/(NREC#^AINV#~1#)
1880 BETA#=((M1#-GAMMA#)/F1#)+GAMMA#
1890 FOR JJ%=1 TO 3
1900 DUM#(JJ%)=0#
1910 Z#=1#+JJ%*AINV#
1920 SUBI#=Z#:GOSUB 5250:DUM#(JJ%)=SUBO#
1930 NEXT JJ%
1940 BA#=1#/SQR(DUM#(2)-DUM#(1)*DUM#(1))
1950 AA#=(1#-DUM#(1))*BA#
1960 M2#=(BETA#-GAMMA#)/BA#
1970 M1#=BETA#-AA#*M2#
1980 M2#=M2#*M2#
1990 SKEW#=(DUM#(3)-3#*DUM#(2)*DUM#(1)+2#*DUM#(1)*DUM#(1)*DUM#(1))*BA#*BA#*BA#
2000 CLS
2010 PRINT:PRINT TAB(B%) TITLE$:PRINT TAB(B%) U$
2020 PRINT TAB(D%) T$:PRINT TAB(D%) V$
2030 PRINT TAB(F%) M$:PRINT TAB(F%) W$
2040 PRINT TAB(19) "Mean is                  ";:PRINT USING FORM$;M1#
2050 PRINT TAB(19) "Variance is              ";:PRINT USING FORM$;M2#
2060 PRINT TAB(19) "Coefficient of skew is   ";:PRINT USING FORM$;SKEW#
2070 PRINT TAB(19) "Parameter Alpha is       ";:PRINT USING FORM$;ALPHA#
2080 PRINT TAB(19) "Parameter Beta is        ";:PRINT USING FORM$;BETA#
2090 PRINT TAB(19) "Parameter Gamma is       ";:PRINT USING FORM$;GAMMA#
2100 IF FLAG% <> 1 GOTO 2180
2110 LPRINT
2120 LPRINT TAB(19) "Mean is                  ";:LPRINT USING FORM$;M1#
2130 LPRINT TAB(19) "Variance is              ";:LPRINT USING FORM$;M2#
2140 LPRINT TAB(19) "Coefficient of skew is   ";:LPRINT USING FORM$;SKEW#
2150 LPRINT TAB(19) "Parameter Alpha is       ";:LPRINT USING FORM$;ALPHA#
2160 LPRINT TAB(19) "Parameter Beta is        ";:LPRINT USING FORM$;BETA#
2170 LPRINT TAB(19) "Parameter Gamma is       ";:LPRINT USING FORM$;GAMMA#
2180 A#=0#:B#=0#:C#=0#:D#=0#:E#=0#:F#=0#:G#=0#:H#=0#:P#=0#:Q#=0#:R#=0#:S#=0#:U#=
0#:V#=0#:W#=0#:Z#=0#
2190 FOR I%=1 TO NREC%
2200 Z#=X#(I%)-GAMMA#
2210 ZZ#=Z#^ALPHA#
2220 A#=A#+ZZ#*LOG(Z#)
2230 B#=B#+ZZ#*LOG(Z#)*LOG(Z#)
2240 C#=C#+ZZ#
2250 D#=D#+(Z#^(ALPHA#~1#))
2260 E#=E#+(Z#^(ALPHA#~2#))
2270 F#=F#+1#/Z#
2280 G#=G#+1#/(Z#*Z#)
```

175

```
2290 W#=W#+(Z#^(ALPHA#-1#))*(1#+ALPHA#*LOG(Z#))
2300 NEXT I%
2310 Z#=BETA#-GAMMA#
2320 H#=Z#^(-ALPHA#)
2330 P#=Z#^(-ALPHA#-1#)
2340 Q#=Z#^(-ALPHA#-2#)
2350 R#=LOG(Z#)
2360 S#=Z#^(ALPHA#-1#)
2370 U#=NREC#/Z#
2380 V#=NREC#/(Z#*Z#)
2390 IM#(1)=-NREC#/(ALPHA#*ALPHA#)-H#*(B#-(R#*R#)*C#)
2400 IM#(2)=-U#+P#*((R#-1#)*C#+A#)
2410 IM#(3)=U#-F#+H#*(W#+ALPHA#*R#*D#)+P#*(1#-ALPHA#*R#)*C#-ALPHA#*S#*A#
2420 TEMP1#=ALPHA#*(ALPHA#-1#)*Q#*C#
2430 TEMP3#=ALPHA#*(ALPHA#-1#)*H#*E#
2440 TEMP2#=ALPHA#*ALPHA#*P#*D#
2450 IM#(5)=ALPHA#*V#-TEMP1#
2460 IM#(6)=-ALPHA#*V#-TEMP1#*TEMP2#
2470 IM#(9)=ALPHA#*V#-(ALPHA#-1#)*G#-TEMP1#+TEMP2#-TEMP3#
2480 IM#(4)=IM#(2)
2490 IM#(7)=IM#(3)
2500 IM#(8)=IM#(6)
2510 FOR KJ%=1 TO 9
2520 IM#(KJ%)=-IM#(KJ%)
2530 NEXT KJ%
2540 DET#=IM#(1)*(IM#(5)*IM#(9)-IM#(6)*IM#(8))-IM#(2)*(IM#(4)*IM#(9)-IM#(6)*IM#(
7))+IM#(3)*(IM#(4)*IM#(8)-IM#(5)*IM#(7))
2550 VARA#=(IM#(5)*IM#(9)-IM#(6)*IM#(8))/DET#
2560 VARB#=(IM#(1)*IM#(9)-IM#(3)*IM#(7))/DET#
2570 VARG#=(IM#(1)*IM#(5)-IM#(2)*IM#(7))/DET#
2580 COVAB#=-(IM#(4)*IM#(9)-IM#(6)*IM#(7))/DET#
2590 COVAG#=(IM#(4)*IM#(8)-IM#(5)*IM#(7))/DET#
2600 COVBG#=-(IM#(1)*IM#(8)-IM#(2)*IM#(7))/DET#
2610 FOR J% = 1 TO 6
2620 Y#=-LOG(1#/T#(J%))
2630 YA#=Y#^AINV#
2640 YB#=(-LOG(1#-1#/T#(J%)))^AINV#
2650 XT#(J%)=GAMMA#+(BETA#-GAMMA#)*YB#
2660 DXDA#=-(BETA#-GAMMA#)*YA#*LOG(Y#)/(ALPHA#*ALPHA#)
2670 DXDB#=YA#
2680 DXDG#=1#-YA#
2690 SX#(J%)=SQR(VARA#*DXDA#*DXDA#+VARB#*DXDB#*DXDB#+VARG#*DXDG#*DXDG#+2#*DXDA#*
DXDB#*COVAB#+2#*DXDA#*DXDG#*COVAG#+2#*DXDB#*DXDG#*COVBG#)
2700 NEXT J%
2710 PRINT:PRINT TAB(2) "T, years  2        5        10        20
   50        100":PRINT TAB(5) ROW$
2720 PRINT TAB(2) "X  " VV$;:FOR J%=1 TO 6:PRINT USING FORM$;XT#(J%);:NEXT J%
2730 PRINT TAB(2) " t " VV$
2740 PRINT TAB(2) "S  " VV$;:FOR J%=1 TO 6:PRINT USING FORM$;SX#(J%);:NEXT J%
2750 PRINT TAB(2) " t " VV$
2760 IF FLAG% <> 1 GOTO 2820
2770 LPRINT:LPRINT TAB(2) "T, years  2        5        10        20
   50        100":LPRINT TAB(5) ROW$
2780 LPRINT TAB(2) "X  " VV$;:FOR J%=1 TO 6:LPRINT USING FORM$;XT#(J%);:NEXT J%
2790 LPRINT TAB(2) " t " VV$
2800 LPRINT TAB(2) "S  " VV$;:FOR J%=1 TO 6:LPRINT USING FORM$;SX#(J%);:NEXT J%
2810 LPRINT TAB(2) " t " VV$
2820 M$="Method of Maximum Likelihood"
2830 E%=LEN(M$)
2840 F%=(80-E%)/2
2850 W$=STRING$(E%,205)
2860 PRINT:PRINT"Press any key to continue"
2870 Q$=INKEY$:IF Q$ ="" GOTO 2870
2880 CLS
2890 PRINT:PRINT TAB(F%) M$:PRINT TAB(F%) W$
2900 PRINT:PRINT TAB(19) "Iteration" TAB(37) "G" TAB(53) "f(G)"
2910 PRINT TAB(19) "---------" TAB(37) "-" TAB(53) "----"
2920 IF FLAG% <> 1 GOTO 2970
2930 LPRINT CHR$(12)
2940 LPRINT:LPRINT TAB(F%) M$:LPRINT TAB(F%) W$
2950 LPRINT:LPRINT TAB(19) "Iteration" TAB(37) "G" TAB(53) "f(G)"
2960 LPRINT TAB(19)"---------" TAB(37) "-" TAB(53) "----"
2970 AL#=ALPHA#
2980 GA=GAMMA#
2990 DELTAX#=XMIN#/20#
3000 GA#=XMIN#+DELTAX#-.00001
3010 KCOUNT%=0
3020 KCOUNT%=KCOUNT%+1
3030 FOR IJK%=1 TO 40
```

176

```
3040 GA#=GA#-DELTAX#
3050 ICOUNT%=0
3060 ICOUNT%=ICOUNT%+1
3070 A#=0#:B#=0#:C#=0#:D#=0#:F#=0#
3080 FOR I%=1 TO NREC%
3090 TEMP1#=X#(I%)-GA#
3100 TEMP2#=TEMP1#^AL#
3110 TEMP3#=TEMP1#^(AL#-1#)
3120 TEMP4#=LOG(TEMP1#)
3130 A#=A#+1#/TEMP1#
3140 B#=B#+TEMP2#
3150 C#=C#+TEMP3#
3160 D#=D#+TEMP3#*TEMP4#
3170 F#=F#+TEMP2#*TEMP4#
3180 NEXT I%
3190 FCN#=((AL#-1#)/(NREC#*AL#))*A#-C#/B#
3200 FPN#=(1#/(NREC#*AL#*AL#))*A#-(B#*D#-C#*F#)/(B#*B#)
3210 BL#=AL#-(FCN#/FPN#)
3220 DELTA#=ABS(.001#*BL#)
3230 IF ABS(BL#-AL#) < DELTA# GOTO 3270
3240 IF ICOUNT% > 24 THEN PRINT:PRINT "Procedure does not converge at 3240":PRIN
T:GOTO 4670
3250 AL#=BL#
3260 GOTO 3060
3270 AL1#(IJK%)=AL#
3280 JCOUNT%=0
3290 JCOUNT%=JCOUNT%+1
3300 A#=0#:B#=0#:C#=0#:D#=0#
3310 FOR I%=1 TO NREC%
3320 TEMP1#=X#(I%)-GA#
3330 TEMP2#=TEMP1#^AL#
3340 TEMP3#=LOG(TEMP1#)
3350 A#=A#+TEMP2#
3360 B#=B#+TEMP2#*TEMP3#
3370 C#=C#+TEMP2#*TEMP3#*TEMP3#
3380 D#=D#+TEMP3#
3390 NEXT I%
3400 FTN#=NREC#+AL#*D#-NREC#*AL#*B#/A#
3410 FUN#=D#-NREC#*AL#*((A#*C#-B#*B#)/(A#*A#))-NREC#*B#/A#
3420 BL#=AL#-(FTN#/FUN#)
3430 DELTA#=.001#
3440 IF ABS(BL#-AL#) < DELTA# GOTO 3480
3450 IF JCOUNT% > 24 THEN PRINT:PRINT "Procedure does not converge at 3450":PRIN
T:GOTO 4670
3460 AL#=BL#
3470 GOTO 3290
3480 AL2#(IJK%)=AL#
3490 DIF#(IJK%)=AL1#(IJK%)-AL2#(IJK%)
3500 IF IJK% = 1 GOTO 3550
3510 DELTA#=ABS(.001#*AL1#(IJK%))
3520 IF ABS(DIF#(IJK%)) < DELTA# GOTO 3640
3530 IF DIF#(IJK%-1) >= 0# AND DIF#(IJK%) < 0# GOTO 3570
3540 IF DIF#(IJK%-1) < 0# AND DIF#(IJK%) > 0# GOTO 3570
3550 NEXT IJK%
3560 PRINT:PRINT"Procedure does not converge at 3560":PRINT:GOTO 4670
3570 GA#=GA#+DELTAX#
3580 AL#=(AL1#(IJK%)+AL1#(IJK%-1)+AL2#(IJK%)+AL2#(IJK%-1))/4#
3590 PRINT TAB(22);:PRINT USING "##";KCOUNT%;:PRINT USING "        ##.###^^^^";A
L#,FTN#
3600 IF FLAG% = 1 THEN LPRINT TAB(22);:LPRINT USING "##";KCOUNT%;:LPRINT USING "
      ##.###^^^^";AL#,FTN#
3610 IF KCOUNT% >24 THEN PRINT:PRINT "Procedure does not converge at 3610":PRINT
:GOTO 4670
3620 DELTAX#=DELTAX#/20#
3630 GOTO 3020
3640 A#=0#:B#=0#
3650 FOR I%=1 TO NREC%
3660 TEMP#=X#(I%)-GA#
3670 B#=B#+TEMP#^AL#
3680 NEXT I%
3690 BE#=GA#+(B#/NREC#)^(1#/AL#)
3700 AINV#=1#/AL#
3710 FOR JJ%=1 TO 3
3720 DUM#(JJ%)=0#:Z#=1#+JJ%*AINV#
3730 SUBI#=Z#:GOSUB 5250:DUM#(JJ%)=SUBO#
3740 NEXT JJ%
3750 BA#=1#/SQR(DUM#(2)-DUM#(1)*DUM#(1))
3760 AA#=(1#-DUM#(1))*BA#
3770 ML2#=(BE#-GA#)/BA#
3780 ML1#=BE#-AA#*ML2#
```

177

```
3790 ML2#=ML2#*ML2#
3800 MLSKEW#=(DUM#(3)-3#*DUM#(2)*DUM#(1)+2#*DUM#(1)*DUM#(1)*DUM#(1))*BA#*BA#*BA#

3810 CLS
3820 PRINT:PRINT TAB(B%) TITLE$:PRINT TAB(B%) U$
3830 PRINT TAB(D%) T$:PRINT TAB(D%) V$
3840 PRINT TAB(F%) M$:PRINT TAB(F%) W$
3850 PRINT TAB(19) "Mean is                         ";:PRINT USING FORM$;ML1#
3860 PRINT TAB(19) "Variance is                     ";:PRINT USING FORM$;ML2#
3870 PRINT TAB(19) "Coefficient of skew is          ";:PRINT USING FORM$;MLSKEW#
3880 PRINT TAB(19) "Parameter Alpha is              ";:PRINT USING FORM$;AL#
3890 PRINT TAB(19) "Parameter Beta is               ";:PRINT USING FORM$;BE#
3900 PRINT TAB(19) "Parameter Gamma is              ";:PRINT USING FORM$;GA#
3910 IF FLAG% <> 1 GOTO 3990
3920 LPRINT
3930 LPRINT TAB(19) "Mean is                        ";:LPRINT USING FORM$;ML1#
3940 LPRINT TAB(19) "Variance is                    ";:LPRINT USING FORM$;ML2#
3950 LPRINT TAB(19) "Coefficient of skew is         ";:LPRINT USING FORM$;MLSKEW
#
3960 LPRINT TAB(19) "Parameter Alpha is             ";:LPRINT USING FORM$;AL#
3970 LPRINT TAB(19) "Parameter Beta is              ";:LPRINT USING FORM$;BE#
3980 LPRINT TAB(19) "Parameter Gamma is             ";:LPRINT USING FORM$;GA#
3990 A#=0#:B#=0#:C#=0#:D#=0#:E#=0#:F#=0#:G#=0#:H#=0#:P#=0#:Q#=0#:R#=0#:S#=0#:U#=
0#:V#=0#:W#=0#:Z#=0#
4000 FOR I%=1 TO NREC%
4010 Z#=X#(I%)-GA#
4020 A#=A#+(Z#^AL#)*LOG(Z#)
4030 B#=B#+(Z#^AL#)*LOG(Z#)*LOG(Z#)
4040 C#=C#+(Z#^AL#)
4050 D#=D#+(Z#^(AL#-1#))
4060 E#=E#+(Z#^(AL#-2#))
4070 F#=F#+1#/Z#
4080 G#=G#+1#/(Z#*Z#)
4090 W#=W#+(Z#^(AL#-1#))*(1#+AL#*LOG(Z#))
4100 NEXT I%
4110 Z#=BE#-GA#
4120 H#=Z#^(-AL#)
4130 P#=Z#^(-AL#-1#)
4140 Q#=Z#^(-AL#-2#)
4150 R#=LOG(Z#)
4160 S#=Z#^(AL#-1#)
4170 U#=NREC#/Z#
4180 V#=NREC#/(Z#*Z#)
4190 IM#(1)=-NREC#/(AL#*AL#)-H#*(B#-(R#*R#)*C#)
4200 IM#(2)=-U#+P#*((R#-1#)*C#+A#)
4210 IM#(3)=U#-F#+H#*(W#+AL#*R#*D#)+P#*(1#-AL#*R#)*C#-AL#*S#*A#
4220 TEMP1#=AL#*(AL#+1#)*Q#*C#
4230 TEMP3#=AL#*(AL#-1#)*H#*E#
4240 TEMP2#=AL#*AL#*P#*D#
4250 IM#(5)=AL#*V#-TEMP1#
4260 IM#(6)=-AL#*V#-TEMP1#*TEMP2#
4270 IM#(9)=AL#*V#-(AL#-1#)*G#-TEMP1#+TEMP2#-TEMP3#
4280 IM#(4)=IM#(2)
4290 IM#(7)=IM#(3)
4300 IM#(8)=IM#(6)
4310 FOR KJ%=1 TO 9
4320 IM#(KJ%)=-IM#(KJ%)
4330 NEXT KJ%
4340 DET#=IM#(1)*(IM#(5)*IM#(9)-IM#(6)*IM#(8))-IM#(2)*(IM#(4)*IM#(9)-IM#(6)*IM#(
7))+IM#(3)*(IM#(4)*IM#(8)-IM#(5)*IM#(7))
4350 VARA#=(IM#(5)*IM#(9)-IM#(6)*IM#(8))/DET#
4360 VARB#=(IM#(1)*IM#(9)-IM#(3)*IM#(7))/DET#
4370 VARG#=(IM#(1)*IM#(5)-IM#(2)*IM#(4))/DET#
4380 COVAB#=-(IM#(4)*IM#(9)-IM#(6)*IM#(7))/DET#
4390 COVAG#=(IM#(4)*IM#(8)-IM#(5)*IM#(7))/DET#
4400 COVBG#=-(IM#(1)*IM#(8)-IM#(2)*IM#(7))/DET#
4410 FLAG2%=0
4420 FOR J% = 1 TO 6
4430 Y#=-LOG(1#/T#(J%))
4440 YA#=Y#^AINV#
4450 YB#=(-LOG(1#-1#/T#(J%)))^AINV#
4460 XT#(J%)=GA#+(BE#-GA#)*YB#
4470 DXDA#=-(BE#-GA#)*YA#*LOG(Y#)/(AL#*AL#)
4480 DXDB#=YA#
4490 DXDG#=1#-YA#
4500 SX#(J%)=VARA#*DXDA#*DXDA#+VARB#*DXDB#*DXDB#+VARG#*DXDG#*DXDG#+2#*DXDA#*DXDB
#*COVAB#+2#*DXDA#*DXDG#*COVAG#+2#*DXDB#*DXDG#*COVBG#
4510 IF SX#(J%) < 0! THEN SX#(J%)=0!:FLAG2%=1:GOTO 4530
4520 SX#(J%)=SQR(SX#(J%))
```

```
4530 NEXT J%
4540 PRINT:PRINT TAB(2) "T, years  2          5          10          20
    50          100":PRINT TAB(5) ROW$
4550 PRINT TAB(2) "X  " VV$;:FOR J%=1 TO 6:PRINT USING FORM$;XT#(J%);:NEXT J%
4560 PRINT TAB(2) " t " VV$
4570 PRINT TAB(2) "S  " VV$;:FOR J%=1 TO 6:PRINT USING FORM$;SX#(J%);:NEXT J%
4580 PRINT TAB(2) " t " VV$
4590 IF FLAG2% = 1 THEN PRINT:PRINT "Standard errors printed as 0 are undefined
in this method"
4600 IF FLAG% <> 1 GOTO 4670
4610 LPRINT:LPRINT TAB(2) "T, years  2          5          10          20
    50          100":LPRINT TAB(5) ROW$
4620 LPRINT TAB(2) "X  " VV$;:FOR J%=1 TO 6:LPRINT USING FORM$;XT#(J%);:NEXT J%
4630 LPRINT TAB(2) " t " VV$
4640 LPRINT TAB(2) "S  " VV$;:FOR J%=1 TO 6:LPRINT USING FORM$;SX#(J%);:NEXT J%
4650 LPRINT TAB(2) " t " VV$
4660 IF FLAG2% = 1 THEN LPRINT:LPRINT "Standard errors printed as 0 are undefine
d in this method"
4670 PRINT:PRINT"Press any key to continue"
4680 Q$=INKEY$:IF Q$ ="" GOTO 4680
4690 CLS
4700 SYSTEM
4710 '
4720 ' Subroutine for standard screen format
4730 '
4740 CLS:PRINT:PRINT
4750 ROW$=STRING$(78,205)
4760 BOXTOP$=CHR$(201)+ROW$+CHR$(187)
4770 PRINT BOXTOP$;
4780 PRINT CHR$(186) TAB(80) CHR$(186);
4790 PRINT CHR$(186) TAB(80) CHR$(186);
4800 C%=LEN(T$)
4810 D%=(80-C%)/2
4820 V$=STRING$(C%,205)
4830 PRINT CHR$(186) TAB(D%) T$ TAB(80) CHR$(186);
4840 PRINT CHR$(186) TAB(D%) V$ TAB(80) CHR$(186);
4850 PRINT CHR$(186) TAB(80) CHR$(186);
4860 PRINT CHR$(186) TAB(80) CHR$(186);
4870 PRINT CHR$(186) "    What is the name of your data file?" TAB(80) CHR$(186);
4880 PRINT CHR$(186) TAB(80) CHR$(186);
4890 PRINT CHR$(186) "    Do you want printer output (Y/N)?" TAB(80) CHR$(186);
4900 PRINT CHR$(186) "    (press Alt Q to quit at this stage)" TAB(80) CHR$(186);
4910 PRINT CHR$(186) TAB(80) CHR$(186);
4920 PRINT CHR$(186) TAB(80) CHR$(186);
4930 BOXBOT$=CHR$(200)+ROW$+CHR$(188)
4940 PRINT BOXBOT$;
4950 LOCATE 24,1,0,0,0
4960 PRINT "G Kite";
4970 LOCATE 10,41,1,0,13
4980 FILE$=""
4990 I$=INPUT$(1)
5000 IF I$ = CHR$(13) GOTO 5100              ' ENTER key
5010 IF I$ = CHR$(8)  GOTO 5050              ' BACKSPACE key
5020 PRINT I$;
5030 FILE$=FILE$+I$
5040 GOTO 4990
5050 H%=POS(0)
5060 LOCATE 10,H%-1,1,0,13
5070 L%=LEN(FILE$)
5080 FILE$=LEFT$(FILE$,L%-1)
5090 GOTO 4990
5100 LOCATE 12,39,1,0,13
5110 YN$=""
5120 I$=INKEY$:IF I$ = "" GOTO 5120
5130 IF LEN(I$) = 1 GOTO 5170
5140 I$=RIGHT$(I$,1)
5150 IF I$ = CHR$(16) THEN IERR%=1:GOTO 5240     ' ALT Q key
5160 GOTO 5100
5170 IF I$ = CHR$(13) GOTO 5210              ' ENTER key
5180 PRINT I$;
5190 YN$=I$
5200 GOTO 5120
5210 IF YN$ = "Y" OR YN$ = "y" OR YN$ = "N" OR YN$ = "n" GOTO 5230
5220 GOTO 5100
5230 LOCATE ,,0,13,13
5240 RETURN
5250 '
```

```
5260 ' Subroutine to calculate the gamma function
5270 '
5280 IF SUBI# <= 57# GOTO 5320
5290 SUBO#=.39909#-.43429#*(SUBI#-1#)+(SUBI#-.5#)*(LOG(SUBI#-1#)/LOG(10#))
5300 SUBO#=10#^SUBO#
5310 GOTO 5560
5320 XXX#=SUBI#
5330 ERRR#=.0000001#:IERR%=0
5340 GX#=1#
5350 IF XXX# <= 2# GOTO 5410
5360 IF XXX# > 2# GOTO 5380
5370 IF XXX# <= 2# GOTO 5510
5380 XXX#=XXX#-1!
5390 GX#=GX#*XXX#
5400 GOTO 5370
5410 IF XXX# = 1# GOTO 5560
5420 IF XXX# > 1# GOTO 5510
5430 IF XXX# > ERRR# GOTO 5480
5440 YYY#=INT(XXX#)-XXX#
5450 IF ABS(YYY#) <= ERRR# GOTO 5550
5460 IF (1#-YYY#) <= ERRR# GOTO 5550
5470 IF XXX# > 1# GOTO 5510
5480 GX#=GX#/XXX#
5490 XXX#=XXX#+1#
5500 GOTO 5470
5510 YYY#=XXX#-1#
5520 SUBO#=1#-.5771017#*YYY#+.985854#*YYY#*YYY#-.8764218#*YYY#*YYY#*YYY#+.832821
2#*YYY#^4-.5684729#*YYY#^5+.2548205#*YYY#^6-.0514993#*YYY#^7
5530 SUBO#=GX#*SUBO#
5540 GOTO 5560
5550 IERR%=1
5560 RETURN
5570 '
5580 ' Subroutine to calculate the psi function
5590 '
5600 DD#=SUBI#+2#
5610 SUBO#=LOG(DD#)-(1#/(2#*DD#))-(1#/(12#*DD#*DD#))+(1#/(120#*DD#*DD#*DD#))
-(1#/(252#*DD#*DD#*DD#*DD#*DD#*DD#))-(1#/(SUBI#+1#))-(1#/SUBI#)
5620 RETURN
```

## Type III Extremal Distribution

### Method of Moments

| | |
|---|---|
| Mean is | 2.174D+00 |
| Variance is | 3.208D+00 |
| Coefficient of skew is | 1.418D+00 |
| Parameter Alpha is | 1.248D+00 |
| Parameter Beta is | 2.337D+00 |
| Parameter Gamma is | -4.730D-02 |

| T, years | 2 | 5 | 10 | 20 | 50 | 100 |
|---|---|---|---|---|---|---|
| $x_t$ | 1.730D+00 | 6.696D-01 | 3.457D-01 | 1.735D-01 | 5.734D-02 | 1.251D-02 |
| $s_t$ | 2.464D-01 | 1.382D-01 | 1.957D-01 | 2.713D-01 | 3.459D-01 | 3.831D-01 |

### Method of Smallest Observed Drought

| Iteration | A | f(A) |
|---|---|---|
| 1 | 1.145D+00 | 2.045D-01 |
| 2 | 1.163D+00 | 1.570D-02 |
| 3 | 1.163D+00 | -1.890D-05 |
| 4 | 1.163D+00 | 1.465D-07 |

| | |
|---|---|
| Mean is | 2.174D+00 |
| Variance is | 3.208D+00 |
| Coefficient of skew is | 1.595D+00 |
| Parameter Alpha is | 1.163D+00 |
| Parameter Beta is | 2.287D+00 |
| Parameter Gamma is | 9.745D-02 |

| T, years | 2 | 5 | 10 | 20 | 50 | 100 |
|---|---|---|---|---|---|---|
| $x_t$ | 1.695D+00 | 7.004D-01 | 4.137D-01 | 2.678D-01 | 1.739D-01 | 1.394D-01 |
| $s_t$ | 2.597D-02 | 1.331D-01 | 2.825D-01 | 4.479D-01 | 6.827D-01 | 8.694D-01 |

### Method of Maximum Likelihood

| Iteration | G | f(G) |
|---|---|---|
| 1 | 1.093D+00 | -1.271D-02 |

| | |
|---|---|
| Mean is | 2.175D+00 |
| Variance is | 3.206D+00 |
| Coefficient of skew is | 1.651D+00 |
| Parameter Alpha is | 1.136D+00 |
| Parameter Beta is | 2.270D+00 |
| Parameter Gamma is | 1.444D-01 |

| T, years | 2 | 5 | 10 | 20 | 50 | 100 |
|---|---|---|---|---|---|---|
| $x_t$ | 1.684D+00 | 7.122D-01 | 4.378D-01 | 3.001D-01 | 2.130D-01 | 1.815D-01 |
| $s_t$ | 0.000D+00 | 8.308D-02 | 2.630D-01 | 4.443D-01 | 6.983D-01 | 8.997D-01 |

Standard errors printed as 0 are undefined in this method

# CHAPTER 12

## COMPARISON OF FREQUENCY DISTRIBUTIONS

### Introduction

Chapters 5-11 have described the use of various continuous probability distributions for estimating events at return periods larger than those of the recorded events. The question naturally arises as to which of these distributions to use for a particular data sample. The methods described in this chapter may be used to test the fit of a distribution to a particular sample. The larger question of whether or not the best fitting distribution is suitable for use in predicting event magnitudes outside the sample range will be discussed in Chapter 15. Tables 12-1 and 12-2 summarize the maximum likelihood event magnitudes and standard errors for several return periods computed for the annual maximum daily flows of the St. Mary's River at Stillwater, Nova Scotia, over the period 1915-1986 for the distributions of Chapters 5-11. In addition to the results taken from Chapters 5-11, calculations were made for the truncated normal distributions.

### Table 12-1
Comparison of T-Year Event Magnitude Using Various
Frequency Distributions, Cubic Metres per Second

| | Cumulative Probability, P, % | | | | | |
| | 50 | 80 | 90 | 95 | 98 | 99 |
| | Corresponding Return Period, T, Years | | | | | |
| Frequency Distribution | 2 | 5 | 10 | 20 | 50 | 100 |
|---|---|---|---|---|---|---|
| Truncated Normal | 413 | 534 | 597 | 649 | 708 | 747 |
| 2-Parameter Lognormal | 390 | 519 | 602 | 680 | 781 | 857 |
| 3-Parameter Lognormal | 387 | 516 | 602 | 687 | 797 | 882 |
| Type I Extremal | 389 | 512 | 593 | 671 | 772 | 848 |
| Pearson Type III | 388 | 519 | 603 | 682 | 782 | 855 |
| Log-Pearson Type III | 397 | 514 | 599 | 683 | 793 | 879 |

### Table 12-2
Comparison of Standard Errors of T-Year Events Using Various
Frequency Distribution, Cubic Metres per Second

| | Cumulative Probability, P, % | | | | | |
| | 50 | 80 | 90 | 95 | 98 | 99 |
| | Corresponding Return Period, T, Years | | | | | |
| Frequency Distribution | 2 | 5 | 10 | 20 | 50 | 100 |
|---|---|---|---|---|---|---|
| Truncated Norma | 16.9 | 19.7 | 22.9 | 26.0 | 29.9 | 32.6 |
| 2-Parameter Lognormal | 15.6 | 24.0 | 32.4 | 41.6 | 54.9 | 65.7 |
| 3-Parameter Lognormal | 16.2 | 24.8 | 35.2 | 49.2 | 72.3 | 93.1 |
| Type I Extremal | 15.0 | 23.0 | 29.5 | 36.1 | 44.9 | 51.6 |
| Pearson Type III | 16.1 | 24.6 | 33.3 | 43.1 | 57.1 | 68.4 |
| Log-Pearson Type III | 16.1 | 24.4 | 34.6 | 49.2 | 75.4 | 101.0 |

Figure 12-1 shows the event magnitudes from each of these
distributions plotted on type I extremal graph paper. The figure
is for illustration only and normally many of these distributions
would be plotted on other types of graph paper. The figure also
shows those recorded events which have return periods greater
than 2 years when computed from the equation (n+1)/m. It will
be noted from the figure that the distributions are very closely
grouped except for the truncated normal. The 2-parameter lognormal and the log-Pearson type III are so close as to be indis-
tinguishable at this scale. Note also that all distributions
predict a 100-year event considerably lower than the maximum
event recorded in the 72-year sample data.

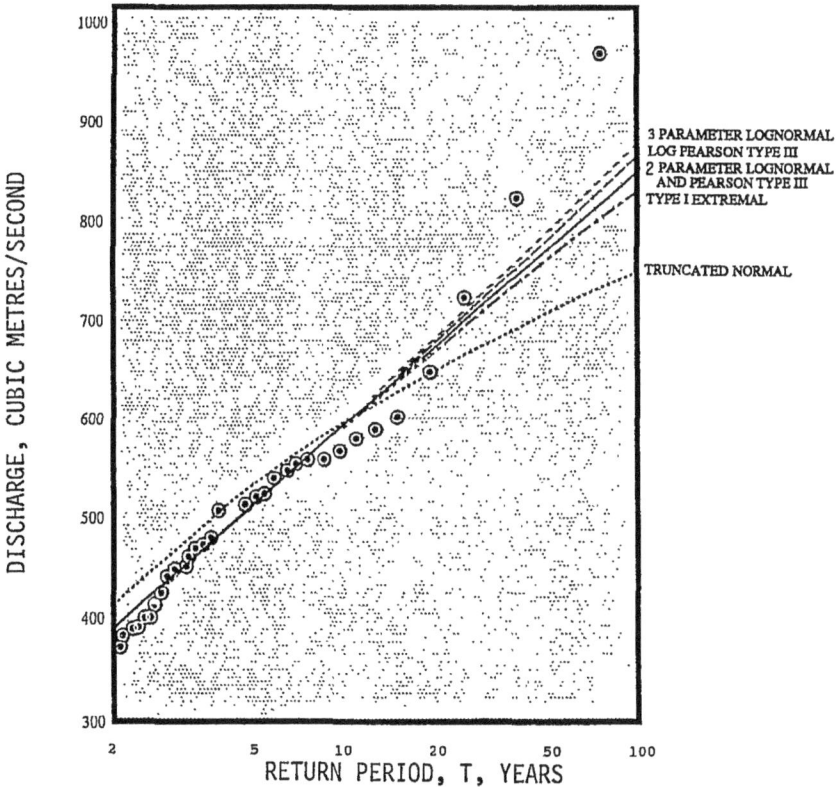

Figure 12-1.   *Comparison of frequency curves from various distributions - St. Mary's River at Stillwater,
1915 - 1986, Station No. 01E0001.*

## Classical Tests

The two most commonly used tests of goodness of fit are the chi-square and Kolmogorov-Smirnov (Yevjevich (6)). An additional check on goodness of fit may be made by computing the sum of squares of differences between observed and computed event magnitudes.

*Chi-Square Test* - The statistic

$$\chi^2 = \sum_{j=i}^{k} \frac{(O_j - E_j)^2}{E_j} \qquad (12\text{-}1)$$

is distributed assymptotically as chi-square with k-1 degrees of freedom where $O_j$ is the observed number of events in the $j_{th}$ class interval and $E_j$ is the number of events that would be expected from the theoretical distribution. If the class intervals are defined such that each interval corresponds to an equal probability then $E_j$ is n/k where n is the sample size and k is the number of class intervals and Equation 12-1 reduces to

$$\chi^2 = \frac{k}{n} \sum_{j=1}^{n} O_j^2 - n \qquad (12\text{-}2)$$

according to Mann and Wald (4). The class intervals were computed for the various distributions as follows:

(a)  Truncated Normal

$$CL = \bar{x} + tS \qquad (12\text{-}3)$$

where $\bar{x}$ and S are the sample mean and standard deviation and t is the standard normal deviate corresponding to the probabilities of exceedence, P, listed in column 2 of Table 12-3.

(b)  2-Parameter Lognormal

$$CL = \exp(\bar{x}_n + tS_n) \qquad (12\text{-}4)$$

where $\bar{x}_n$ and $S_n$ are the mean and standard deviation of the logarithms of the recorded events.

(c)  3-Parameter Lognormal

$$CL = a + \exp(\bar{x}_{na} + tS_{na}) \qquad (12\text{-}5)$$

where a is the lower boundary of the distribution and $\bar{x}_{na}$ and $S_{na}$

Table 12-3

Comparison of Class Limits (in Cubic Metres per Second), Chi-Square and Kolmogorov - Smirnov Statistics and Standard Error for Various Frequency Distributions.

St. Mary's River at Stillwater

| Class Interval | Probability | Truncated Normal | 2-Parameter Lognormal | 3-Parameter Lognormal | Type I Extremal | Pearson Type III | Log-Pearson Type III |
|---|---|---|---|---|---|---|---|
| 0 | 0 | 0 | 0 | 0 | 0 | 0 | 0 |
| 1 | 0.14286 | 263 | 277 | 277 | 264 | 277 | 278 |
| 2 | 0.28571 | 332 | 325 | 322 | 318 | 316 | 323 |
| 3 | 0.42857 | 387 | 369 | 365 | 366 | 355 | 365 |
| 4 | 0.57143 | 439 | 415 | 411 | 416 | 400 | 410 |
| 5 | 0.71429 | 494 | 471 | 468 | 478 | 457 | 466 |
| 6 | 0.85714 | 563 | 552 | 554 | 572 | 543 | 551 |
| 7 | 1.0 | $\infty$ | $\infty$ | $\infty$ | $\infty$ | $\infty$ | $\infty$ |
| Chi-Square | | 8.11 | 3.44 | 3.05 | 6.16 | 1.88 | 2.08 |
| D | | 0.099 | 0.040 | 0.040 | 0.046 | 0.036 | 0.026 |
| SE | | 42.2 | 24.9 | 21.5 | 24.8 | 23.5 | 21.9 |

are the mean and standard deviation of the logarithms of the distribution x-a.

(d) Type I Extremal

$$CL = \bar{x} + [\frac{y_m - \mu}{\sigma}] \, S \qquad (12-6)$$

where $y_m$ is $- \ln (- \ln P)$ and $\mu$ and $\sigma$ are the mean and standard deviation of the plotting positions.

(e) Pearson Type III

$$CL = \bar{x} + [\frac{x^2 \gamma_1}{4} - \frac{2}{\gamma_1}] \, S \qquad (12-7)$$

where $x^2$ is the value of chi-square at probability P and $8/\gamma_1^2$ degrees of freedom, $\gamma_1$ is the sample coefficient of skew.

185

(f) Log-Pearson Type III

$$CL = \exp \left( \bar{x}_n + \left[ \frac{x^2 \gamma_n}{4} - \frac{2}{\gamma_n} \right] S_n \right) \qquad (12\text{-}8)$$

where $\gamma_n$ is the coefficient of skew of ln x.

The recorded events were then sorted in order of magnitude and the numbers of events within each class interval determined for each distribution.

Table 12-3 lists the computed class limits for each distribution together with the derived chi-square values. All the computed values of chi-square are less than the tabulated value at 95% and k-m-1 degrees of freedom, where m is the number of parameters estimated from the sample for the particular distribution. While no distinction can be made between the distributions on the basis of this test they may be listed in order of increasing magnitude of chi-square as:

> Pearson type III
> log-Pearson type III
> 3-parameter lognormal
> 2-parameter lognormal
> type I extremal
> truncated normal

*Kolmogorov-Smirnov Test* - In order to avoid the loss of information due to grouping suffered by the chi-square test, other tests of goodness of fit have been developed such as the Neyman-Barton "smooth" tests, and the Cramer-von Mises $W^2$ test. The most important of these alternatives to chi-square is the Kolmogorov statistic, D, which is based on deviations of the sample distribution function P(x) from the completely specified continuous hypothetical distribution function $P_o(x)$, such that:

$$D_n = \max |P(x) - P_o(x)| \qquad (12\text{-}9)$$

Developed by Kolmogorov (3) in 1933, the test requires that the value of $D_n$ computed from the sample distribution be less than the tabulated value of $D_n$ at the required confidence level.

Table 12-3 shows, on the next to last line, the calculated value of $D_n$, the Kolmogorov-Smirnov statistic, for each distribution. In practice the values P(x) are obtained as $n_j/n$ where $n_j$ is the cumulative number of sample events at class limit j and n is the total number of events. $P_o(x)$ is then 1/k, 2/k, etc. where k is the number of class intervals. Given the 95% and 90% critical values of 0.16 and 0.14, for a sample size of 72, it can be seen that all distributions are well within the acceptance limits. On the basis of the Komogorov-Smirnov test the preferred order of distributions would be:

log-Pearson type III
Pearson type III
3-parameter lognormal
2-parameter lognormal
type I extremal
truncated normal

*Least Squares* - A third method of comparing the fit of different distributions to a data sample is to compute the sum of squares of the differences between calculated and observed discharges. Bobée and Robitaille (1) used this method to compare calculated discharges at 2, 5, 10, 20, 50, and 100-year return periods with discharges at the same return periods interpolated (or extrapolated) from the recorded data. A difficulty with this method is that the interpolation itself is a form of distribution with which all the other distributions are then compared. Deininger and Westfield (2) compared the recorded events at their given return periods, (n+1)/m, to computed events from the distribution being tested at the same return periods (i.e. at 61 years for the maximum event in a 60-year record, at 30.5 years for the second largest event, etc.). The disadvantage of this procedure is the large number of computations needed.

The least squares method also has the disadvantage of being dependent on the plotting position equation used to compute the return periods of the observed events. Although this dependence will govern the absolute value of the sum of squares of differences for each distribution it will not affect the relative ranking of the distributions.

The program at the end of this chapter was used to compute the standard error of each of the six distributions as

$$SE_j = \left[ \frac{\sum\limits_{i=1}^{n} (x_i - y_i)^2}{n - m_j} \right]^{\frac{1}{2}} \tag{12-10}$$

where $x_i$, $i=1,\ldots,n$ are the recorded events, $y_i$, $i-1,\ldots,n$ are the event magnitudes computed from the jth probability distribution at probabilities computed from the sorted ranks of $x_i$, $i=1$, $\ldots,n$, and $m_j$ is the number of parameters estimated for the jth distribution. The different distributions are represented in the program by subroutines abstracting the maximum likelihood procedure from each of the programs in Chapters 6-10 plus a simple procedure for the truncated normal distribution. The ranking of the distributions by order of least standard error is as follows:

> 3-parameter lognormal
> log-Pearson type III
> Pearson type III
> type I extremal
> 2-parameter lognormal
> truncated normal

The results of the three tests of goodness of fit are not identical but indicate that the best-fitting distributions are the Pearson type III, log-Pearson type III and 3-parameter lognormal and the worst fitting is, as would be expected, the truncated normal.

## Bayesian Analysis

It has been suggested (5) that Bayesian analysis may provide a means of choosing one distribution from a set of distributions. By assuming a set of prior probabilities of choosing each distribution (all would be equal in the absence of any additional information) $p'(f_i)$, where $f_i$ represents the ith probability distribution function of the set, then the posterior probabilities can be derived by Bayesian analysis as

$$p''(f_i) = \frac{K_i}{K_*} \ p'(f_i) \qquad (12\text{-}11)$$

where $K_i$ is the marginal likelihood function of the observed data coming from the ith distribution and $K_*$ is a normalizing constant. By combining the probability distributions $f_i(q)$ and the posterior probabilities, $p''(f_i)$, as

$$f(q) = \Sigma \ p''(f_i) \ f_i(q) \qquad (12\text{-}12)$$

the composite probability density function, $f(q)$, is the one which should be used in making inferences about future flood discharges. A Bayesian analyst would also require that $f_i(q)$ $i=1,\ldots,n$ all be Bayesian distributions so that $f(q)$ is a composite Bayesian distribution and accounts for both uncertainty in the choice of model and uncertainty in the parameter values of each model. As an example of this latter procedure, Wood and Rodriguez-Iturbe (5) have shown that using three Bayesian distributions, the normal, lognormal and exponential distributions, with equal prior probabilities of 1/3 the composite Bayesian distribution became

$$\hat{f}(q) = 0.996 \ f_1(q) + 0.004 \ f_2(q) \qquad (12\text{-}13)$$

where $f_1(q)$ and $f_2(q)$ represent the exponential and lognormal pdf's and 0.996 and 0.004 are the posterior probabilities of these distributions. The posterior probability of the normal distribution was zero.

## References

1.  Bobée, B., and R. Robitaille, 1976, The Use of the Pearson Type 3 and log-Pearson Type 3 Distributions Revisited, Wat. Res. Res., Vol. 13, No. 2, April, pp. 427-443.

2.  Deininger, R. A. and J. D. Westfield, 1969, Estimation of the Parameters of Gumbel's Third Asymptotic Distribution by Different Methods, Wat. Res. Res., Vol. 5, No. 6, pp. 1238-1243.

3.  Kolmogorov, A., 1933, Sulla Determinazione Empirica di una Legge di Distribuzione, G. Ist. Ital. Attuari, 4, 83.

4.  Mann, H. B. and A. Wald, 1942, On the Choice of the Number of Intervals in the Application of the Chi-Square Test, Ann. Math. Stat., 13, 306.

5.  Wood, E. F. and I. Rodriguez-Iturbe, 1975, Bayesian Inference and Decision Making for Extreme Hydrologic Events, Wat. Res. Tes., Vol. 11, No. 4, pp. 533-542.

6.  Yevjevich, V., 1972, Probability and Statistics in Hydrology, Water Resources Publications, Fort Collins, Colorado.

```
      PROGRAM SER(INPUT,OUTPUT,TAPE5=INPUT,TAPE6=OUTPUT)              A    1
C                                                                     A    2
C                                                                     A    3
C     PROGRAM TO COMPUTE THE STANDARD ERRORS OF EVENTS COMPUTED       A    4
C     FROM VARIOUS PROBABILITY DISTRIBUTIONS COMPARED TO THE          A    5
C     OBSERVED EVENT MAGNITUDES                                       A    6
C     INPUT                                                           A    7
C     TITLE                                                           A    8
C     N NUMBER OF ANNUAL MAXIMUM EVENTS                               A    9
C     X SERIES OF EVENTS                                              A   10
C                                                                     A   11
C                                                                     A   12
      DIMENSION X(100), Y(100), Z(100), P(100), RP(100), T(100), TITLE(8  A   13
     10)                                                              A   14
      REAL L1,L2,L3,LG                                                A   15
      REAL K,M1,M2,M3                                                 A   16
      E=2.515517                                                      A   17
      C1=0.802853                                                     A   18
      C2=0.010328                                                     A   19
      D1=1.432788                                                     A   20
      D2=0.189269                                                     A   21
      D3=0.001308                                                     A   22
      READ (5,6) TITLE                                                A   23
      READ (5,7) N                                                    A   24
      XN=N                                                            A   25
      READ (5,8) (X(I),I=1,N)                                         A   26
      WRITE (6,9) TITLE                                               A   27
      A=0.0                                                           A   29
      B=0.0                                                           A   30
      C=0.0                                                           A   31
      DO 1 I=1,N                                                      A   32
      A=A+X(I)                                                        A   33
      B=B+X(I)**2                                                     A   34
      C=C+X(I)**3                                                     A   35
1     CONTINUE                                                        A   36
      M1=A/XN                                                         A   37
      M2=(B/XN)-(A/XN)**2                                             A   38
      M3=(C/XN)+2.0*M1**3-3.0*M1*(B/XN)                               A   39
      M2=M2*XN/(XN-1.0)                                               A   40
      G=M3/(M2**1.5)                                                  A   41
      WRITE (6,11) M1                                                 A   42
      WRITE (6,12) M2                                                 A   43
      WRITE (6,13) G                                                  A   44
      CALL SORTX (N,X)                                                A   45
      WRITE (6,10)                                                    A   28
      WRITE (6,19) (X(I),I=1,N)                                       A   46
      WRITE (6,17)                                                    A   47
      A=0.0                                                           A   48
      B=0.0                                                           A   49
      C=0.0                                                           A   50
      DO 2 I=1,N                                                      A   51
      Y(I)=ALOG(X(I))                                                 A   52
      A=A+Y(I)                                                        A   53
      B=B+Y(I)**2                                                     A   54
      C=C+Y(I)**3                                                     A   55
2     CONTINUE                                                        A   56
      L1=A/XN                                                         A   57
      L2=(B/XN)-(A/XN)**2                                             A   58
      L3=(C/XN)+2.0*L1**3-3.0*L1*(B/XN)                               A   59
      L2=L2*XN/(XN-1.0)                                               A   60
      LG=L3/(L2**1.5)                                                 A   61
      WRITE (6,14) L1                                                 A   62
      WRITE (6,15) L2                                                 A   63
      WRITE (6,16) LG                                                 A   64
      DO 3 I=1,N                                                      A   65
      XI=I                                                            A   66
      P(I)=1.0-XI/(XN+1.0)                                            A   67
      RP(I)=1.0/(1.0-P(I))                                            A   68
      D=P(I)                                                          A   69
      IF (D.GT.0.5) D=1.0-D                                           A   70
      W=SQRT(ALOG(1.0/(D**2)))                                        A   71
      T(I)=W-(E+C1*W+C2*W**2)/(1.0+D1*W+D2*W**2+D3*W**3)              A   72
      IF (P(I).LT.0.5) T(I)=-T(I)                                     A   73
3     CONTINUE                                                        A   74
      WRITE (6,18)                                                    A   75
      CALL IN (N,M1,M2,T,Z)                                           A   76
      WRITE (6,19) (Z(I),I=1,N)                                       A   77
      CALL SSQ (N,X,Z)                                                A   78
      WRITE (6,20)                                                    A   79
      CALL IN (N,L1,L2,T,Z)                                           A   80
```

```
          DO 4 I=1,N                                      A  81
    4     Z(I)=EXP(Z(I))                                  A  82
          WRITE (6,19) (Z(I),I=1,N)                       A  83
          CALL SSQ (N,X,Z)                                A  84
          WRITE (6,21)                                    A  85
          CALL LN3 (N,X,1,Z)                              A  86
          WRITE (6,19) (Z(I),I=1,N)                       A  87
          CALL SSQ (N,X,Z)                                A  88
          WRITE (6,22)                                    A  89
          CALL TIE (N,X,M1,M2,RP,Z)                       A  90
          WRITE (6,19) (Z(I),I=1,N)                       A  91
          CALL SSQ (N,X,Z)                                A  92
          WRITE (6,23)                                    A  93
          CALL PT3 (N,X,T,Z)                              A  94
          WRITE (6,19) (Z(I),I=1,N)                       A  95
          CALL SSQ (N,X,Z)                                A  96
          WRITE (6,24)                                    A  97
          CALL PT3 (N,Y,T,Z)                              A  98
          DO 5 I=1,N                                      A  99
    5     Z(I)=EXP(Z(I))                                  A 100
          WRITE (6,19) (Z(I),I=1,N)                       A 101
          CALL SSQ (N,X,Z)                                A 102
          STOP                                            A 103
    C                                                     A 104
    6     FORMAT (80A1)                                   A 105
    7     FORMAT (I5)                                     A 106
    8     FORMAT (8F10.0)                                 A 107
    9     FORMAT (1H1,/,80A1,/)                           A 108
   10     FORMAT (3X,22HSORTED RECORDED EVENTS)           A 109
   11     FORMAT (20X,9HMEAN OF X,10X,E12.5)              A 110
   12     FORMAT (20X,13HVARIANCE OF X,12X,E12.5)         A 111
   13     FORMAT (20X,9HSKEW OF X,10X,E12.5,/)            A 112
   14     FORMAT (20X,15HMEAN OF LN(X)  ,10X,E12.5)       A 113
   15     FORMAT (20X,19HVARIANCE OF LN(X)  ,6X,E12.5)    A 114
   16     FORMAT (20X,15HSKEW OF LN(X)  ,10X,E12.5,/)     A 115
   17     FORMAT (///)                                    A 116
   18     FORMAT (3X,23HTRUNCATED NORMAL EVENTS)          A 117
   19     FORMAT (3X,6E12.5)                              A 118
   20     FORMAT (3X,28H2 PARAMETER LOGNORMAL EVENTS)     A 119
   21     FORMAT (3X,28H3 PARAMETER LOGNORMAL EVENTS)     A 120
   22     FORMAT (3X,22HTYPE 1 EXTREMAL EVENTS)           A 121
   23     FORMAT (3X,21HPEARSON TYPE 3 EVENTS)            A 122
   24     FORMAT (3X,25HLOG-PEARSON TYPE 3 EVENTS)        A 123
          END                                             A 124


          SUBROUTINE SSQ (N,X,Z)                          D   1
          DIMENSION X(1), Z(1)                            D   2
          SUM=0.0                                         D   3
          DO 1 I=1,N                                      D   4
    1     SUM=SUM+(Z(I)-X(I))**2                          D   5
          XN=N                                            D   6
          SUM=SQRT(SUM/XN)                                D   7
          WRITE (6,2) SUM                                 D   8
          RETURN                                          D   9
    C                                                     D  10
    2     FORMAT (3X,17HSTANDARD ERROR IS,E12.5,//)       D  11
          END                                             D  12


          SUBROUTINE TN (N,XBAR,XVAR,T,X)                 E   1
          DIMENSION X(1), T(1)                            E   2
          XSTD=SQRT(XVAR)                                 E   3
          DO 1 I=1,N                                      E   4
    1     X(I)=XBAR+T(I)*XSTD                             E   5
          RETURN                                          E   6
          END                                             E   7


          SUBROUTINE SORTX (N,X)                          F   1
    C     SORTS IN DECREASING ORDER, X(1)=LARGEST         F   2
          DIMENSION X(1)                                  F   3
          K=N-1                                           F   4
          DO 2 L=1,K                                      F   5
          M=N-L                                           F   6
          DO 2 J=1,M                                      F   7
          IF (X(J)-X(J+1)) 1,1,2                          F   8
    1     XT=X(J)                                         F   9
          X(J)=X(J+1)                                     F  10
          X(J+1)=XT                                       F  11
    2     CONTINUE                                        F  12
          RETURN                                          F  13
          END                                             F  14
```

191

```
      SUBROUTINE PT3 (N,X,SND,XT)                                    B    1
      DIMENSION X(1), SND(1), XT(1)                                  B    2
      XN=N                                                           B    3
      ICOUNT=0                                                       B    4
      XMIN=10000000.                                                 B    5
      DO 1 I=1,N                                                     B    6
1     IF (X(I).LT.XMIN) XMIN=X(I)                                    B    7
      GML=XMIN*0.99                                                  B    8
2     ICOUNT=ICOUNT+1                                                B    9
      A=0.0                                                          B   10
      B=0.0                                                          B   11
      C=0.0                                                          B   12
      R=0.0                                                          B   13
      DO 3 I=1,N                                                     B   14
      A=A+1.0/(X(I)-GML)                                             B   15
      B=B+(X(I)-GML)                                                 B   16
      C=C+ALOG(X(I)-GML)                                             B   17
      R=R+1.0/((X(I)-GML)**2)                                        B   18
3     CONTINUE                                                       B   19
      BETA=A/(A-(XN**2)/B)                                           B   20
      ALPHA=B/(XN*BETA)                                              B   21
      D=BETA+2.0                                                     B   22
      PSI=ALOG(D)-(1.0/(2.0*D))-(1.0/(12.0*D**2))+(1.0/(120.0*D**4))-(1. B   23
     10/(252.0*D**6))-(1.0/(BETA+1.0))-(1.0/BETA)                    B   24
      FCN=-XN*PSI+C-XN*ALOG(ALPHA)                                   B   25
      TRI=(1.0/D)+(1.0/(2.0*D**2))+(1.0/(6.0*D**3))-(1.0/(30.0*D**5))+(1 B   26
     1.0/(42.0*D**7))-(1.0/(30.0*D**9))+(1.0/((BETA+1.0)**2))+(1.0/(BETA B   27
     2**2))                                                          B   28
      V=A-(XN**2)/B                                                  B   29
      U=A                                                            B   30
      W=(B/XN)-(XN/A)                                                B   31
      DU=R                                                           B   32
      DV=R-(XN**3)/(B**2)                                            B   33
      DW=-1.0+(XN*R)/(A**2)                                          B   34
      FPN=-XN*TRI*((V*DU-U*DV)/(V**2))-A-XN*DW/W                     B   35
      AS=GML-(FCN/FPN)                                               B   36
      DELTA=ABS(0.00000001*AS)                                       B   37
      IF (ABS(AS-GML).LT.DELTA) GO TO 4                             B   38
      IF (ICOUNT.GT.25) GO TO 6                                     B   39
      GML=AS                                                         B   40
      GO TO 2                                                        B   41
4     CONTINUE                                                       B   42
      GAMMA=AS                                                       B   43
      DO 5 J=1,N                                                     B   44
      T=SND(J)                                                       B   45
      E=BETA**(1./3.)-1.0/(9.0*BETA**(2./3.))+T/(3.0*BETA**(1./6.))  B   46
      XT(J)=GAMMA+ALPHA*E**3                                         B   47
5     CONTINUE                                                       B   48
6     CONTINUE                                                       B   49
      RETURN                                                         B   50
      END                                                            B   51.

      SUBROUTINE TIE (N,X,M1,M2,T,XT)                                C    1
      DIMENSION X(1), T(1), XT(1)                                    C    2
      REAL M1,M2                                                     C    3
      XN=N                                                           C    4
      ICOUNT=0                                                       C    5
      ALPHA=1.2825/(SQRT(M2))                                        C    6
      AML=ALPHA                                                      C    7
1     ICOUNT=ICOUNT+1                                                C    8
      A=1.0/(AML**2)                                                 C    9
      B=M1-1.0/AML                                                   C   10
      C=0.0                                                          C   11
      D=0.0                                                          C   12
      E=0.0                                                          C   13
      DO 2 I=1,N                                                     C   14
      TEMP=EXP(-AML*X(I))                                            C   15
      C=C+TEMP                                                       C   16
      D=D+TEMP*X(I)                                                  C   17
      E=E+TEMP*X(I)**2                                               C   18
2     CONTINUE                                                       C   19
      FCN=D-B*C                                                      C   20
      FPN=B*D-E-A*C                                                  C   21
      AS=AML-(FCN/FPN)                                               C   22
      DELTA=ABS(0.0000001*AS)                                        C   23
      IF (ABS(AS-AML).LT.DELTA) GO TO 3                             C   24
      IF (ICOUNT.GT.25) GO TO 5                                     C   25
      AML=AS                                                         C   26
      GO TO 1                                                        C   27
3     CONTINUE                                                       C   28
```

```
      ALPHA=AS                                                   C  24
      HETA=(1.0/ALPHA)*ALOG(XN/C)                                C  31
      DO 4 J=1,N                                                 C  32
      YM=-ALOG(-ALOG(1.0-1.0/T(J)))                              C  33
      XT(J)=HETA+YM/ALPHA                                        C  33
4     CONTINUE                                                   C  34
5     CONTINUE                                                   C  35
      RETURN                                                     C  36
      END                                                        C  37

      SUBROUTINE LN3 (N,X,SND,XT)                                G   1
      DIMENSION X(1), SND(1), XT(1)                              G   2
      REAL MU                                                    G   3
      XN=N                                                       G   4
      XMIN=10000000.                                             G   5
      DO 1 I=1,N                                                 G   6
1     IF (X(I).LT.XMIN) XMIN=X(I)                                G   7
      AML=XMIN*0.80                                              G   8
      ICOUNT=0                                                   G   9
2     ICOUNT=ICOUNT+1                                            G  10
      A=0.0                                                      G  11
      B=0.0                                                      G  12
      C=0.0                                                      G  13
      D=0.0                                                      G  14
      E=0.0                                                      G  15
      F=0.0                                                      G  16
      P=0.0                                                      G  17
      DO 3 I=1,N                                                 G  18
      A=A+ALOG(X(I)-AML)                                         G  19
      B=B+(ALOG(X(I)-AML))**2                                    G  20
      P=P+(ALOG(X(I)-AML))**3                                    G  21
      C=C+1.0/((X(I)-AML))                                       G  22
      D=D+1.0/((X(I)-AML)**2)                                    G  23
      E=E+(1.0/((X(I)-AML)))*ALOG(X(I)-AML)                      G  24
      F=F+(1.0/((X(I)-AML)**2))*ALOG(X(I)-AML)                   G  25
3     CONTINUE                                                   G  26
      G=(B/XN)-(A/XN)**2-(A/XN)                                  G  27
      H=(-2.0*E/XN)+(2.0*A/XN)*(C/XN)+(C/XN)                     G  28
      FCN=C*G+E                                                  G  29
      FPN=C*H+D*G+F-D                                            G  30
      AS=AML-(FCN/FPN)                                           G  31
      DELTA=ABS(0.0000001*AS)                                    G  32
      IF (ABS(AS-AML).LT.DELTA) GO TO 4                          G  33
      IF (ICOUNT.GT.25) GO TO 6                                  G  34
      AML=AS                                                     G  35
      GO TO 2                                                    G  36
4     CONTINUE                                                   G  37
      AML=AS                                                     G  38
      MU=A/XN                                                    G  39
      VAR=(B/XN)-(A/XN)**2                                       G  40
      VAR=VAR*XN/(XN-1)                                          G  41
      SKEW=(P/XN)+2.0*MU**3-3.0*MU*(B/XN)                        G  42
      SD=SQRT(VAR)                                               G  43
      DO 5 J=1,N                                                 G  44
      T=SND(J)                                                   G  45
      Z=EXP(MU+T*SD)                                             G  46
      XT(J)=AML+Z                                                G  47
5     CONTINUE                                                   G  48
6     CONTINUE                                                   G  49
      RETURN                                                     G  50
      END                                                        G  51.

10 '
20 ' Program SER
30 ' Copyright G.W. Kite 1986
40 '
50 ' Compute the standard errors of events using various
60 ' probability distributions compared to the observed
70 ' event magnitudes
80 '
90 DIM X#(250),Y#(250),Z#(250),P#(250),RP#(250),T#(250)
100 DIM XS#(250),TN#(250),LN2#(250),LN3#(250),T1E#(250),T3E#(250),PT3#(250),LP3#
(250)
110 DIM ST#(6)
120 E1#=2.515517#
130 C1#=.802853#
140 C2#=.010328#
150 D1#=1.432788#
160 D2#=.189269#
170 D3#=.001308#
180 FORM$=" #.####^^^^"
```

```
190 VV$=CHR$(179)
200 IERR%=0
210 GOSUB 3060
220 IF IERR% = 1 GOTO 3041
230 FLAG%=0:IF YN$ = "Y" OR YN$ = "y" THEN FLAG%=1
240 NREC%=0
250 OPEN FILE$ FOR INPUT AS #1
260 LINE INPUT#1,TITLE$
270 IF EOF(1) GOTO 330
280 NREC%=NREC%+1
290 LINE INPUT#1,I$
300 X#(NREC%)=VAL(MID$(I$,5,12))
310 XS#(NREC%)=X#(NREC%)
320 GOTO 270
330 CLOSE#1
340 NREC#=NREC%
350 A%=LEN(TITLE$)
360 B%=(80-A%)/2
370 U$=STRING$(A%,205)
380 CLS
390 PRINT TAB(B%) TITLE$:PRINT TAB(B%) U$
400 PRINT TAB(D%) T$:PRINT TAB(D%) V$
410 IF FLAG% <> 1 GOTO 450
420 LPRINT CHR$(12):LPRINT:LPRINT
430 LPRINT TAB(B%) TITLE$:LPRINT TAB(B%) U$
440 LPRINT:LPRINT TAB(D%) T$:LPRINT TAB(D%) V$
450 A#=0#:B#=0#:C#=0#
460 FOR I%=1 TO NREC%
470 A#=A#+X#(I%)
480 B#=B#+X#(I%)*X#(I%)
490 C#=C#+X#(I%)*X#(I%)*X#(I%)
500 NEXT I%
510 M1#=A#/NREC#
520 M2#=(B#/NREC#)-M1#*M1#
530 M3#=(C#/NREC#)+2#*M1#*M1#*M1#-3#*M1#*(B#/NREC#)
540 M2#=M2#*NREC#/(NREC#-1#)
550 G#=M3#/(M2#^1.5#)
560 PRINT
570 PRINT TAB(20)    "Mean is                             ";:PRINT USING FORM$;M1#
580 PRINT TAB(20)    "Variance is                         ";:PRINT USING FORM$;M2#
590 PRINT TAB(20)    "Coefficient of skew is              ";:PRINT USING FORM$;G#
600 IF FLAG% <> 1 GOTO 650
610 LPRINT
620 LPRINT TAB(20)    "Mean is                            ";:LPRINT USING FORM$;M1#
630 LPRINT TAB(20)    "Variance is                        ";:LPRINT USING FORM$;M2#
640 LPRINT TAB(20)    "Coefficient of skew is             ";:LPRINT USING FORM$;G#
650 A#=0#:B#=0#:C#=0#
660 GOSUB 3610
670 FOR I%=1 TO NREC%
680 Y#(I%)=LOG(XS#(I%))
690 A#=A#+Y#(I%)
700 B#=B#+Y#(I%)*Y#(I%)
710 C#=C#+Y#(I%)*Y#(I%)*Y#(I%)
720 NEXT I%
730 L1#=A#/NREC#
740 L2#=(B#/NREC#)-L1#*L1#
750 L3#=(C#/NREC#)+2#*L1#*L1#*L1#-3#*L1#*(B#/NREC#)
760 L2#=L2#*NREC#/(NREC#-1#)
770 LG#=L3#/(L2#^1.5#)
780 PRINT
790 PRINT TAB(20) "Mean of logs is                ";:PRINT USING FORM$;L1#
800 PRINT TAB(20) "Variance of logs is            ";:PRINT USING FORM$;L2#
810 PRINT TAB(20) "Coefficient of skew of logs is ";:PRINT USING FORM$;LG#
820 IF FLAG% <> 1 GOTO 880
830 LPRINT
840 LPRINT TAB(20) "Mean of logs is                ";:LPRINT USING FORM$;L1#
850 LPRINT TAB(20) "Variance of logs is            ";:LPRINT USING FORM$;L2#
860 LPRINT TAB(20) "Coefficient of skew of logs is ";:LPRINT USING FORM$;LG#
870 LPRINT CHR$(12)
880 GOSUB 3750
890 '
900 ' truncated normal
910 '
920 XSTD#=SQR(M2#)
930 FOR I%=1 TO NREC%
940 I#=I%
950 P#(I%)=1#-I#/(NREC#+1#)
960 RP#(I%)=1#/(1#-P#(I%))
961 D#=P#(I%)
```

194

```
970 IF D# > .5# THEN D#=1#-D#
980 W#=SQR(LOG(1#/(D#*D#)))
990 T#(I%)=W#-(E1#+C1#*W#+C2#*W#*W#)/(1#+D1#*W#+D2#*W#*W#+D3#*W#*W#*W#)
1000 IF P#(I%) < .5 THEN T#(I%)=-T#(I%)
1010 TN#(I%)=M1#+T#(I%)*XSTD#
1020 TEMP#=TN#(I%)-XS#(I%)
1030 SUM#=SUM#+TEMP#*TEMP#
1040 NEXT I%
1050 ST#(1)=SQR(SUM#/NREC#)
1060 '
1070 ' 2 parameter lognormal
1080 '
1090 SUM#=0#
1100 LSTD#=SQR(L2#)
1110 FOR I%=1 TO NREC%
1120 LN2#(I%)=L1#+T#(I%)*LSTD#
1130 LN2#(I%)=EXP(LN2#(I%))
1140 TEMP#=LN2#(I%)-XS#(I%)
1150 SUM#=SUM#+TEMP#*TEMP#
1160 NEXT I%
1170 ST#(2)=SQR(SUM#/NREC#)
1180 '
1190 ' 3 parameter lognormal
1200 '
1210 AML#=XS#(NREC%)*.8#
1220 COUNT%=0
1230 COUNT%=COUNT%+1
1240 A#=0#:B#=0#:C#=0#:D#=0#:E#=0#:F#=0#:P#=0#
1250 FOR I%=1 TO NREC%
1260 TEMP1#=XS#(I%)-AML#
1270 TEMP2#=LOG(TEMP1#)
1280 A#=A#+TEMP2#
1290 B#=B#+TEMP2#*TEMP2#
1300 P#=P#+TEMP2#*TEMP2#*TEMP2#
1310 C#=C#+1#/TEMP1#
1320 D#=D#+1#/(TEMP1#*TEMP1#)
1330 E#=E#+(1#/TEMP1#)*TEMP2#
1340 F#=F#+(1#/(TEMP1#*TEMP1#))*TEMP2#
1350 NEXT I%
1360 TEMP3#=A#/NREC#
1370 G#=(B#/NREC#)-TEMP3#*TEMP3#-TEMP3#
1380 H#=(-2#*E#/NREC#)+(2#*A#/NREC#)*(C#/NREC#)+(C#/NREC#)
1390 FCN#=C#*G#+E#
1400 FPN#=C#*H#+D#*G#+F#-D#
1410 AAS#=AML#-(FCN#/FPN#)
1420 DELTA#=ABS(.0000001#*AAS#)
1430 IF ABS(AAS#-AML#) < DELTA# GOTO 1470
1440 IF COUNT% > 25 THEN PRINT:PRINT "Three parameter lognormal procedure does n
ot converge":PRINT:ST#(3)=99999#:GOTO 1640
1450 AML#=AAS#
1460 GOTO 1230
1470 AML#=AAS#
1480 MU#=TEMP3#
1490 VAR#=(B#/NREC#)-TEMP3#*TEMP3#
1500 VAR#=VAR#*NREC#/(NREC#-1#)
1510 SKEW#=(P#/NREC#)+2#*MU#*MU#*MU#-3#*MU#*(B#/NREC#)
1520 SD#=SQR(VAR#)
1530 SUM#=0
1540 FOR I% = 1 TO NREC%
1550 Z#=EXP(MU#+T#(I%)*SD#)
1560 LN3#(I%)=AML#+Z#
1570 TEMP#=LN3#(I%)-XS#(I%)
1580 SUM#=SUM#+TEMP#*TEMP#
1590 NEXT I%
1600 ST#(3)=SQR(SUM#/NREC#)
1610 '
1620 ' type 1 extremal
1630 '
1640 ALPHA#=1.2825/XSTD#
1650 COUNT%=0
1660 AML#=ALPHA#
1670 COUNT%=COUNT%+1
1680 A#=1#/(AML#*AML#)
1690 B#=M1#-1#/AML#
1700 C#=0#:D#=0#:E#=0#
1710 FOR I%=1 TO NREC%
1720 TEMP#=EXP(-AML#*XS#(I%))
1730 C#=C#+TEMP#
1740 D#=D#+TEMP#*XS#(I%)
```

```
1750 E#=E#+TEMP#*XS#(I%)*XS#(I%)
1760 NEXT I%
1770 FCN#=D#-B#*C#
1780 FPN#=B#*D#-E#-A#*C#
1790 AAS#=AML#-FCN#/FPN#
1800 DELTA#=ABS(.0000001#*AAS#)
1810 IF ABS(AAS#-AML#) < DELTA# GOTO 1850
1820 IF COUNT% > 24 THEN PRINT:PRINT "Type I extremal procedure does not converg
e":PRINT:ST#(4)=99999#:GOTO 1980
1830 AML#=AAS#
1840 GOTO 1670
1850 ALPHA#=AAS#
1860 BETA#=(1#/ALPHA#)*LOG(NREC#/C#)
1870 SUM#=0#
1880 FOR I%=1 TO NREC%
1890 YM#=-LOG(-LOG(1#-1#/RP#(I%)))
1900 T1E#(I%)=BETA#+YM#/ALPHA#
1910 TEMP#=T1E#(I%)-XS#(I%)
1920 SUM#=SUM#+TEMP#*TEMP#
1930 NEXT I%
1940 ST#(4)=SQR(SUM#/NREC#)
1950 '
1960 ' Pearson type 3
1970 '
1980 COUNT%=0
1990 GML#=XS#(NREC%)*.99#
2000 COUNT%=COUNT%+1
2010 A#=0#:B#=0#:C#=0#:R#=0#
2020 FOR I%=1 TO NREC%
2030 TEMP#=XS#(I%)-GML#
2040 A#=A#+1#/TEMP#
2050 B#=B#+TEMP#
2060 C#=C#+LOG(TEMP#)
2070 R#=R#+1#/(TEMP#*TEMP#)
2080 NEXT I%
2090 BETA#=A#/(A#-(NREC#*NREC#/B#))
2100 ALPHA#=B#/(NREC#*BETA#)
2110 D#=BETA#+2#
2120 PSI#=LOG(D#)-(1#/(2#*D#))-(1#/(12#*D#*D#))+(1#/(120#*D#^4#))-(1#/(252#*D#^6
#))-(1#/(BETA#+1#))-(1#/BETA#)
2130 FCN#=-NREC#*PSI#+C#-NREC%*LOG(ALPHA#)
2140 TRI#=(1#/D#)+(1#/(2#*D#*D#))+(1#/(6#*D#*D#*D#))-(1#/(30#*D#^5#))+(1#/(42#*D
#^7#))-(1#/(30#*D#^9#))+(1#/((BETA#+1#)*(BETA#+1#)))+(1#/(BETA#*BETA#))
2150 V#=A#-(NREC#*NREC#)/B#
2160 U#=A#
2170 W#=(B#/NREC#)-(NREC#/A#)
2180 DU#=R#
2190 DV#=R#-(NREC#*NREC#*NREC#)/(B#*B#)
2200 DW#=-1#+(NREC#*R#)/(A#*A#)
2210 FPN#=-NREC#*TRI#*((V#*DU#-U#*DV#)/(V#*V#))-A#-NREC#*DW#/W#
2220 AAS#=GML#-(FCN#/FPN#)
2230 DELTA#=ABS(.00000001#*AAS#)
2240 IF ABS(AAS#-GML#) < DELTA# GOTO 2280
2250 IF COUNT% > 24 THEN PRINT:PRINT "Pearson type III procedure does not conver
ge":PRINT:ST#(5)=99999#:GOTO 2410
2260 GML#=AAS#
2270 GOTO 2000
2280 GAMMA#=AAS#
2290 SUM#=0#
2300 FOR I%=1 TO NREC%
2310 E#=BETA#^(1#/3#)-1#/(9#*BETA#^(2#/3#))+T#(I%)/(3#*BETA#^(1#/6#))
2320 PT3#(I%)=GAMMA#+ALPHA#*E#*E#*E#
2330 TEMP#=PT3#(I%)-XS#(I%)
2340 SUM#=SUM#+TEMP#*TEMP#
2350 NEXT I%
2360 ST#(5)=SQR(SUM#/NREC#)
2370 '
2380 '
2390 ' Log-Pearson type 3
2400 '
2410 COUNT%=0
2420 GML#=Y#(NREC%)*.99#
2430 COUNT%=COUNT%+1
2440 A#=0#:B#=0#:C#=0#:R#=0#
2450 FOR I%=1 TO NREC%
2460 TEMP#=Y#(I%)-GML#
2470 A#=A#+1#/TEMP#
2480 B#=B#+TEMP#
2490 C#=C#+LOG(TEMP#)
```

```
2500 R#=R#+1#/(TEMP#*TEMP#)
2510 NEXT I%
2520 BETA#=A#/(A#-(NREC#*NREC#/B#))
2530 ALPHA#=B#/(NREC#*BETA#)
2540 D#=BETA#+2#
2550 PSI#=LOG(D#)-(1#/(2#*D#))-(1#/(12#*D#*D#))+(1#/(120#*D#^4#))-(1#/(252#*D#^6
#))-(1#/(BETA#+1#))-(1#/BETA#)
2560 FCN#=-NREC#*PSI#+C#-NREC%*LOG(ALPHA#)
2570 TRI#=(1#/D#)+(1#/(2#*D#*D#))+(1#/(6#*D#*D#*D#))-(1#/(30#*D#^5#))+(1#/(42#*D
#^7#))-(1#/(30#*D#^9#))+(1#/((BETA#+1#)*(BETA#+1#)))+(1#/(BETA#*BETA#))
2580 V#=A#-(NREC#*NREC#)/B#
2590 U#=A#
2600 W#=(B#/NREC#)-(NREC#/A#)
2610 DU#=R#
2620 DV#=R#-(NREC#*NREC#*NREC#)/(B#*B#)
2630 DW#=-1#+(NREC#*R#)/(A#*A#)
2640 FPN#=-NREC#*TRI#*((V#*DU#-U#*DV#)/(V#*V#))-A#-NREC#*DW#/W#
2650 AAS#=GML#-(FCN#/FPN#)
2660 DELTA#=ABS(.00000001#*AAS#)
2670 IF ABS(AAS#-GML#) < DELTA# GOTO 2710
2680 IF COUNT% > 24 THEN PRINT:PRINT "Log Pearson type III procedure does not co
nverge":PRINT:ST#(6)=99999#:GOTO 3750
2690 GML#=AAS#
2700 GOTO 2430
2710 GAMMA#=AAS#
2720 SUM#=0#
2730 FOR I%=1 TO NREC%
2740 E#=BETA#^(1#/3#)-1#/(9#*BETA#^(2#/3#))+T#(I%)/(3#*BETA#^(1#/6#))
2750 LP3#(I%)=GAMMA#+ALPHA#*E#*E#*E#
2760 LP3#(I%)=EXP(LP3#(I%))
2770 TEMP#=LP3#(I%)-XS#(I%)
2780 SUM#=SUM#+TEMP#*TEMP#
2790 NEXT I%
2800 ST#(6)=SQR(SUM#/NREC#)
2810 PRINT TAB(B%)  TITLE$:PRINT TAB(B%) U$
2820 PRINT TAB(D%)  T$:PRINT TAB(D%) V$
2830 IF FLAG% <> 1 GOTO 2870
2840 LPRINT CHR$(12):LPRINT:LPRINT
2850 LPRINT TAB(B%)  TITLE$:LPRINT TAB(B%) U$
2860 LPRINT:LPRINT TAB(D%)  T$:LPRINT TAB(D%) V$
2870 FORM2$="#.###^^^^ #.###^^^^ #.###^^^^ #.###^^^^ #.###^^^^ #.###^^
^^ #.###^^^^"
2880 FORM3$="              #.###^^^^ #.###^^^^ #.###^^^^ #.###^^^^ #.###^^
^^ #.###^^^^"
2890 ICOUNT%=0
2900 FOR I%=1 TO NREC%
2910 I#=I%
2920 ICOUNT%=ICOUNT%+1
2930 IF I%=1 THEN FLAG2%=0:GOSUB 3790:FLAG2%=1
2940 IF ICOUNT%=15 THEN ICOUNT%=0:GOSUB 3750:GOSUB 3790
2950 PRINT USING FORM2$; (NREC#+1#)/I#,XS#(I%),TN#(I%),LN2#(I%),LN3#(I%),T1E#(I%
),PT3#(I%),LP3#(I%)
2960 IF FLAG%<>1 GOTO 2980
2970 LPRINT USING FORM2$; (NREC#+1#)/I#,XS#(I%),TN#(I%),LN2#(I%),LN3#(I%),T1E#(I
%),PT3#(I%),LP3#(I%)
2980 NEXT I%
2990 PRINT:PRINT TAB(26) "Standard Errors of Estimate":PRINT TAB(26) "----------
------------------"
3000 PRINT:PRINT USING FORM3$;ST#(1),ST#(2),ST#(3),ST#(4),ST#(5),ST#(6)
3010 IF FLAG%<>1 GOTO 3040
3020 LPRINT:LPRINT TAB(26) "Standard Errors of Estimate":LPRINT TAB(26) "-------
-------------------"
3030 LPRINT:LPRINT USING FORM3$;ST#(1),ST#(2),ST#(3),ST#(4),ST#(5),ST#(6)
3040 GOSUB 3750
3041 CLS
3050 SYSTEM
3060 '
3070 ' Subroutine for standard screen format
3080 '
3090 CLS:PRINT:PRINT
3100 ROW$=STRING$(78,205)
3110 BOXTOP$=CHR$(201)+ROW$+CHR$(187)
3120 PRINT BOXTOP$;
3130 PRINT CHR$(186) TAB(80) CHR$(186);
3140 PRINT CHR$(186) TAB(80) CHR$(186);
3150 T$="Comparison of Various Probability Distributions"
3160 C%=LEN(T$)
3170 D%=(80-C%)/2
3180 V$=STRING$(C%,205)
3190 PRINT CHR$(186) TAB(D%) T$ TAB(80) CHR$(186);
```

197

```
3200 PRINT CHR$(186) TAB(D%) V$ TAB(80) CHR$(186);
3210 PRINT CHR$(186) TAB(80) CHR$(186);
3220 PRINT CHR$(186) TAB(80) CHR$(186);
3230 PRINT CHR$(186) "   What is the name of your data file?" TAB(80) CHR$(186);

3240 PRINT CHR$(186) TAB(80) CHR$(186);
3250 PRINT CHR$(186) "   Do you want printer output (Y/N)?" TAB(80) CHR$(186);
3260 PRINT CHR$(186) "   (press Alt Q to quit at this stage)" TAB(80) CHR$(186);

3270 PRINT CHR$(186) TAB(80) CHR$(186);
3280 PRINT CHR$(186) TAB(80) CHR$(186);
3290 BOXBOT$=CHR$(200)+ROW$+CHR$(188)
3300 PRINT BOXBOT$;
3310 LOCATE 24,1,0,0,0
3320 PRINT "G Kite";
3330 LOCATE 10,41,1,0,13
3340 FILE$=""
3350 I$=INPUT$(1)
3360 IF I$ = CHR$(13) GOTO 3460                        ' ENTER key
3370 IF I$ = CHR$(8)  GOTO 3410                        ' BACKSPACE key
3380 PRINT I$;
3390 FILE$=FILE$+I$
3400 GOTO 3350
3410 H%=POS(0)
3420 LOCATE 10,H%-1,1,0,13
3430 L%=LEN(FILE$)
3440 FILE$=LEFT$(FILE$,L%-1)
3450 GOTO 3350
3460 LOCATE 12,39,1,0,13
3470 YN$=""
3480 I$=INKEY$:IF I$ = "" GOTO 3480
3490 IF LEN(I$) = 1 GOTO 3530
3500 I$=RIGHT$(I$,1)
3510 IF I$ = CHR$(16) THEN IERR%=1:GOTO 3600           ' ALT Q key
3520 GOTO 3460
3530 IF I$ = CHR$(13) GOTO 3570                        ' ENTER key
3540 PRINT I$;
3550 YN$=I$
3560 GOTO 3480
3570 IF YN$ = "Y" OR YN$ = "y" OR YN$ = "N" OR YN$ = "n" GOTO 3590
3580 GOTO 3460
3590 LOCATE ,,0,13,13
3600 RETURN
3610 '
3620 ' sort an array in descending order
3630 '
3640 KK%=NREC%-1
3650 FOR LL%=1 TO KK%
3660 MM%=NREC%-LL%
3670 FOR JJ%=1 TO MM%
3680 IF XS#(JJ%) > XS#(JJ%+1) GOTO 3720
3690 XTEMP#=XS#(JJ%)
3700 XS#(JJ%)=XS#(JJ%+1)
3710 XS#(JJ%+1)=XTEMP#
3720 NEXT JJ%
3730 NEXT LL%
3740 RETURN
3750 PRINT:PRINT"Press any key to continue"
3760 Q$=INKEY$:IF Q$ ="" GOTO 3760
3770 CLS
3780 RETURN
3790 ' titles
3800   PRINT " Return    Sorted                         Predicted Magnitudes"
3810   PRINT " Period   Recorded  Truncated  2 Param.  3 Param.  Type I    Pearso
n  L-Pearson"
3820   PRINT " (years)   Event     Normal  Lognormal Lognormal Extremal  Type I
II  Type III"
3830   PRINT " --------------------------------------------------------------------
--------------"
3840 IF FLAG2% = 1 GOTO 3910
3850 IF FLAG%<>1 GOTO 3910
3860 LPRINT:LPRINT " Return    Sorted                         Predicted Magnitude
s"
3870 LPRINT " Period   Recorded  Truncated  2 Param.  3 Param.  Type I    Pearson
  L-Pearson"
3880 LPRINT " (years)   Event     Normal  Lognormal Lognormal Extremal  Type II
I  Type III"
3890 LPRINT " --------------------------------------------------------------------
--------------"
3900 LPRINT
3910 RETURN
```

St. Mary's River at Stillwater, Nova Scotia. Station No. 01E0001. m**3/s.

## Comparison of Various Probability Distributions

| | |
|---|---|
| Mean is | 0.4132D+03 |
| Variance is | 0.2067D+05 |
| Coefficient of skew is | 0.1195D+01 |
| | |
| Mean of logs is | 0.5969D+01 |
| Variance of logs is | 0.1088D+00 |
| Coefficient of skew of logs is | 0.1662D+00 |

| Return Period (years) | Sorted Recorded Event | Truncated Normal | 2 Param. Lognormal | 3 Param. Lognormal | Type I Extremal | Pearson Type III | L-Pearson Type III |
|---|---|---|---|---|---|---|---|
| 0.730D+02 | 0.974D+03 | 0.730D+03 | 0.810D+03 | 0.844D+03 | 0.814D+03 | 0.822D+03 | 0.840D+03 |
| 0.365D+02 | 0.324D+03 | 0.689D+03 | 0.737D+03 | 0.759D+03 | 0.738D+03 | 0.748D+03 | 0.755D+03 |
| 0.243D+02 | 0.725D+03 | 0.663D+03 | 0.694D+03 | 0.710D+03 | 0.693D+03 | 0.704D+03 | 0.706D+03 |
| 0.183D+02 | 0.651D+03 | 0.643D+03 | 0.663D+03 | 0.676D+03 | 0.661D+03 | 0.672D+03 | 0.672D+03 |
| 0.146D+02 | 0.603D+03 | 0.627D+03 | 0.639D+03 | 0.649D+03 | 0.636D+03 | 0.647D+03 | 0.645D+02 |
| 0.122D+02 | 0.593D+03 | 0.613D+03 | 0.619D+03 | 0.626D+03 | 0.616D+03 | 0.626D+03 | 0.623D+03 |
| 0.104D+02 | 0.583D+03 | 0.601D+03 | 0.602D+03 | 0.608D+03 | 0.598D+03 | 0.608D+03 | 0.604D+03 |
| 0.913D+01 | 0.569D+03 | 0.590D+03 | 0.587D+03 | 0.591D+03 | 0.583D+03 | 0.592D+03 | 0.588D+03 |
| 0.811D+01 | 0.565D+03 | 0.580D+03 | 0.573D+03 | 0.577D+03 | 0.569D+03 | 0.578D+03 | 0.574D+03 |
| 0.730D+01 | 0.564D+03 | 0.570D+03 | 0.561D+03 | 0.564D+03 | 0.557D+03 | 0.566D+03 | 0.561D+03 |
| 0.664D+01 | 0.564D+03 | 0.562D+03 | 0.550D+03 | 0.552D+03 | 0.546D+03 | 0.554D+03 | 0.549D+03 |
| 0.608D+01 | 0.552D+03 | 0.554D+03 | 0.540D+03 | 0.541D+03 | 0.536D+03 | 0.543D+03 | 0.538D+03 |
| 0.562D+01 | 0.544D+03 | 0.546D+03 | 0.530D+03 | 0.530D+03 | 0.526D+03 | 0.533D+03 | 0.528D+03 |
| 0.521D+01 | 0.527D+03 | 0.538D+03 | 0.521D+03 | 0.521D+03 | 0.517D+03 | 0.524D+03 | 0.519D+03 |
| 0.487D+01 | 0.524D+03 | 0.531D+03 | 0.513D+03 | 0.512D+03 | 0.509D+03 | 0.515D+03 | 0.510D+03 |
| 0.456D+01 | 0.518D+03 | 0.525D+03 | 0.505D+03 | 0.504D+03 | 0.501D+03 | 0.507D+03 | 0.502D+03 |
| 0.429D+01 | 0.515D+03 | 0.518D+03 | 0.498D+03 | 0.496D+03 | 0.493D+03 | 0.499D+03 | 0.494D+03 |
| 0.406D+01 | 0.514D+03 | 0.512D+03 | 0.490D+03 | 0.488D+03 | 0.486D+03 | 0.491D+03 | 0.487D+03 |
| 0.384D+01 | 0.510D+03 | 0.505D+03 | 0.484D+03 | 0.481D+03 | 0.479D+03 | 0.484D+03 | 0.480D+03 |
| 0.365D+01 | 0.487D+03 | 0.499D+03 | 0.477D+03 | 0.474D+03 | 0.473D+03 | 0.477D+03 | 0.473D+03 |
| 0.348D+01 | 0.479D+03 | 0.494D+03 | 0.471D+03 | 0.468D+03 | 0.467D+03 | 0.470D+03 | 0.466D+03 |
| 0.332D+01 | 0.479D+03 | 0.488D+03 | 0.464D+03 | 0.461D+03 | 0.461D+03 | 0.464D+03 | 0.460D+03 |
| 0.317D+01 | 0.464D+03 | 0.482D+03 | 0.458D+03 | 0.455D+03 | 0.455D+03 | 0.458D+03 | 0.454D+03 |
| 0.304D+01 | 0.456D+03 | 0.477D+03 | 0.453D+03 | 0.449D+03 | 0.449D+03 | 0.452D+03 | 0.448D+03 |
| 0.292D+01 | 0.456D+03 | 0.471D+03 | 0.447D+03 | 0.443D+03 | 0.444D+03 | 0.446D+03 | 0.443D+03 |
| 0.281D+01 | 0.453D+03 | 0.466D+03 | 0.442D+03 | 0.438D+03 | 0.438D+03 | 0.440D+03 | 0.437D+03 |
| 0.270D+01 | 0.442D+03 | 0.461D+03 | 0.436D+03 | 0.432D+03 | 0.433D+03 | 0.435D+03 | 0.432D+03 |
| 0.261D+01 | 0.428D+03 | 0.456D+03 | 0.431D+03 | 0.427D+03 | 0.428D+03 | 0.429D+03 | 0.427D+03 |
| 0.252D+01 | 0.411D+03 | 0.451D+03 | 0.426D+03 | 0.422D+03 | 0.423D+03 | 0.424D+03 | 0.422D+03 |
| 0.243D+01 | 0.405D+03 | 0.445D+03 | 0.421D+03 | 0.417D+03 | 0.419D+03 | 0.419D+03 | 0.417D+02 |
| 0.235D+01 | 0.405D+03 | 0.440D+03 | 0.416D+03 | 0.412D+03 | 0.414D+03 | 0.414D+03 | 0.412D+03 |
| 0.228D+01 | 0.394D+03 | 0.435D+03 | 0.412D+03 | 0.407D+03 | 0.409D+03 | 0.409D+03 | 0.407D+03 |
| 0.221D+01 | 0.394D+03 | 0.430D+03 | 0.407D+03 | 0.403D+03 | 0.405D+03 | 0.404D+03 | 0.403D+03 |
| 0.215D+01 | 0.394D+03 | 0.425D+03 | 0.402D+03 | 0.398D+03 | 0.400D+03 | 0.399D+03 | 0.398D+03 |
| 0.209D+01 | 0.385D+03 | 0.421D+03 | 0.398D+03 | 0.393D+03 | 0.396D+03 | 0.395D+03 | 0.394D+03 |
| 0.203D+01 | 0.378D+03 | 0.416D+03 | 0.393D+03 | 0.389D+02 | 0.391D+03 | 0.390D+03 | 0.389D+03 |
| 0.197D+01 | 0.377D+03 | 0.411D+03 | 0.389D+03 | 0.385D+03 | 0.387D+03 | 0.385D+03 | 0.385D+03 |
| 0.192D+01 | 0.371D+03 | 0.406D+03 | 0.385D+03 | 0.380D+03 | 0.383D+03 | 0.381D+03 | 0.381D+03 |
| 0.187D+01 | 0.371D+03 | 0.401D+03 | 0.380D+03 | 0.376D+03 | 0.379D+03 | 0.376D+03 | 0.376D+03 |
| 0.183D+01 | 0.368D+02 | 0.396D+03 | 0.376D+03 | 0.372D+03 | 0.375D+03 | 0.372D+03 | 0.372D+03 |
| 0.178D+01 | 0.368D+03 | 0.391D+03 | 0.372D+03 | 0.367D+03 | 0.370D+03 | 0.368D+03 | 0.368D+03 |
| 0.174D+01 | 0.365D+03 | 0.386D+03 | 0.367D+03 | 0.363D+03 | 0.366D+03 | 0.363D+03 | 0.364D+03 |
| 0.170D+01 | 0.360D+03 | 0.381D+03 | 0.363D+03 | 0.359D+03 | 0.362D+03 | 0.359D+03 | 0.360D+03 |
| 0.166D+01 | 0.351D+03 | 0.376D+03 | 0.359D+03 | 0.355D+03 | 0.358D+03 | 0.355D+03 | 0.356D+03 |
| 0.162D+01 | 0.348D+03 | 0.371D+03 | 0.355D+03 | 0.351D+03 | 0.354D+03 | 0.351D+03 | 0.352D+03 |
| 0.159D+01 | 0.348D+03 | 0.365D+03 | 0.351D+03 | 0.347D+03 | 0.350D+03 | 0.346D+03 | 0.348D+03 |
| 0.155D+01 | 0.345D+03 | 0.360D+03 | 0.346D+03 | 0.343D+03 | 0.346D+03 | 0.342D+03 | 0.344D+03 |
| 0.152D+01 | 0.342D+03 | 0.355D+03 | 0.342D+03 | 0.339D+03 | 0.342D+03 | 0.338D+03 | 0.340D+03 |
| 0.149D+01 | 0.337D+03 | 0.349D+03 | 0.338D+03 | 0.335D+03 | 0.338D+03 | 0.334D+03 | 0.336D+03 |
| 0.146D+01 | 0.337D+03 | 0.344D+03 | 0.334D+03 | 0.330D+03 | 0.334D+03 | 0.329D+03 | 0.332D+03 |
| 0.143D+01 | 0.334D+03 | 0.338D+03 | 0.330D+03 | 0.326D+03 | 0.330D+03 | 0.325D+03 | 0.327D+03 |
| 0.140D+01 | 0.334D+03 | 0.333D+03 | 0.325D+03 | 0.322D+03 | 0.326D+03 | 0.321D+03 | 0.323D+03 |
| 0.138D+01 | 0.328D+03 | 0.327D+03 | 0.321D+03 | 0.318D+03 | 0.322D+03 | 0.317D+03 | 0.319D+03 |
| 0.135D+01 | 0.318D+03 | 0.321D+03 | 0.317D+03 | 0.314D+03 | 0.317D+03 | 0.312D+03 | 0.315D+03 |
| 0.133D+01 | 0.311D+03 | 0.315D+03 | 0.312D+03 | 0.310D+03 | 0.313D+03 | 0.308D+03 | 0.311D+03 |
| 0.130D+01 | 0.303D+03 | 0.308D+03 | 0.308D+03 | 0.305D+03 | 0.309D+03 | 0.304D+03 | 0.307D+03 |
| 0.128D+01 | 0.303D+03 | 0.302D+03 | 0.303D+03 | 0.301D+03 | 0.304D+03 | 0.299D+03 | 0.302D+03 |

```
0.126D+01 0.294D+03 0.295D+03 0.298D+03 0.297D+03 0.300D+03 0.295D+03 0.298D+03
0.124D+01 0.292D+03 0.283D+03 0.293D+03 0.292D+03 0.295D+03 0.290D+03 0.293D+03
0.122D+01 0.292D+03 0.281D+03 0.289D+03 0.288D+03 0.290D+03 0.286D+03 0.289D+03
0.120D+01 0.289D+03 0.273D+03 0.283D+03 0.283D+03 0.286D+03 0.281D+03 0.284D+03
0.118D+01 0.289D+03 0.265D+03 0.278D+03 0.278D+03 0.280D+03 0.276D+03 0.279D+03
0.116D+01 0.280D+03 0.256D+03 0.273D+03 0.273D+03 0.275D+03 0.271D+03 0.274D+03
0.114D+01 0.255D+03 0.247D+03 0.267D+03 0.268D+03 0.270D+03 0.265D+03 0.269D+03
0.112D+01 0.255D+03 0.237D+03 0.261D+03 0.262D+03 0.264D+03 0.260D+03 0.263D+03
0.111D+01 0.238D+03 0.225D+03 0.254D+03 0.256D+03 0.257D+03 0.254D+03 0.257D+03
0.109D+01 0.232D+03 0.213D+03 0.247D+03 0.250D+03 0.250D+03 0.248D+03 0.251D+03
0.107D+01 0.232D+03 0.199D+03 0.239D+03 0.243D+03 0.243D+03 0.241D+03 0.244D+03
0.106D+01 0.232D+03 0.183D+03 0.231D+03 0.235D+03 0.234D+03 0.234D+03 0.236D+03
0.104D+01 0.228D+03 0.163D+03 0.220D+03 0.226D+03 0.224D+03 0.225D+03 0.226D+03
0.103D+01 0.202D+03 0.137D+03 0.208D+03 0.215D+03 0.211D+03 0.215D+03 0.215D+03
0.101D+01 0.190D+03 0.960D+02 0.189D+03 0.199D+03 0.192D+03 0.201D+03 0.198D+03
```

Standard Errors of Estimate
----------------------------

```
0.422D+02 0.249D+02 0.215D+02 0.248D+02 0.235D+02 0.219D+02
```

# CHAPTER 13
## REGIONAL ANALYSIS

Introduction

Chapters 5-11 have described some of the probability distributions that can be used to carry out a frequency analysis
on a set of observed or computed data. Using any one of those
techniques the event magnitude corresponding to a given probability of occurrence can be determined. This event magnitude will,
however, apply only to the exact location at which the original
observations were made. Frequently in hydrology it is necessary
to estimate event magnitudes at sites where no observations have
been taken. As an example, the design of a highway culvert may
require the estimation of a design flood for a small ungauged
catchment area. Regional analysis is the term given to techniques which make this estimation possible.

In addition, as noted in Chapter 1, the use of more than
one set of data tends to reduce the sampling error and, even for
a gauged site, will produce more reliable event estimates than
a single station frequency analysis. Indeed, as discussed later,
the Bayesian approach to frequency analysis might be to use distributions of parameters obtained from a regional analysis in
conjunction with the point values to account for the uncertainty
in the estimates.

The earliest approach to the regionalization problem was to
use empirical equations relating floodflow, Q, to drainage area,
A, within a particular region (3) such as:

$$Q = cA^n \qquad (13-1)$$

where c and n are constants. Other types of empirical equation
(some still in use, such as the Rational Formula) related floodflow to rainfall intensity and area, as:

$$Q = ciA \qquad (13-2)$$

where c is a runoff coefficient, i is the rainfall intensity and
A is the area. The objective of all these equations was to
extrapolate from gauged basins to ungauged basins by means of
parameters which could be estimated (rainfall intensity) or
measured from maps (area).

Other methods in use include that designed by Coulson (10)
for Southern Ontario. From the records of 59 gauging stations
in the area, together with an isohyetal map of mean annual precipitation, a map of lines of equal mean annual runoff was drawn.
Rewriting the standard frequency equation (see Chapter 3) as

$$Q_T = \bar{Q} (K z + 1) \qquad (13-3)$$

where $Q_T$ is the event magnitude at the required return period, T, $\bar{Q}$ is the mean annual runoff, K is a frequency factor depending on the probability distribution used and z is the coefficient of variation. For an ungauged drainage basin for which an estimate of $Q_T$ is required, Coulson obtained $\bar{Q}$ by planimetering the mean annual runoff isoline map for the particular basin. Using a Pearson type III distribution the value of K, the frequency factor, depends upon the return period, T, and the coefficient of skew of the distribution, $\gamma_1$. It was found in Southern Ontario that the coefficient of skew could best be derived from the coefficient of variation as:

$$\gamma_1 = 2z \qquad (13-4)$$

while the coefficient of variation could be obtained from the drainage area, A, as (14):

$$z = 0.35 - 0.03 \log (A + 1) \qquad (13-5)$$

It should be noted that this method was used only for annual flows and not for instantaneous maxima or minima.

One of the methods derived by the USGS is described as an "index-flood" technique (12). There are two major parts to this method. Firstly, basic dimensionless frequency curves are drawn representing the ratios of the floods at various frequencies to the mean annual flood for each gauged basin. Secondly, relationships are developed between the characteristics of drainage areas and the mean annual flood. Combining the mean annual flood with a regional frequency curve enables flood magnitudes to be estimated at any location within the region.

Many modifications of the original index-flood method have been made (3, 11), mainly to try and increase the number of independent variables used to transfer the hydrologic information. In general there are two types of variables used: (a) physiographic characteristics such as drainage area, elevation, slope, percent of basin covered by lakes, swamps, etc. and (b) hydrometeorologic variables such as mean annual precipitation and mean annual temperature.

One of the major developments since the index-flood technique is the "square-grid" method (26). Originally programmed for mean annual runoff this method has been adapted for frequency analyses, simulation and modelling (16, 25, 18).

Alternate methods include the use of standard single-station frequency distributions modified for use as regional distributions and the regional record maxima technique.

Many of the methods used in regional analysis depend upon inter-station correlation of streamflows in order to produce time series with a uniform period of record. The final section of this chapter discusses information transfer together with the concept of effective number of years of record and number of stations.

## Index-Flood Method

The basic idea behind the index-flood method (12) is to increase the reliability of the frequency characteristics within a region. If, within a hydrologically homogeneous area, a number of hydrometric stations have been operating and recording the effects of the same meteorologic factors then a combination of these records will provide, not a longer record, but a more reliable record. The following brief description of the index-flood method does not include all the computational details of the procedure. These can be obtained, if required, from Dalrymple (12).

Firstly, the data sets available within a region are listed, unsuitable stations eliminated, and a common period of record selected. Generally stations having less than 5 years of record of gauging and all regulated or controlled streams are excluded. Since streamflows are not pure random but contain trends and periodicities the period over which measurements are made becomes important when records are combined. A bar graph showing the period of record of each gauge is useful in determining which base period to use. The base period should be planned so as to include the maximum information content i.e. maximum number of station-years. Missing data points may be filled in by inter-station correlations (see section on regional record maxima). Data points filled in in this way are not used directly but only as aids in assigning representative return periods to the recorded events.

The index-flood method next computes return periods, T, for each recorded event for each station in the region using the equation:

$$T = \frac{n + 1}{m} \qquad (13-6)$$

where n is the sample size and m is the order number of an event; m = 1 for the maximum event and m = n for the minimum event. For each station a graph of T versus event magnitude is plotted and a smooth curve drawn through the points. No attempt is made to force a straight line fit or to fit any mathematical distribution. The mean annual event for the station is then picked off the smooth curve at the point T = 2.33. This is a theoretical result taken from the type I extremal distribution (see Chapter 8). Benson (2) has confirmed experimentally that the mean annual event (i.e. the mean of all the observed annual maxima) does occur with a return period of 2.33 years. It is preferred in the index-flood method to derive the mean annual event graphically rather than arithmetically.

Dalrymple (12) has described a test which should be used at this stage of the index-flood procedure to check for regional hydrologic homogeneity. If the standard error of estimate of the reduced variable, y, in a type I extremal distribution is given by:

$$\sigma_y = \frac{e^y}{\sqrt{n}} \sqrt{\frac{1}{T-1}} \qquad (13\text{-}7)$$

then, assuming a normal distribution of the estimates, 95% of the estimates will lie within $\pm 2\sigma_y$ of the most probable value. If T, the return period of the estimate, is taken as 10 years, then

$$2\sigma_y = \frac{0.666e^y}{\sqrt{n}} \qquad (13\text{-}8)$$

Since for T = 10 the reduced variable in a type I extremal distribution is 2.25 (see Chapter 8) then the confidence limits are given by

$$2.25 \pm 6.33/\sqrt{n} \qquad (13\text{-}9)$$

Table 13-1. after Dalrymple (12), gives the upper and lower confidence limits with the corresponding return periods for various values of n.

Table 13-1

Confidence Limits for Index-Flood Homogeneity Test[1]

| Sample Size n | Lower Limit $y-2\sigma_y$ | $T_L$ | Upper Limit $y+2\sigma_y$ | $T_U$ |
|---|---|---|---|---|
| 5 | -0.59 | 1.2 | 5.09 | 160 |
| 10 | 0.25 | 1.8 | 4.25 | 70 |
| 20 | 0.83 | 2.8 | 3.67 | 40 |
| 50 | 1.35 | 4.4 | 3.15 | 24 |
| 100 | 1.62 | 5.6 | 2.88 | 18 |
| 200 | 1.80 | 6.5 | 2.70 | 15 |
| 500 | 1.97 | 7.7 | 2.53 | 13 |
| 1000 | 2.05 | 8.3 | 2.45 | 12 |

[1]From Dalrymple (12)

The procedure used for the test is to first of all plot $T_L$ and $T_U$ from Table 13-1 versus n on probability scale graph paper. Then, for each station in the region to be tested, the ratio of the 10-year event to the mean annual event is computed and an average ratio for the region calculated. Then the average ratio for the region is multiplied by the mean annual event for each station to give a modified 10-year event magnitude for each station. The return periods corresponding to these modified 10-year events are then found for each station

204

from the individual station frequency curves, say $T_E$. The effective period of record of each gauging station is determined as the number of recorded annual events plus one half the number of events computed for that station by inter-station correlation, say $N_E$. Next, the coordinate pairs $(T_E, N_E)$ for each station are plotted on the test graph showing curves of $T_L$ and $T_U$. Any station for which the plotted point is outside the confidence limit curves is then excluded from the homogeneous region. Figure 13-1 is a base graph for use in this test.

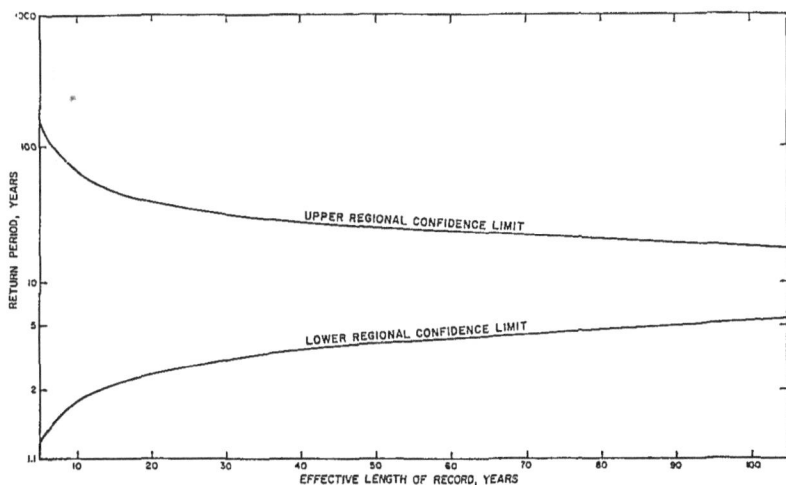

Figure 13-1. Plot to be used in regional homogeneity test.

For each station which remains in the hydrologically homogeneous region, ratios of events of different return periods to the mean annual event are computed for T values of say 1.1, 1.5, 5, 10, 20, 50 and median values of these ratios are determined for the region. A plot of these median ratios versus return period is then the regional frequency curve and represents the most likely relationship for all parts of the region.

The next major step in the index-flood analysis is to plot drainage area versus mean annual event for those stations within the homogeneous region and graphically fit a smooth curve through the points. An alternative available at this stage is to develop a multiple regression equation between mean annual event and basin characteristics.

To define a frequency curve at any location within the homogeneous region the mean annual event is determined from the curve relating this event to drainage area. The mean annual event is then multiplied by the median ratios for the various return periods required, as determined from the regional frequency curve.

Regional index-flood studies have been carried out for most of the states in the United States and for the South Saskatchewan River Basin (6), New Brunswick - Gaspé (7) and Nova Scotia (9) areas in Canada.

Benson (3) has noted three deficiencies found in the index-flood method:

(a)  The index-flood (mean annual flood) for stations with short periods of record may not be typical.  This means that the ratios of floods of different return periods to the index-flood may vary widely between stations.

(b)  The homogeneity test is used to determine whether the differences in slopes of frequency curves are greater than may be attributed to chance alone.  This test uses the ratio of the 10-year flood because many individual records are too short to adequately define the frequency curve at higher levels.  It has been found in some studies that although homogeneity is apparently established at the 10-year level, the individual curves show wide and sometimes systematic differences at higher levels.

(c)  In the use of the index-flood method, it has been accepted that within a flood-frequency region, frequency curves may be combined for all sizes of drainage areas, excluding only the largest.  Although the variation in the slope of the frequency curve with drainage area had been investigated at the time of each study, it was studied at the 10-year point where the effect is small.  The error of neglecting this drainage-area effect has been reduced by giving separate and special treatment to large streams.  Recent studies by the USGS, for which ratios of less frequent floods were used, have shown, in all regions where such data are available, that the ratios of any specified flood to the mean annual flood will vary inversely with the drainage area. In general, the larger the drainage area, the flatter the frequency curve.  The effect of drainage area is relatively greater for floods of higher recurrence intervals.

In applying the index-flood technique in Canada, Collier (6) has provided comments on some of the problems he encountered. It has been found in the foothills area of Alberta that normally the annual flood is due to snowmelt in the high regions usually combined with rainfall from Pacific air masses moving over the mountains from the west.  Occasionally, however, moist tropical air is sucked up from the southeast and heavy precipitation occurs on the foothills.  Simple flood frequency analysis does not distinguish between these two types of events and, as a result, a plot of the floods on arithmetic-normal probability paper shows a distinct S-shape.

Other comments made by Collier (6) included a discussion of the steps to take when a group of stations fails the regional homogeneity test as well as a discussion of the validity of the index-flood approach.

## Multiple Regression Techniques

*Estimating Events at Specific Return Periods* - Rather than plotting drainage area versus mean annual flood, as in the index-flood method, many investigators (see Benson (3) for a partial list) have studied the relationships between discharges at specified return periods and basin characteristics. In general, the relationships take the form

$$Q_T = f(A^a, B^b, C^c, ..., Z^z) \qquad (13\text{-}10)$$

where A, B, C,...,Z are independent variables and a, b, c,..., z are constants derived by multiple regression analysis. Since a similar relationship, but with different exponents, will be found for the mean annual flood, the ratio of $Q_T$ to the mean annual flood will not be a dimensionless constant as assumed in the index-flood method, but will be proportionate to the basin characteristics. A further advantage of this multiple regression modification to the index-flood method is that it obviates the necessity of assuming any underlying distribution for the flood peaks.

Many different procedures are available for determining the relevant parameters in Equation 13-10 including the simple linear regression, multiple linear regression, forward, backward and stepwise procedures. Packages of computer programs for these procedures are commonly available as library functions at computer centres. Wampler (31) has tested and compared more than 20 linear least squares computer programs.

De Coursey (13) used a canonical correlation procedure to select watershed characteristics and then derived a multiple regression matrix relationship

$$Q = aA + b \qquad (13\text{-}11)$$

where Q is a column vector of peak flows at various return periods, A is a column vector of watershed characteristics and a and b are respectively a matrix and a vector of regression coefficients. This approach preserves the intercorrelation between the dependent variables in vector Q.

*Estimating Regional Distribution Parameters* - An alternate approach is to regress the basin characteristics not with flood

magnitudes at given return periods, but with the parameters of
a chosen probability distribution. The relevant parameters are
first derived for each of the individual sites in the region
using single-station frequency analysis techniques and then
regression equations are developed to estimate the distribution
parameters at the required location. As an example, if the lognormal probability distribution is used (16) then the mean and
standard deviation of the annual events should be used in the
correlation.

Thomas and Benson (28) examined both techniques of estimating specific return periods and regional distribution parameters
using over 70 parameters of streamflow including low flow peaks
and flood volumes, parameters of annual and monthly means and
flow distribution statistics.

These streamflow characteristics were each regressed against
up to 31 basin characteristics including such parameters as
drainage area, slopes, storage indices, precipitation totals
and intensity, evaporation and temperature. The study showed
that the most important parameters were drainage areas and mean
annual precipitation. As might be expected the equations for
characteristics of the mid-range of flows were more accurate than
for low flows. No significant relationships were found for
coefficients of skew or serial correlation coefficients.

## Square-Grid Method

The definition of areal runoff in geographical areas exhibiting basically similar hydrological characteristics can be
facilitated by employing the square-grid technique as proposed
by Solomon et al. (26). As implied in the term square-grid this
technique entails dividing the study area into a uniform grid,
the squares of which are identified in cartesian coordinates.
Each square-grid has associated with it a set of parameters such
as the elevations of each corner and centre point, the percentage
areas of the square covered by lakes, swamps, forests, urban
areas, etc., soil types, indicator of bedrock, etc., derived
from topographic maps and other information sources such as soil
surveys. From these basic data other physiographic characteristics such as mean slope, azimuth of slope, barrier height in
different directions, distance to the sea in different directions
can be easily derived for each square. Those square-grids containing meteorologic or hydrometric stations will also have
associated with them data on mean annual precipitation and temperature and streamflow respectively. Each grid-square also
carries identification of up to four streamflow directions into
and out of adjacent squares so that drainage basin data can be
accumulated and averaged automatically.

The grid interval determines to a large extent the accuracy
of the results since the finer the grid the more basic data are

available. Nevertheless, for a given set of conditions, the gain in accuracy obtained by further decreasing the grid interval diminishes greatly beyond a certain value of the interval, and a further increase of the number of squares is not warranted. In general, the optimum grid interval is determined by the size of the area, the size of the individual drainage basins considered, the details of the available data, the computer characteristics, the purpose and budget of the study, etc.. For usual problems, grid intervals between about 1 and 10 km can be considered. In Canada grid sizes of 10 km, 5 km and 4 km have been used in different studies. For the mountainous areas of British Columbia, physiographic data have been abstracted using a 2 km grid interval.

The steps involved in estimating the distribution of mean annual runoff using the square-grid method are as follows:

(a) Using data at selected meteorological stations a regression equation is established between mean annual temperature at the stations and corresponding square-grid physiographic characteristics. This equation is used with the data file of square-grid physiographic characteristics to estimate the mean annual temperature in the remaining squares. A similar analysis is used to make a preliminary estimate of mean annual precipitation in each square.

(b) Preliminary evaporation is estimated for each square using temperature, preliminary precipitation and Turc's equation (29)

$$E = P/(0.9 + P^2/L^2)^{\frac{1}{2}} \qquad (13-12)$$

where E is evaporation in mm, P is precipitation in mm, T is temperature in degrees Centigrade, and L is defined as

$$L = 300 + 25T + 0.5T^2 \qquad (13-13)$$

Using this equation, $E = P$ for $P^2/L^2$ less than 0.1 $mm^2$.

(c) The preliminary estimates of precipitation and evaporation as described above are used to calculate a preliminary value of mean annual runoff for each square of the watershed having flow data as:

$$\text{Runoff} = \text{Precipitation} - \text{Evaporation} \qquad (13-14)$$

(d) The square-grid runoff established under step (c) is used to compute preliminary average runoff for each basin having flow records and ratios (K) between the recorded and computed average flows are calculated.

(e) A new precipitation value is computed for each square of the basins having flow data, using K as a correction factor:

Precipitation (corrected) = K x Runoff + Evaporation

$$(13-15)$$

The entire error is attributed to precipitation.

(f) Using the corrected values of the precipitation in each square and the precipitation data at rain gauging stations, a new regression equation is established between precipitation and physiographic factors. Data at rain gauging stations is weighted 10 times larger than the precipitation estimates in ungauged squares.

(g) The procedure is then repeated as often as is required to obtain K values as close to 1 as is considered reasonable.

(h) Once square-grid runoff, precipitation and evaporation values are finalized in gauged basin areas, the runoff and precipitation distributions in ungauged areas are estimated using regression equations between the final precipitation and runoff in gauged areas and physiographic data.

The end result of the square-grid technique is a data file of mean annual temperature, precipitation and runoff for each grid square. Thus estimates of these parameters have been transferred from gauged to ungauged basins using physiographic characteristics as the transfer media. Solomon et al. (26) originally applied the square-grid method to Newfoundland but it has since been extended to cover all of Canada except for northern Ontario and the Arctic Archipelago (27).

Pentland and Cuthbert (25) have described a method by which the square-grid technique was extended for the generation of synthetic streamflow traces. The operational hydrology model proposed by Young and Pisano (35) was used because of the small number of statistical parameters required. This model uses only estimates of a single variance-covariance matrix and a single lag 1 covariance matrix whereas most other models require monthly matrices.

Young and Pisano's model (35) is set up as follows:

(a) The available streamflow data are logarithmically transformed.

(b) The transformed data are standardized on a monthly basis to eliminate the annual periodicity.

(c) The generating equation

$$X_{i+1} = AX_i + Be \qquad (13-16)$$

is then used.

For m stations, $X_{i+1}$ and $X_i$ are standardized flows in successive time periods (m x 1 matrices). A and B are m x m matrices to be defined and e is an m x 1 matrix of random components.

(d) The matrix $M_0$ is the variance-covariance matrix, and $M_1$ is the covariance matrix with a lag of 1 time period.

(e) The matrix A can be defined from the equation

$$M_1 = AM_0 \tag{13-17}$$

(f) The matrix B can be calculated by solving the equation:

$$B B^T = M_0 - M_1 M_0^{-1} M_1^T \tag{13-18}$$

(g) After having generated standardized variables, the data are destandardized, and the inverse logarithmic transform applied.

If one or more of the stations has no recorded data, monthly means and standard deviations can be estimated by regression analysis with basin physiographic characteristics derived from the square-grid data.

$$\mu = K_1 A^{a_1} B^{b_1} C^{c_1} \dots Z^{z_1} \tag{13-19}$$

$$\sigma = K_2 A^{a_2} B^{b_2} C^{c_2} \dots Z^{z_2} \tag{13-20}$$

where $\mu$ and $\sigma$ are the monthly mean streamflow and standard deviation, A, B,..., Z are physiographic characteristics and $K_1$, $K_2$, $a_1$, $b_1$,...,$z_1$ $a_2$,$b_2$,..., $z_2$ are regression constants. In order to estimate covariances for the streams with no recorded data, a multiple regression equation is established between covariances (representing cross correlations between stations) and the differences in physiographic characteristics for all pairs of gauged streams in the region.

In the covariance matrix with a lag of one time period, elements of the diagonal (representing serial correlation for each stream) for gauged stations can be calculated directly, and can be estimated for ungauged stations by regression analysis. The remainder of this matrix can be estimated as the product of its diagonal and the variance covariance matrix calculated earlier.

Pentland and Cuthbert (25) tested the generation procedure on five streams in the northeast of the Province of New Brunswick. Comparisons of simulated and recorded monthly means, standard deviations, serial and cross correlations and firm flows showed good agreements.

A procedure for using the square-grid technique to estimate events at required return periods on ungauged streams has been described by Kite (16) for the Mackenzie River area, Northwest Territories. Basically, equations similar to 13-19 and 13-20 were developed relating the mean and standard deviation of the annual maximum instantaneous flows to the square-grid physiographic data for those streams which are gauged. The relationships developed were then extended to ungauged streams and, assuming a lognormal probability distribution of annual extremes, event magnitudes at any required return period were calculated. Any other probability distribution thought suitable could have been used in place of the lognormal.

Kouwen (18) has described an advanced model, based upon square-grid techniques, used for the simulation of complete watersheds. The model allows hydrographs based on weather forecasts to be incorporated, for prediction of flood peaks. The basic input to the model consists of topographic data such as streambed elevations and landslope, drainage channel directions, watershed boundary coordinates, precipitation records, streamflow records and a soil permeability index.

The program is also set up in such a way that precipitation data from radar, and snow pack and soil moisture measurements from satellite can be included. The principal characteristic of the simulation is that runoff passes through successive 1 km x 1 km square elements from higher to lower elevations. For each element there exists relationships between channel capacity and drainage area, surface storage and channel inflow, surface storage and infiltration, subsurface storage and channel inflow, and channel storage and channel discharge. Recent developments have incorporated sediment transport into the list of parameters modelled by the square-grid procedures.

## Use of Standard Frequency Distributions

In order to remove the necessity for personal judgement in drawing the preliminary frequency curves and to provide a means of computing confidence limits on the regional frequency curve in the index-flood method, Collier (6) produced an alternative regional analysis procedure. This procedure is recommended for regions where the Gumbel (type I extremal) distribution produces reasonably reliable individual flood frequency curves. The procedure is described briefly below:

(a) All stations in the region with 10 or more years of record (either natural flow or with minor regulation only) are selected. Stations with less than 10 years of record would usually be discarded, and most of the selected stations should have at least 15 years of record.

(b)   A frequency curve covering the range up to the 100-year flood is constructed by the Gumbel (type I extremal) method for each of the individual stations.  For the purpose of this discussion these will be called the preliminary curves.  Confidence limits are constructed on each of the preliminary curves using the degree of confidence required in the regional curve.

(c)   A homogeneity test is carried out exactly as in the index-flood method.  For the purpose of this discussion it will be considered that each station passes the test and the region has therefore been demonstrated to be homogeneous.

(d)   The preliminary curves are considered to be a sample composed of a number of different estimates of the same regional curve.  The estimates are averaged by the method described in the next paragraph to obtain the required estimate of the regional curve.

(e)   The averaging procedure can be carried out only if the preliminary curves are reduced to dimensionless terms (to remove the effect of the different sizes of the drainage basins concerned).  This is accomplished by computing a set of flood ratios (ratio of flood to mean annual flood) for each of the stations over a range of arbitrarily selected recurrence intervals.  The data for computing the ratios are read from the preliminary curves.

(f)   For each of the selected recurrence intervals, the mean of the ratios from all the stations is computed.  The resulting means are the flood ratios for the regional curve. These are plotted on Gumbel paper (with arithmetic ordinate scale) and the best-fit straight line drawn through them.  The resulting line is taken as the required regional frequency curve.

(g)   To compute confidence limits for the regional curve, a recurrence interval (say 50 years) is selected arbitrarily and the width of the confidence band at this interval is read off each of the preliminary curves.  The width is taken as the vertical distance between the preliminary curve and the upper (or lower) band and it is expressed in cfs.  The resulting figures are divided by the appropriate mean annual floods to produce a set of ratios, which are defined as the "errors" in the individual curves.  The errors are combined by computing the square root by the number of stations.  The resulting ratio is taken as the "error" in the regional curve or, in other words it is the width of the confidence bands for the regional curve at the selected recurrency interval (50 years in this case).  The procedure is repeated at another recurrence interval (say 5 years). The "errors" from the two sets of computations are plotted on the regional curve by laying them off at the appropriate recurrence intervals in a vertical direction either side of the main curve.  The resulting points are joined by straight lines to

produce the required confidence bands for the regional curve.
Note that these bands represent the same degree of confidence
as was used in computing the confidence limits for the preliminary curves.

(h)  Having obtained the dimensionless regional frequency
curve, complete with confidence limits, it is necessary to
introduce a relationship between mean annual flood and basin
characteristics so that estimates may be made for ungauged
drainage basins.  In a study of the Province of Nova Scotia
using Collier's procedure, Coulson (9) ran stepwise linear
regressions of mean annual flood versus drainage area, size and
position of lakes and swamps, main channel slope, average basin
elevation, mean barrier elevation and mean annual precipitation.
He ended up with an equation of the form

$$\bar{Q} = f(A_u + \lambda^k A_c) \qquad (13-21)$$

where $A_u$ and $A_c$ are the drainage areas uncontrolled and controlled by lakes and swamps respectively,

$$\lambda = \frac{A_c - A_L}{A_c} \qquad (13-22)$$

where $A_L$ is the total surface area of major lakes and swamps,
and k is a constant optimized by minimizing the standard error
of $\bar{Q}$, the mean annual flood.

Collier and Nix (7) used a similar approach in a flood
frequency study of the New Brunswick-Gaspé region.

The principal advantage of Collier's alternative procedure
for the regional frequency curve is that since no personal judgement is involved, the entire procedure can be programmed for
computer.  Although Collier (6) described the alternative procedure utilizing a type I extremal distribution there is no
reason why any other type of distribution thought suitable could
not be used.

Cruff and Rantz (11) have described the adaptations made
by United States agencies to use the lognormal, extremal type I
(Gumbel) and Pearson type III distributions in regional analysis.
Basically the procedures used consist of the following steps:
(a)  The mean and standard deviation of the peak discharge
data at each gauging station are computed for the available
periods of record.  In the case of the lognormal distribution
the means and standard deviations of the logarithms of the peak
discharge data are computed in the procedure described by Cruff
and Rantz (11) but this is not strictly necessary (see Chapter
6).

(b) The computed statistical parameters are then adjusted to a standard base period by computing linear correlations between concurrent peak discharges for a long-term station and the short-term stations. Then

$$\sigma_{1b} = \sigma_{1a} + (\sigma_{2b} - \sigma_{2a}) \, R^2 \, \sigma_{1a}/\sigma_{2a} \qquad (13\text{-}23)$$

and

$$\mu_{1b} = \mu_{1a} + (\mu_{2b} - \mu_{2a}) \, R^2 \, \sigma_{1b}/\sigma_{2b} \qquad (13\text{-}24)$$

where $\mu$ and $\sigma$ are the means and standard deviations respectively; subscript 1 refers to the short-term station, 2 to the long-term station, subscript a refers to the short-term period and b refers to the base period; and R is the coefficient of correlation between 1 and 2.

If these equations are derived by standardizing the means and standard deviations of the two periods of record at each station then the $R^2$ is not statistically correct and should be omitted.

In the case of the Pearson type III distribution, for which skews are needed, Equations 13-23 and 13-24 are used to generate events at the short-term stations to complete the record for all years of the base period by using the relationship:

$$x_1 = \mu_{1b} + R (x_2 - \mu_{2b}) \, \sigma_{2b}/\sigma_{1b} \qquad (13\text{-}25)$$

where $x_1$ is the peak discharge to be estimated at a short-term station, $x_2$ is the peak discharge measured at the long-term station and the other parameters are as previously defined. When the full number of annual events is available for each of the short-term stations the coefficients of skew are computed for each station.

(c) The parameters of the distribution (mean, standard deviation and, for Pearson type III, skewness) are then related to the basin and climatologic characteristics by multiple linear regression equations as explained in section 13-3.

(d) For any site the mean, standard deviation (and, if necessary, coefficient of skew) can then be determined from the derived regression equations, and the event magnitude at return period T can be obtained from

$$x_T = \mu + K \sigma \qquad (13\text{-}26)$$

where K is the frequency factor. As explained in Chapter 3 the frequency factor can be developed in terms of T for each distribution and tables are commonly available.

## Regional Record Maxima

Suppose that there exists a set of n independent identically distributed concurrent series each containing k extreme events $x_{ij}$ i=1,n; j=1,k. If the maximum event of each of the n series is abstracted and ordered from highest to lowest in a new series, $y_i$, i=1,n then the probability, $P(y > y_i)$, that another event y exceeds the ith event in the series of maxima $y_i$ is given by Conover and Benson (8) as:

$$P(y > y_i) = \sum_{m=0}^{i-1} n!/[(n-m)! \ k \prod_{j=0}^{m} (n+1/k-j)] \quad (13\text{-}27)$$

As an example, Carrigan (5) has shown that for the three series of four events, $x_{ij}$,

$$
\begin{array}{ccc}
3 & 69 & 3 \\
38 & 24 & 48 \\
17 & 61 & 60 \\
32 & 30 & 83
\end{array}
$$

the series of maximum events, $y_i$, (83, 69, 38) and the probability, $P(y > 69)$, that another event y exceeds the second largest event in the series of maxima is

$$P(y > 69) = \sum_{m=0}^{1} 3!/[(3-m)! \ 4 \prod_{j=0}^{m} (3+1/4-j)] \quad (13\text{-}28)$$

$$P(y > 69) = 0.180 \quad (13\text{-}29)$$

An analogy which can be made here is with annual maximum streamflows recorded at a set of gauging stations within a hydrologically homogeneous region. Within this homogeneous region it is a reasonable assumption that the same probability distribution is applicable to the records of maximum events on each stream. Just as in the index-flood method the different records could be reduced to an identical distribution by normalizing with the computed mean annual floods. The procedure outlined above then takes the n independent samples of k events and forms a sample size nk; thus probabilities can be computed associated with return periods of nk years instead of only n years. The catch is that streamflow records are not independent but are quite strongly cross-correlated. This reduces the maximum return period available from nk to f(R)nk where f(R) is some function of the correlation coefficient, R, between streamflow records. The expression f(R) varies between 1 when R=0 (independent records) to 1/n when R=1 (identical records).

The probability of another random event, y, exceeding one of the ordered record maxima, $y_i$, cannot be determined analytically when the records are not independent. By assuming that the exceedence probability is independent of the identical distribution of the records, Carrigan (5) has derived the probability by data generation. Using the normal distribution for simplicity the generation model used by Carrigan is

$$X = B\varepsilon \qquad (13\text{-}30)$$

where X is an n x k matrix of generated events, B is an n x n principal component matrix and $\varepsilon$ is an n x k matrix of independent normally distributed random numbers with zero mean and unit variance. The principal component matrix B is derived from the correlation matrix R of the n records of k hydrologic events as follows:

$$B = E\lambda \qquad (13\text{-}31)$$

where E is an n x n matrix of eigenvectors obtained from R, and $\lambda$ is an n x n matrix for which the diagonal elements are the square roots of the eigenvalues of R and the off-diagonal elements are zero.

After generation of X, the n maxima are selected and put in order of magnitude and the exceedence probabilities computed using a digital approximation to the normal distribution.

In summary, the method extracts from the records of a series of gauging stations a matrix of the inter-station correlation. This correlation is then incorporated into a large number of generated events from which extreme probabilities can be measured. In effect the method converts the spatially-distributed information into time-distributed information on extreme events.

## Bayesian Analysis

A further method of combining a single station analysis with regional analysis information is the use of Bayesian analysis. Bayes theorem states that the posterior probability distribution function (pdf) is proportional to the product of the pdf containing the prior information (e.g. from regional analysis) and the likelihood function of the sample pdf. In order to provide a tractable posterior pdf it is necessary to choose a prior pdf and a likelihood function which will conveniently combine. Vicens et al. (30) used a normal distribution for the mean annual flows of a New Hampshire river and a normal inverted 2-parameter gamma distribution for the prior pdf. For extreme events a better distribution to use might be the lognormal. Given a sample of n annual maximum events $x_i$, i=1, n then the sufficient statistics of the normal distribution of $y = \log_{10} x$ are the mean $\bar{y}$, the variance $s^2$, the sample size n, and the number of degrees of freedom, $v = n-1$. As an example, Table 13-2, lists the sufficient statistics for the Tobique River at Plaster Rock, New Brunswick, for the period 1955-1966. Wood and Rodriguez-Iturbe (32) have shown other examples.

The same statistics can then be defined for the prior distribution as:

$$\bar{y} = E[\mu] \qquad (13\text{-}32)$$

$$S'^2 = [(v'-2)/v'] \, E[\sigma^2] \qquad (13\text{-}33)$$

$$n' = E[\sigma^2]/V[\mu] \qquad (13\text{-}34)$$

$$v' = 2E^2[\sigma^2]/V[\sigma^2] \qquad (13\text{-}35)$$

where $E[\mu]$, $V[\mu]$, $E[\sigma^2]$, $V[\sigma^2]$ are the expected value and variance of the mean and variance of the prior distribution. These statistics may be obtained by regressions against physiographic or meteorologic variables. For example, using data from Collier and Nix (7), a regional analysis was carried out using 19 basins in Gaspé, New Brunswick and northern Maine with periods of record varying from 12-40 years. Independent variables used included drainage area, area of lakes and mean slope of basin. The resulting statistics are shown in Table 13-2. Note that $n'$ and $v'$ from the prior pdf can be interpreted as the number of years of data-equivalent for the mean and variance contained in the prior pdf (Vicens et al. (30)).

The regional information in the prior distribution and the sample information in the single station record are then combined in Bayes theorem to yield a posterior pdf. The statistics of this posterior pdf are given by

$$\bar{y}'' = (n'\bar{y}' + n\bar{y})/(n' + n) \qquad (13\text{-}36)$$

$$S''^2 = (v's'^2 + n'\bar{y}'^2 + vs^2 + n\bar{y}^2 - n''\bar{y}''^2)/v'' \qquad (13\text{-}37)$$

$$n'' = n' + n \qquad (13\text{-}38)$$

$$v'' = v' + v + 1 \qquad (13\text{-}39)$$

Table 13-2

Bayesian Analysis, Tobique River, New Brunswick

| Statistic | Sample Value | Prior Value | Posterior Value |
|-----------|-------------|-------------|-----------------|
| n, years | 12 | 19 | 31 |
| v, years | 11 | 21 | 33 |
| $\bar{y}$ | 4.0674 | 4.1500 | 4.1180 |
| $s^2$ | 0.0579 | 0.0312 | 0.0417 |

The statistics for the posterior pdf of the Tobique river are also given in Table 13-2. Figure 13-2 shows the frequency curves from the sample and from the posterior distribution.

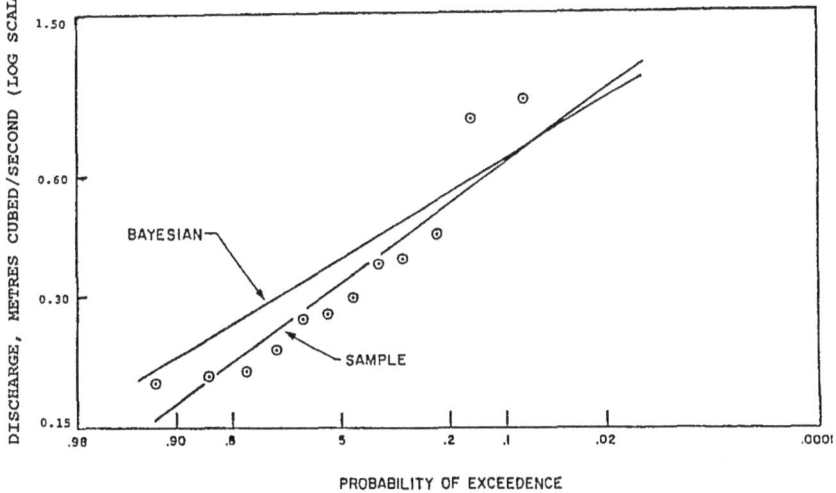

Figure 13-2. Comparison of sample and Bayesian frequency curves, Tobique River at Plaster Rock.

The prior information does not necessarily have to come from a regional analysis; Vicens et al. (30) also give an example of prior information obtained from a simple watershed model. Given the four sufficient parameters of the total annual precipitation distribution (assumed to be independent normal) and a simple model such as:

$$q_i = (1-b) \, x_i \qquad (13\text{-}40)$$

where $q_i$ is the runoff in year i, b is the "loss" and $x_i$ is the annual precipitation in the same units as $q_i$, then the sufficient parameters of the runoff distribution are

$$E[\mu'] = (1 - E[b]) \, E[\mu_x] \qquad (13\text{-}41)$$

$$V[\mu'] = E^2[\mu_x] \, V[b] + (1 - E[b]^2) \, V[\mu_x] \qquad (13\text{-}42)$$

$$E[\sigma'^2] = \{(1 - E[b])^2 + V[b]\} \, E[\sigma_x^2] \qquad (13\text{-}43)$$

$$V[\sigma'^2] = 4(1 - E[b])^2 \, E^2[\sigma_x^2] \, V[b] + (1 - E[b])^4 \, V[\sigma_x^2] \qquad (13\text{-}44)$$

assuming that all covariances are zero. The hydrologist must then estimate E[b] and V[b] before the Bayesian analysis can be used to combine the information from the model with the information from the sample to give the posterior information.

219

Vicens et al. have compared the posterior variance of the mean, $V[\mu'']$, resulting from combining in Bayesian analysis (a) prior information from regional analysis with sample data of different length and (b) prior information from a model (subjective experience) with sample data of different length. The variance of the mean was clearly reduced by including the prior information from either source but the effect died out at record lengths greater than about 40 events.

One problem with the use of Bayesian analysis in this manner is that no account is taken of the uncertainty in the prior information. For example, some of the regression equations used in the regional analysis for the Tobique River had coefficients of multiple correlation as low as 0.3 and so the prior pdf statistics are not very reliable. Perhaps a weighting factor based on the coefficients of determination of the regressions could be introduced into Equations 13-36 to 13-39.

## Single Station and Regional Information Content

*General* - A time series may or may not consist of observed outcomes which are independent of one another. Streamflow is a hydrologic variable whose observations, equally spaced in time, are not necessarily randomly distributed. Because of natural storage such as groundwater, lakes, swamps and annual persistance as well as manmade factors such as reservoirs, the stochastic precipitation variable becomes modulated and serially correlated. This means that each unit of streamflow data does not contain totally new and independent information; units tend to repeat some of the previously obtained information. A data set of N units may therefore only contain a lesser number, $N_e$, of effective data units.

The early stage of any regional analysis procedure calls for the examination of available data. It is usual that some gauging station records will be longer than others within the required region and often, after selection of a base period of record, it will be necessary to fill in gaps in some records and extend other records to the full base period. This provision of missing data can have two general purposes: (a) To provide estimates of event magnitudes in order to better obtain plotting positions of recorded events. This procedure is used in the index-flood regional analysis technique where the estimated event magnitudes are not themselves used at all, they are there merely to improve estimates for the recorded events. (b) To improve estimates of the parameters of a theoretical distribution of recorded events such as the mean and standard deviation.

Data are not only serially correlated, but because streamflows in rivers within a region are affected by the same conditions of precipitation and radiation, simultaneous observations of streamflow in different rivers will not be independent

observations but will contain an overlap of information. Thus
a region of n gauging stations may reduce to a much smaller
number, $n_e$, of effective stations or equivalent independent
stations.

*Single Station Information Content* - The purpose of maintaining records of precipitation, stage, streamflow, etc., is to
extract from the recorded observations information on the parameters of the underlying distribution. Matalas and Langbein (21)
defined the amount of information given by a statistical estimate, I, as the reciprocal of the variance of the estimate. Considering the mean, $\mu$, of a random series of N events, $x_i$, i=1...n,

$$\mu = \sum_{i=1}^{N} x_i/N \tag{13-45}$$

an estimate of the variance of $\mu$ is given by

$$\text{var } \mu = \sigma^2/N \tag{13-46}$$

where $\sigma^2$ is an estimate of the population variance of the random
time series. Defining the random series as the standard, the
relative information content about the mean of any other time
series with variance of the estimated mean, $\sigma_\mu^2$, is

$$I_\mu = (\sigma^2/N)/\sigma_\mu^2 \tag{13-47}$$

referred to the random series. Since variances are always positive $I_\mu$ can vary from zero to plus infinity. If $I_\mu$ is less than
unity the time series being tested conveys less information about
the mean than a random series of the same length. If $I_\mu$ is
greater than unity then the time series contains more information about the mean than an equal length random series.

Many time series exhibiting persistance, such as streamflow,
can be described by a simple first order linear Markov model (34)
such as:

$$x_{i+1} = R_1 x_i + \varepsilon_{i+1} \tag{13-48}$$

where $x_i$ and $x_{i+1}$ are the variable values at time i and i+1
respectively, $\varepsilon_{i+1}$ is a random component independent of x, and
$R_1$ is the first order serial or autocorrelation coefficient
where, in general, the kth order autocorrelation coefficient is
defined (33) as:

$$R_k = \frac{\sum_{i=1}^{n-k} (x_i-\mu)(x_{i+k}-\mu)}{(N-k)\sigma^2} \tag{13-49}$$

where N is the number of observations and $\mu$ and $\sigma$ are the sample estimates of the mean and standard deviation of the time series.

For a first order linear Markov model the variance of the mean, $\sigma_\mu^2$, is given (21) by

$$\sigma_\mu^2 = \frac{\sigma^2}{N}\left[\frac{1+R_1}{1-R_1} - \frac{2}{N}\frac{R_1\,(1-R_1^N)}{(1-R_1)^2}\right] \qquad (13\text{-}50)$$

From Equation 13-47 the relative information content on the mean is

$$I_\mu = \left[\frac{1+R_1}{1-R_1} - \frac{2}{N}\frac{R_1\,(1-R_1^N)}{(1-R_1)^2}\right]^{-1} \qquad (13\text{-}51)$$

which is less than unity for $R_1 > 0$.

If a number of independent observations of a random time series, $N_e$, contain the same information content about the mean as the number N observations of the Markov model then

$$I_\mu = N_e/N \qquad (13\text{-}52)$$

and

$$N_e = N\left[\frac{1+R_1}{1-R_1} - \frac{2}{N}\frac{R_1\,(1-R_1^N)}{(1-R_1^2)}\right]^{-1} \qquad (13\text{-}53)$$

*Two-Station Transfer of Information* - Inter-station transfer of information is commonly used in regional analysis to fill in missing data or to extend short time series to a longer common base period. The method used in the index-flood method of regional analysis (12) is as follows:

A graph is drawn of the flow at one station versus the flow at the other station for each year of the common period of record. A straight line is fitted by eye through the coordinate points and this line is then used to extend the shorter period of record. This process is a simplification of the least squares fitting of a linear regression equation of the type

$$y = mx + c \qquad (13\text{-}54)$$

where x and y are the annual maximum flows at the two stations, m is the slope and c is the intercept of the straight line. The missing event magnitudes are then estimated from this regression line and the total events, recorded and estimated are placed in order of decreasing magnitude. Plotting positions (see Chapter 2) are assigned to the recorded events on the basis of the total number of events and the estimated events are then discarded and used no further.

Many investigators (19, 24) have found that the logarithms of hydrologic events are better correlated than the recorded events and have used equations such as

$$\ln y = m_1 \ln x + c_1 \tag{13-55}$$

where x and y may be the recorded events or the deviations of the recorded events from some mean value.

The question arises as to whether the estimated events actually increase the information content of the shorter time-series, i.e. does the extended data provide better estimates of the population distribution parameters than the originally recorded series? Langbein (19) has shown that to improve the significance of the mean of a time series the effective period of record, $N_e$, of a combined recorded and estimated record must be greater than $N_1$, the number of years of recorded data, where

$$N_e = \frac{N_1 + N_2}{1 + \dfrac{N_2}{N_1 - 2}(1 - R^2)} \tag{13-56}$$

$N_2$ is the number of years of estimated data and $R^2$ is the coefficient of determination of the simple linear regression used to provide the estimated data.

If two random normally distributed time series x, of length $N_1 + N_2$, and y, of length $N_1$, are linearly related with a simple linear correlation coefficient R, and the time series x is used to extend time series y by $N_2$ data points, then the variance, $\sigma_\mu^2$, of the weighted mean of series, y, $\mu_y$, where

$$\mu_y = \frac{N_1 \mu_1 + N_2 \mu_2}{N_1 + N_2} \tag{13-57}$$

is given (21) by

$$\sigma_\mu^2 = \frac{\sigma_y^2}{N_1} \left[ 1 - \frac{N_2}{N_1 + N_2} \left\{ R^2 - \frac{(1 - R^2)}{(N_1 - 3)} \right\} \right] \tag{13-58}$$

where $\mu_1$ and $\mu_2$ are the sample means of time series y based on $N_1$ observations and $N_2$ regression estimates respectively and $\sigma_y^2$ is the variance of time series y based on the $N_1$ observations.

Referring back to Equation 13-47 the information content on the mean of the extended time series y is seen to be:

$$I_\mu = \left[ 1 - \frac{N_2}{N_1 + N_2} \left\{ R^2 - \frac{(1 - R^2)}{(N_1 - 3)} \right\} \right]^{-1} \qquad (13\text{-}59)$$

and the effective number of observations, $N_e$, is given by:

$$N_e = (N_1 + N_2) \left[ 1 - \frac{N_2}{N_1 + N_2} \left\{ R^2 - \frac{(1 - R^2)}{(N_1 - 3)} \right\} \right]^{-1} \qquad (13\text{-}60)$$

For the cross-correlation to provide additional information on the mean, $I_\mu > 1$ and from Equation 13-59

$$\frac{- N_2 R^2}{N_1 + N_2} + \frac{N_2}{(N_1 + N_2)(N_1 - 3)} - \frac{N_2 R^2}{(N_1 - 3)} < 0 \qquad (13\text{-}61)$$

from which

$$R^2 > \frac{1}{N_1 - 2} \qquad (13\text{-}62)$$

Similarly, Fiering (14) concluded that correlation should not be used to augment time series for estimation of the variance unless the computed information content on the variance, $I_{\sigma^2}$, is greater than unity, where

$$I_{\sigma^2} = \left\{ 1 + \frac{N_2}{2(N_1 + N_2 - 1)^2} \left[ 2A(N_1 - 1) + (N_2 + 2)(N_1 - 1)B \right. \right.$$

$$\left. \left. + (N_1 + N_2 - 1)(N_1 - 1)C + (N_1 + 1)(2N_1 + N_2 - 2) \right] \right\}^{-1} \qquad (13\text{-}63)$$

in which

$$A = (N_1 - 1)R^4 + (N_1 + 4)R^2 (1 - R^2) + \frac{N_1 + 1}{N_1 - 3} (1 - R^2)^2 \qquad (13\text{-}64)$$

$$B = R^4 + \frac{6R^2(1 - R^2)}{N_1 - 3} + \frac{3(1 - R^2)}{(N_1 - 3)(N_1 - 5)} \qquad (13\text{-}65)$$

and

$$C = \frac{2(N_1 - 4)(1 - R^2)}{N_1 - 3} \qquad (13\text{-}66)$$

Fiering (14) concluded that, in general, the estimate of the population variance will be improved if $R > 0.85$.

If the two time series x and y are not random but are serially correlated then for an equal number of observations, N, the effective number of data points, $N_e$, has been given by Yevjevich (34) as:

$$N_e = N/(1 + 2R_1R_1' + 2R_2R_2' + \ldots + 2R_{n-1}R_{n-1}') \quad (13\text{-}67)$$

where $R_k$ and $R_k'$ are the kth order autocorrelation coefficients of the x and y time series respectively. This equation is only useful if all periodicities have been removed from both time series. If the equal-length time series x and y can be described by first order linear Markov models then the effective number of observations is given by:

$$N_e = N \left[ \frac{1 - R_1R_1'}{1 + R_1R_1'} \right] \quad (13\text{-}68)$$

If x and y are two first order linear Markov models of length $N_1 + N_2$ and $N_1$ respectively and x is correlated with y to provide $N_2$ regression estimates for y then the relative information content for the mean of the augmented time series varies with $N_1$, $N_2$, $R_1$, $R_1'$ and R in a complex fashion. Assuming that $R_1 = R_1'$, Matalas and Langbein (21) have tabulated values of $I_\mu$ for different values of these variables.

Equations such as 13-54 and 13-55 yield event magnitudes on the regression line and, although this does not affect the estimate of the mean, it does induce a bias in the estimate of the variance. To overcome this bias it is necessary to introduce into the generating equation a random variable with mean zero and variance $(1 - R^2)\sigma^2$ where $\sigma^2$ is the variance of the recorded series. The true regression equation thus becomes:

$$y = \mu_y + m(x - \mu_x) + (1 - R^2)^{\frac{1}{2}} \sigma_y \varepsilon \quad (13\text{-}69)$$

where $\varepsilon$ is a normally distributed $(0,1)$ random number. The term $(1 - R^2)^{\frac{1}{2}} \sigma_y \varepsilon$ is the random component of the generated time series and in communications theory is referred to as noise.

Under the assumptions that (a) events are independently distributed in time, (b) the concurrent events for the two sequences have a joint normal distribution, (c) the relation between the concurrent events is defined by a linear regression, and (d) no changes occur in the hydrologic regimes with which the sequences are associated, Matalas and Jacobs (22) have evaluated the reliability of estimates of population parameters under conditions of noise and no-noise.

Matalas and Jacobs (22) recommended the use of the following equations to compute the mean, $\mu_y$, and variance, $\sigma_y^2$, of an extended time series:

$$\mu_y = \bar{y}_1 + \frac{N_2}{N_1 + N_2} m(\bar{x}_2 - \bar{x}_1) \qquad (13\text{-}70)$$

and

$$\sigma_y^2 = \frac{1}{N_1 + N_2 + 1} \left\{ (N_1 - 1) S_{y_1}^2 + (N_2 - 1) m^2 S_{x_2}^2 \right.$$

$$\left. + (N_2 - 1) \alpha^2 (1 - R^2) S_{y_1}^2 + \frac{N_1 N_2}{(N_1 + N_2)} m^2 (\bar{x}_2 - \bar{x}_1)^2 \right\} \qquad (13\text{-}71)$$

where $\bar{y}_1$ and $S_{y_1}^2$ are the mean and variance of the recorded events in the augmented series, $\bar{x}_1$ is the mean of the concurrent augmenting series, $\bar{x}_2$ and $S_{x_2}^2$ are the mean and variance of the total number of events in the augmenting series, and

$$\alpha^2 = \frac{N_2 (N_1 - 4)(N_1 - 1)}{(N_2 - 1)(N_1 - 3)(N_1 - 2)} \qquad (13\text{-}72)$$

Equations 13-70 and 13-71 should only be used if the inter-station correlation coefficient, R, is greater than the critical values given in Tables 13-3, 13-4 and 13-5.

Table 13-3

Critical Minimum Values of R for Estimation of the Mean[1]

| N | 10 | 15 | 20 | 25 | 30 |
|---|----|----|----|----|----|
| R | 0.35 | 0.28 | 0.24 | 0.21 | 0.19 |

[1]From Matalas and Jacobs (22)

Regional Transfer of Information - As well as considering the transfer of information from a long-term station to an adjacent short-term station, hydrologists are often interested in studying how an hydrologic variable, such as river discharge, varies with the physical parameters describing the drainage area. The variation may be studied by assembling the data for many gauging stations and using regression analysis to define a relationship between the hydrologic variable and the physiographic characteristics.

In a given region, however, rivers may rise in response to a rain storm that affects all the rivers in the region and, at another time, may be low due to a common lack of rainfall. Thus the flows of different streams are affected by common causes and are therefore not independent but cross-correlated.

Table 13-4

Critical Minimum Values of R for Estimation of the Variance Including Noise Component[1]

| $N_2$ | $N_1$ | | | | |
|---|---|---|---|---|---|
| | 10 | 15 | 20 | 25 | 30 |
| 10 | 0.65 | 0.54 | 0.52 | 0.42 | 0.38 |
| 15 | 0.65 | 0.54 | 0.51 | 0.42 | 0.39 |
| 20 | 0.65 | 0.54 | 0.51 | 0.42 | 0.39 |
| 25 | 0.65 | 0.54 | 0.50 | 0.42 | 0.39 |
| 30 | 0.65 | 0.54 | 0.50 | 0.42 | 0.39 |

Table 13-5

Critical Minimum Values of R for Estimation of the Variance Excluding Noise Component[1]

| $N_2$ | $N_1$ | | | | |
|---|---|---|---|---|---|
| | 10 | 15 | 20 | 25 | 30 |
| 10 | 0.73 | 0.63 | 0.70 | 0.76 | 0.76 |
| 15 | 0.75 | 0.77 | 0.79 | 0.80 | 0.80 |
| 20 | 0.76 | 0.79 | 0.81 | 0.81 | 0.82 |
| 25 | 0.78 | 0.80 | 0.84 | 0.83 | 0.81 |
| 30 | 0.77 | 0.80 | 0.82 | 0.83 | 0.84 |

[1] From Matalas and Jacobs (22)

where $R_{ij}$ is the cross-correlation coefficient between stations i and j.

If a number n of hydrometric gauging station records within a hydrologically homogeneous region are intercorrelated, then the effective number of stations or equivalent number of independent gauging stations, $n_e$ can be derived as follows (34): If the mean and variance of the observations at the jth gauging station are given by:

$$\mu_j = \frac{1}{N} \sum_{i=1}^{N} x_{ij} \qquad (13\text{-}72)$$

$$\sigma_j^2 = \frac{1}{N-1} \sum_{i=1}^{N} (x_{ij} - \mu_j)^2 \qquad (13\text{-}73)$$

where N is the number of observations at each station, then the estimates of the regional mean, $\mu$, and the variance of the regional mean, $\sigma_\mu^2$, are (33)

$$\mu = \frac{1}{N} \sum_{j=1}^{n} \mu_j \qquad (13\text{-}74)$$

$$\sigma_\mu^2 = \frac{1}{n} \sum_{j=1}^{n} \sigma_j^2 + \frac{2}{n^2} \sum_{j=1}^{n-1} \sum_{i=j+1}^{n} R_{ij}\sigma_i\sigma_j \qquad (13\text{-}75)$$

Equation 13-75 can be simplified by defining the regional mean cross-correlation coefficient, $\bar{R}$, as:

$$\bar{R} = \frac{2 \sum_{j=1}^{n-1} \sum_{i=j+1}^{n} R_{ij}}{n(n-1)} \qquad (13\text{-}76)$$

If the time-series are standardized to a common mean of zero and variance $\sigma^2$ then Equation 13-75, incorporating Equation 13-76 becomes:

$$\sigma_m^2 = \frac{\sigma^2}{n} [1 + \bar{R}(n - 1)] \qquad (13\text{-}77)$$

227

The relative information content on the regional mean is therefore given by:

$$I_\mu = [1 + \bar{R}(n - 1)]^{-1} \qquad (13\text{-}78)$$

and the effective number of stations, $n_e$, or equivalent number of independent stations is:

$$n_e = n/[1 + \bar{R}(n - 1)] \qquad (13\text{-}79)$$

In a study of regional flood frequency relations for 164 basins in New England, Benson (4) found that using Equation 13-79 the effective number of gauging stations or equivalent number of independent stations was 3.8. In a later study, Matalas and Benson (20) point out that, assuming the same value of $\bar{R}$, if n were only 20 stations, $n_e$ would be 3.4 and if n were 500, $n_e$ would be 3.8. This illustrates the rapid arrival at the limiting number of independent records. No appreciable increase in information is attained by using 500 stations instead of 20, if they are within the same region. As a further example, if there are an infinite number of stations and $\bar{R} = 0.1$, the effective number of stations is only 10.

The theory of regression analysis is based on the assumption, amongst others, that the values of the dependent variable are mutually independent. It is apparent, then, that in equations such as:

$$x = a_0 + a_1 A + a_2 B + a_3 C + \ldots \qquad (13\text{-}80)$$

where x is some streamflow characteristic and A, B, C, ... are basin physiographic characteristics, the regression theory is impaired since the x values are not independent. Matalas and Benson (20) have investigated this problem very thoroughly and have concluded that the estimation of the regression constants, $a_0$, $a_1$, ... and the subsequent estimated value $\hat{x}$ are not affected by inter-station correlation. If inter-station correlation is present, however, the variance of $a_0$ will be larger than if there were no correlation, the variances of $a_1$, $a_2$,...,$a_n$ will be smaller and the variance of $\hat{x}$ may be larger or smaller.

Considering a set of n gauging stations each of N observations but which are both serially and cross-correlated then Equations 13-67 and 13-79 can be combined to give a total effective number of station-events defined as:

$$N_e n_e = Nn/\left\{\left[1 + \bar{R}(n - 1)\right](1 + 2\bar{R}_1 + 2\bar{R}_2 + \ldots 2\bar{R}_n)\right\}$$

$$(13\text{-}81)$$

where $\bar{R}_1$, $\bar{R}_2$ are the average serial correlation coefficients of the n time series.

Developing this analysis, the procedure can be used as a means of defining homogeneous hydrologic regions (33, 17).

## References

1.  Alexander, G. N., 1954, Some Aspects of Time Series in Hydrology, J. Inst. Engineers (Australia) p. 196.

2.  Benson, M. A., 1960, Characteristics of Frequency Curves Based on a Theoretical 1,000-Year Record, USGS Water Supply Paper 1543-A, pp. 51073.

3.  Benson, M. A., 1962, Evolution of Methods for Evaluating the Occurrence of Floods, USGS Water Supply Paper 1580-A.

4.  Benson, M. A., 1962, Factors Influencing the Occurrence of Floods in a Humid Region of Diverse Terrain, USGS Water Supply Paper 1580-B.

5.  Carrigan, P. H., Jr., 1971, A Flood-Frequency Relation Based On Regional Record Maxima, USGS Professional Paper No. 434-F.

6.  Collier, E. P., 1963, Regional Flood Frequency Analysis, unpublished paper, Water Resources Branch, Ottawa.

7.  Collier, E. P. and G. A. Nix, 1967, Flood Frequency Analysis for the New Brunswick-Gaspé Region, Technical Bulletin No. 9, Inland Waters Branch, Ottawa.

8.  Conover, W. J. and M. A. Benson, 1963, Long-Term Flood Frequencies Based on Extremes of Short-Term Records, USGS Professional Paper No. 450-E, pp. E159-E160.

9.  Coulson, A., 1967, Flood Frequencies of Nova Scotia Streams, Technical Bulletin No. 4, Water Resources Branch, Ottawa.

10. Coulson, A., 1967, Estimating Runoff in Southern Ontario, Technical Bulletin No. 7, Inland Waters Branch, Ottawa.

11. Cruff, R. W. and S. E. Rantz, 1965, A Comparison of Methods Used in Flood Frequency Studies for Coastal Basins in California, USGS Water Supply Paper 1580-E.

12. Dalrymple, T., 1960, Flood Frequency Analyses, USGS Water Supply Paper 1543-A.

13. De Coursey, D. G., 1973, Objective Regionalization of Peak Flow Rates, Proceedings of Second International Symposium in Hydrology, pp. 395-405, Water Resources Publications, Fort Collins, Colorado.

14. Fiering, M. B., 1963, Use of Correlation to Improve Estimates of the Mean and Variance, USGS Professional Paper No. 434-C.

15. Kalinin, G. P., 1960, Calculation and Forecasts of Streamflow from Scanty Hydrometric Readings, Trans. Interregional Seminar on Hydrologic Networks and Methods, Bangkok, 1959, WMO Flood Control Series No. 15, pp. 42-52.

16. Kite, G. W., 1974, Case Study of Regional Analysis Techniques for Design Flood Estimation, Can. J. Earth Sciences, Vol. 11, No. 6, pp. 801-808.

17. Kite, G. W., 1973, Serial Correlation as a Measure of Regional Uniformity, unpublished notes, Water Resources Branch, Ottawa.

18. Kouwen, N., 1973, Watershed Modelling Using a Square-Grid Technique, Proceedings of the 9th Canadian Hydrology Symposium, Edmonton, Alberta.

19. Langbein, W. B., 1960, Hydrologic Data Networks and Methods of Extrapolating or Extending Available Hydrologic Data, Trans. Interregional Seminar on Hydrologic Networks and Methods, Bangkok, 1959, WMO Flood Control Series No. 15, pp. 13-41.

20. Matalas, N. C. and M. A. Benson, 1961, Effect of Interstation Correlation on Regression Analysis, Journal of Geophysical Research, Vol. 66, No. 10, pp. 3285-3293.

21. Matalas, N. C. and W. B. Langbein, 1962, Information Content of the Mean, Journal of Geophysical Research, Vol. 67, No. 9, pp. 344-348.

22. Matalas, N. C. and B. Jacobs, 1964, A Correlation Procedure for Augmenting Hydrologic Data, USGS Professional Paper No. 434-E.

23. Matalas, N. C., 1967, Time Series Analysis, Water Resources Research, Vol. 3, No. 3, pp. 817-830.

24. Pentland, R. L., 1967, Extending Streamflow Records, Program No. 20, unpublished paper, Water Resources Branch, Ottawa.

25. Pentland, R. L. and D. R. Cuthbert, 1971, Operational Hydrology for Ungauged Streams by the Grid Square Technique, Water Resources Research, Vol. 7, No. 2, pp. 283-291.

26. Solomon, S. I., T. P. Denouvilliez, C. Cadou and E. J. Chart, 1968, The Use of a Square-Grid System for Computer Estimation of Precipitation, Temperature and Runoff in a Sparsely Gauged Area, Water Resources Research, Vol. 4, No. 5, pp. 919-930.

27. Solomon, S. I. and A. S. Qureshi, 1972, Hydrologic Data Banks - Present Status and Potential, Engineering Journal, Vol. 55, No. 1/2, pp. 9-14.

28. Thomas, D. M. and M. A. Benson, 1969, Generalization of Streamflow Characteristics from Drainage Basin Characteristics, USGS Open-file Report, Washington, D. C., 45 p.

29. Turc, L., 1954, Le Bilan D'Eau des Sols: Relations entre les Precipitations, L'évaporation et L'écoulement, Annales Agronomiques, Vol. 4, pp. 491-595.

30. Vicens, G. J., Rodriguez-Iturbe, I. and J. C. Schaake, Jr., 1975, A Bayesian Framework for the Use of Regional Information in Hydrology, Water Resources Research, Vol. 11, No. 3, pp. 405-414.

31. Wampler, R. H., 1969, An Evaluation of Linear Least Squares Computer Programs, Journal of Research of the National Bureau of Standards - B. Mathematical Sciences, Vol. 73B, No. 2, pp. 59-90.

32. Wood, E. F. and I. Rodriguez-Iturbe, 1975, Bayesian Inference and Decision Making for Extreme Hydrologic Events, Water Resources Research, Vol. 11, No. 4, pp. 533-542.

33. Yevjevich, V. M., 1964, Fluctuations of Wet and Dry Years, Part II, Analysis by Serial Correlation, Hydrology Paper No. 4, Colorado State University, Fort Collins, Colorado.

34. Yevjevich, V. M., 1972, Probability and Statistics in Hydrology Water Resources Publications, Fort Collins, Colorado.

35. Young, G. K. and Pisano, W. C., 1968, Operational Hydrology Using Residuals, Proc. ASCE, Vol. 94, No. HY4, pp. 909-923.

# CHAPTER 14
## RISK

The Need for Risk Analysis

The most important question facing the designer of any hydrologic structure is: what is the risk of failure? The price of failure of a major dam is high and the risk of this occurrence must be minimized. A study of over 1600 dams (in (8)) has shown the following causes of failure:

| | |
|---|---|
| Foundation problems | 40% |
| Inadequate spillway | 23% |
| Poor construction | 12% |
| Uneven settlement | 10% |
| High pore pressure | 5% |
| Acts of war | 3% |
| Embankment slips | 2% |
| Defective materials | 2% |
| Incorrect operation | 2% |
| Earthquakes | 1% |

In a more recent study of over 300 dam disasters (8) it was found that roughly 35% of the failures were due to inadequate spillway design. Also of importance here is the study of dam failures noted by the AWWA (3). Inadequate spillway design is usually caused by inadequate design flood analysis and this is the direct concern of the hydrologist. Design floods are estimated either from frequency techniques or as the Probable Maximum Flood.

As noted in Chapter 1, the technique of Probable Maximum Flood, despite its name, is a totally deterministic concept and as such has no risk associated with it. Because there is no proof of the existence of extreme boundaries in the meteorological factors which cause floods (19) the concepts of maximum probable precipitation, maximum probable flood and other similarly named imaginary events may be considered as arbitrary. They are concepts of expediency.

Frequency analysis, on the other hand, accepts events of any magnitude as being possible although as the magnitude increases so the probability of occurrence decreases.

The simplest procedure in the frequency analysis estimation of spillway design floods is to select a return period and use either graphical techniques or a mathematical distribution to derive the corresponding event magnitude. Some of the return periods commonly used for different types of structure are (in (7)):

Major dams with probable loss of life

| | |
|---|---|
| Earth dam | 1000 years |
| Masonry or concrete dam | 500 years |
| Costly dams with no likeli hood of loss of life | 500 years |
| Moderately costly dams | 100 years |
| Minor dams | 20 years |

In addition, McCaig and Erickson (12) note that in the past it has been common practice to design major dams for floods having theoretical return periods of up to 10,000 years. The ASCE Hydraulics Division Committee on Hydrometeorology (2) has suggested that the Probable Maximum Flood is perhaps equivalent to a design return period of 10,000 years. This elementary procedure takes no account of the increase of risk with increasing project life or of the economically optimum design.

## Economic Design

A second procedure sometimes used in the design of hydraulic structures relates the design spillway capacity not only to the magnitude and frequency of possible floods but also to the monetary value of the dam, the unit cost of the spillway and the value placed upon the lives and property of the people downstream of the dam. McCaig and Erickson (12) have provided a very clear description of this method of design using in their example lognormal distributions of fall and spring floods.

If the average annual losses for a particular structure can be expressed as:

$$C_1 = \Sigma \Delta L \ P \qquad (14-1)$$

where $\Delta L$ is the incremental average loss for a particular design flood, x, in dollars and P is the exceedence probability of that design flood; and if the average annual cost of the spillway is given by:

$$C_2 = \Delta x \ Q \qquad (14-2)$$

where $\Delta x$ is the incremental cost, in dollars per cumec, of providing spillway capacity for flow Q cumec; then the optimum structure design will occur when

$$C = C_1 + C_2 \qquad (14-3)$$

is at a minimum. That is to say, for a particular structure and a set of flood flows there will result a particular value of C. By repeating the same set of flood flows with different structure capacities a graph of C versus capacity or design flood can be obtained.

233

McCaig and Erickson (12) assumed a 2-parameter lognormal probability distribution for flood events so that:

$$p = \frac{1}{\sqrt{2\pi}\ \sigma_y} \int_y^\infty e^{-\frac{(y-\mu_y)^2}{2\sigma_y^2}}\ dy \qquad (14\text{-}4)$$

where $y$ is the logarithm of the flood event, $x$, and $\mu_y$ and $\sigma_y$ are respectively the population mean and standard deviation estimated from the logarithms of the recorded flood events. Substituting Equations 14.4, 14.1 and 14.2 into Equation 14.3, differentiating and equating to zero, the optimum design capacity, $Q_d$, can be obtained.

The ASCE (2) has recently described a similar procedure to McCaig and Erickson but designed for the re-evaluation of the spillway capacity of existing dams. A series of alternate project designs are identified by their spillway design floods, e.g. the 500 year design project, the 1000 year design project, etc.. This series would include the existing project. For each of the possible projects the costs associated with an array of floods with return periods varying from very low to very high are determined.

Damages caused by the various floods to each of the alternate project designs should include upstream damages (in the event of overtopping and subsequent failure of the dam) to recreation, piers, boats, buildings, loss of power, loss of water supply; to the structure itself including dam fill eroded, repair time, powerhouse losses, switchyard losses, etc., and damage downstream of the dam including deaths, injuries, property damage, compensation for loss of water supply, power supply, telephone, road access and lost employment. It is instructive to note that in the ASCE example (2) death was valued at $150,000, permanent disabling injury at $200,000 and a non-disabling injury at $10,000. The property damages should be determined by carrying out a stage-damage analysis using measured flood profiles.

For each project the average annual risk can be calculated by arithmetic strip integration of the area beneath the return period-damage curve. The cost of each alternate project design is known and can be converted to an average annual cost. This cost, sometimes known as the "operating rate" (12), may include items for interest, taxes, depreciation, etc., and normally ranges between 8 and 10 percent of the total capital cost. Curves of the type shown in Figure 14-1 can then be drawn and the optimum project design determined.

Note that the series of alternate projects might consist of one dam design with floods of successively longer return

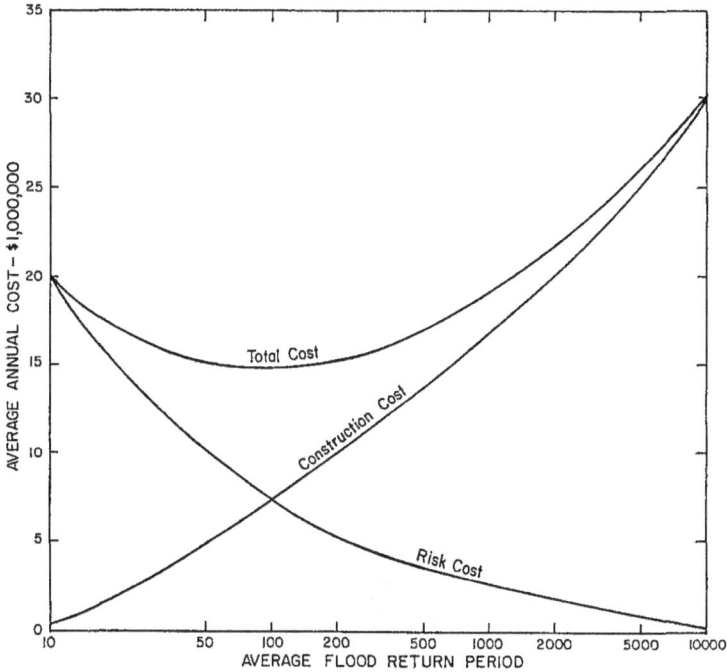

Figure 14-1.    Average annual costs for different designs
                (example only).

period being accommodated by a longer spillway, by downstream
flood protection work, by paving the dam top and downstream dam
surface to reduce erosion from overtopping, by construction of
an upstream reservoir to reduce inflows or by other similar
means.

## Risk Design

Neither of the two techniques described so far include the
concept of total risk.  For any hydraulic structure there is a
total risk of failure which can be broken down into the risk of
failure of each project component i.e. hydrologic, hydraulic
and structural.  The risk within any component can then be broken
down into true risk and uncertainty.  Yen and Ang (18) have used
the terms objective risk and subjective risk.

For the hydrologic component, risk is the calculable probability of failure e.g. occurrence of a certain flood, occurrence of a drought, etc..  The calculation of risk is based on
the assumption that the underlying event distribution is known.
As an example, if it is known that flood magnitudes in a particular river valley location follow the lognormal distribution
and that the time-distribution of the floods follow a Poisson

distribution then the risk that the flood of a certain magnitude will occur in the next five years can be computed exactly.

Uncertainty occurs because the basic data available contain random measurement and computation errors, systematic errors, non-homogeneity in time, loss of information in changing from a continuous record to a discrete data set and so on. These imperfect data are then used to estimate the parameters of the assumed population distribution. Uncertainty generally increases as the variance of the sample data increases and decreases as the sample length increases.

Thomas (16) has evaluated the errors in streamflow estimates made from a continuous stage record while Moss (13) has related the standard error of discharge estimates to the number of streamflow measurements made per year and the associated costs of maintaining the station.

The effect of uncertainty on the parameters of the population distribution can be included in an analysis by computing the standard error of estimate of the particular distribution at the required probability level. Confidence limits around the expected event magnitude can then be calculated.

To summarize this concept, hydrologic risk is made up of basic risk and uncertainty both of which can be evaluated. What cannot be evaluated is the error caused by selecting the wrong distribution to fit the sample data. It is true that the goodness of fit of a distribution, once chosen, can be measured using the Chi-Square or Kolmogorov-Smirnov or similar tests and thus the best-fitting distribution can be selected. Generally, however, the sample data will occupy the central portion of a frequency distribution while the event magnitudes which it is required to compute will be in the extremes so that the best-fitting distribution may not necessarily be the best to use.

The computation of standard errors of estimates for various common distributions has been described earlier. The remainder of this chapter will cover the calculation of basic risk, the assumption being made that the underlying distribution is known.

Suppose that for a time invariant hydrologic system the probability of occurrence of an event, x, greater than the design event, $x_0$, during a period of n years is P. Then the probability of non-occurrence, Q, is 1-P.

If this design event has a return period of T-years and a corresponding annual probability of exceedence of p then:

$$p = \frac{1}{T} \tag{14-5}$$

the probability of non-occurrence in any one year is:

$$q = 1 - \frac{1}{T} \qquad (14\text{-}6)$$

the probability of non-occurrence in n years is:

$$Q = (1 - \frac{1}{T})^n \qquad (14\text{-}7)$$

So that, finally, the probability that x will occur at least once in the n years is:

$$P = 1 - (1 - \frac{1}{T})^n \qquad (14\text{-}8)$$

This is the risk of failure and is based on the assumption of independence of annual events. Yen (17) has tabulated values of T, the required design return period, for various expected project lives, n, and permissible risks of failure, P. Table 14-1 is adapted from Yen. Figure 14-2 is based on the solution of Equation 14-8.

Table 14-1

Design Return Period for Various
Project Lives and Risks of Failure[1]

| Permissible risk of failure | Expected Project Life, n, in years | | | | | | | |
|---|---|---|---|---|---|---|---|---|
| | 1 | 2 | 5 | 10 | 20 | 25 | 50 | 100 |
| 0.99 | 1.01 | 1.11 | 1.66 | 2.71 | 4.86 | 5.95 | 11.4 | 22.2 |
| 0.95 | 1.05 | 1.29 | 2.22 | 3.86 | 7.16 | 8.85 | 17.2 | 33.9 |
| 0.90 | 1.11 | 1.46 | 2.71 | 4.86 | 9.19 | 11.4 | 22.2 | 43.9 |
| 0.75 | 1.33 | 2.00 | 4.13 | 7.73 | 14.9 | 18.6 | 36.6 | 72.6 |
| 0.50 | 2.00 | 3.41 | 7.73 | 14.9 | 29.4 | 36.6 | 72.6 | 145. |
| 0.33 | 3.00 | 5.45 | 12.9 | 25.2 | 49.9 | 62.1 | 124. | 247. |
| 0.25 | 4.00 | 7.46 | 17.9 | 35.3 | 70.0 | 87.3 | 174. | 348. |
| 0.20 | 5.00 | 9.47 | 22.9 | 45.3 | 90.1 | 113. | 225. | 449. |
| 0.10 | 10.0 | 19.5 | 48.0 | 95.4 | 190. | 238. | 475. | 950. |
| 0.05 | 20.0 | 39.5 | 98.0 | 195. | 390. | 488. | 975. | 1,950. |
| 0.02 | 50.0 | 99.0 | 248. | 495. | 990. | 1,238. | 2,476. | 4,951. |
| 0.01 | 100. | 199.5 | 498. | 995. | 1,990. | 2,488. | 4,977. | 9,953. |

[1] From Yen (17)

If the series of recorded or measured events are not an annual series but a partial duration series with an average of K observations per year then the probability that the T-year event will be equalled or exceeded in n consecutive years is:

$$P = 1 - [1 - 1/T \ K]^{nK} \qquad (14\text{-}9)$$

Figures similar to Figure 14-2 (for which K = 1) can be drawn for all values of K, (1).

If failure is associated not with exceedence of the design event but with failure to reach the design event, e.g. a drought design, then the return period must be redefined.

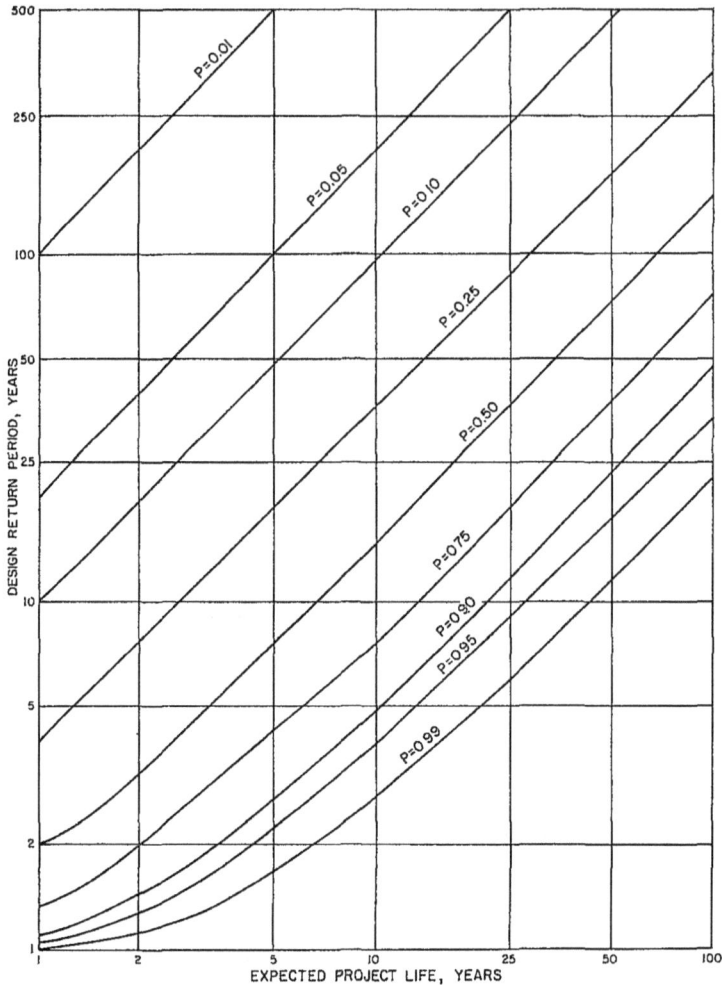

Figure 14-2.  Theoretical probability of failure for given project life and design return period.

In the event that the design return period is made equal to the expected project life there is a 63.4% chance of failure of the project.  This can be shown in Equation 14-8 by putting $T = n$:

$$P = 1 - (1 - \frac{1}{n})^n \qquad (14\text{-}10)$$

In the limit as $n \to \infty$

$$(1 - \frac{1}{n})^n \to \frac{1}{e} = 0.368 \qquad (14\text{-}11)$$

and so, for large n, P tends to 63%.

Similarly, supposing that a project has been designed against a hydrologic event of return period T years then the risk of failure after completion of n' years of the expected project life of n years can be calculated (14).

Writing Equation 14-7 as

$$Q = [(1 - \frac{1}{T})^T]^{T/n} \qquad (14\text{-}12)$$

and using the same asymptotic approximation as in Equation 14-11, Gill (10) has shown that for a given value of P or Q there is a linear relationship between T and n, as:

$$Q = (\frac{1}{e})^{T/n} \qquad (14\text{-}13)$$

$$n = T \ln (1/Q) \qquad (14\text{-}14)$$

A frequently used approximation resulting from Equation 14-8 is:

$$T \approx n/P \qquad (14\text{-}15)$$

Gumbel (11) termed this the "design quotient".

The probabilities referred to above are all probabilities of occurrence of an event of a certain magnitude. Also of interest is the average probability of occurrence of all events above that certain magnitude. For example, in a series of n annual events the number, m, of events which equal or exceed the T-year event is $(n+1)/T$. The annual probability of occurrence of the maximum event is $1/(n+1)$, of the second largest event is $2/(n+1)$, of the third largest event is $3/(n+1)$, etc. so that, the average probability $\bar{p}$ of the n' events which exceed the T-year event is given by:

$$\bar{p} = (\frac{1}{n+1} + \frac{2}{n+1} + \frac{3}{n+1} + \ldots + \frac{n'}{n+1})/n' \qquad (14\text{-}16)$$

Benson (6) has shown that this expression reduces to:

$$\bar{p} = (n+T)/2T(n+1) \qquad (14\text{-}17)$$

which, as n approaches infinity, becomes

$$\bar{p} \approx 1/2T \qquad (14\text{-}18)$$

Thus, in general, the average probability of occurrence of all events above the T-year event is approximated by the probability of the 2T-year event. For example, the average probability of occurrence of all events greater than the 100-year event is approximately 0.005, which corresponds to the 200-year event.

The expressions developed so far in this chapter have all been distribution-free, that is, no assumptions have been made regarding the underlying event distribution. If it is required to estimate the event magnitude corresponding to the design return period computed from, for example, Equation 14-8 then a probability distribution must be assumed.

To show the wide variation possible in the results of this assumption of a distribution, Figure 14-3 is adapted from Gumbel (11). This figure shows the relationship between design quotient and the reduced variable, z, where

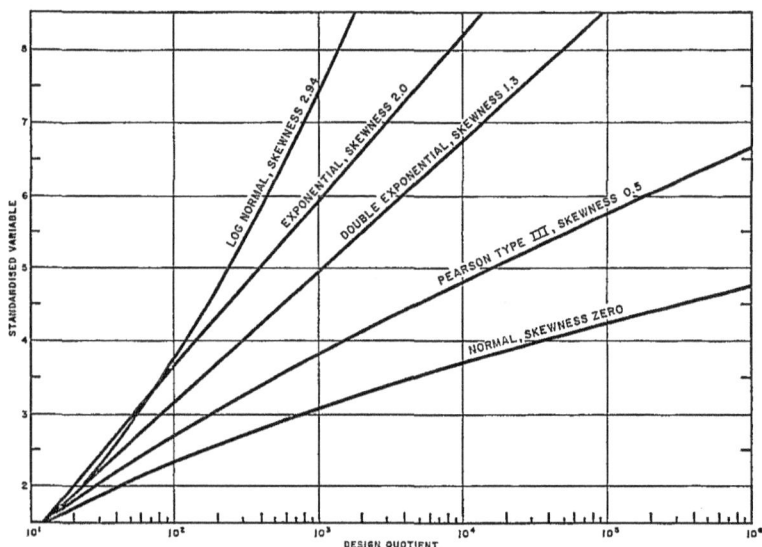

Figure 14-3. Standardized variable vs design quotient.

$$Z = (x - \mu)/\sigma \qquad (14-19)$$

for several commonly used distributions.

Those distributions shown on Figure 14-3 which have fixed coefficient of skew are the normal with $\gamma_1 = 0$ and the double exponential (or type I extremal) with $\gamma_1 = 1.3$. The coefficients of skew have been arbitrarily chosen for the other distributions shown on the figure.

Alternatively, if the assumptions are made that events are independent and the mean number of events in unit time is constant then the Binomial and Poisson distributions can be used to evaluate risk, as described in Chapter 4. For a Poisson distribution of event occurrences and an extremal type I distribution of event magnitudes, Shane (15) has defined the design event, x, as:

$$x = v + \gamma \ln (\lambda F) \qquad (14\text{-}20)$$

where v is a base flow, $\gamma$ is a parameter of the extremal distribution, $\lambda$ is the expected rate of occurrence of events, $\lambda = np$, in the Poisson process and F is the risk factor. The maximum likelihood estimates of $\gamma$ and $\lambda$ are given (15) as:

$$\hat{\gamma} = \bar{x} - v \qquad (14\text{-}21)$$

and

$$\hat{\lambda} = n_e/n \qquad (14\text{-}22)$$

where $n_e$ is the number of events recorded and n is the period of record.

Benson (5) investigated the variation which occurs when small samples are used to estimate a frequency distribution for which the parameters are known exactly. Starting with a known frequency curve Benson (5) constructed short random data sets, drew best-fitting curves and estimated events at different return periods from those curves. From a basic set of 1000 points, 100 records of ten points, 40 records of 25 points, 20 records of 50 points and 10 records of 100 points each were drawn. It was found that records of up to 25 points could not define satisfactorily even short-term events. Long-term records (40 to 50 points) were found to define event magnitudes up to the length of the records with reasonable accuracy.

Yen and Ang (18) have described a procedure for designing hydraulic structures on the basis of a risk analysis. Using as an example the design of an urban sewer system, an overall project risk was chosen on the basis of possible property damage. The hydraulic and hydrologic risks are combined as $\alpha_h$ and are related to the structural risk, $\alpha_s$, and overall risk, $\alpha$

$$(1 - \alpha) = (1 - \alpha_s)(1 - \alpha_h) \qquad (14\text{-}23)$$

Yen and Ang then defined the combined hydraulic and hydrologic risks as:

$$\alpha_h = P(x > Q_c) \, P(N > v) \qquad (14\text{-}24)$$

where $P(x > Q_c)$ is the probability of an event X exceeding a design event, $Q_c$, (the hydrologic risk) and $P(N > v)$ is the probability that N, a random variable, will exceed $v$, a safety factor (hydraulic risk), where

$$v = Q_b/Q_c \qquad (14\text{-}25)$$

and $Q_b$ is the discharge actually used in design. N was assumed to be distributed lognormally with unit mean and a variance, $\sigma_N^2$, equal to the total of the variances of the uncertainties such as inaccuracy of measurement, systematic errors in computation, etc.

$$\sigma_N^2 = \sigma_1^2 + \sigma_2^2 + \ldots \qquad (14\text{-}26)$$

as discussed earlier in the chapter.

If $\alpha$ and $\alpha_s$ are known, then $\alpha_h$ can be determined from Equation 14-23 and for various values of $\nu$, the safety factor, corresponding values of $P(X > Q_c)$, the hydrologic risk, can be found. The equivalent design return period can be found from Equation 14-8 and, assuming a probability distribution to fit the observed data, the corresponding event magnitude, $Q_c$, is found. Yen and Ang (18) used a type I extremal distribution although any other suitable distribution could equally well have been used. By plotting values of $Q_c$ versus $\alpha$ (or $Q_b = \nu \ Q_c$ vs $\alpha$) the optimum discharge can be found. Thus, by defining rigorously the hydrologic risk, the common hydraulic practice of using a safety factor to include the effects of hydraulic risk is provided with a scientific basis.

In the event that no streamflow records are available at the design site, Davis et al. (9) have described a method of evaluating uncertainty by considering the distribution of rainfall events. If the number of rainfall events per season, N, is Poisson distributed with mean $\lambda$, i.e.

$$P_N(x|\lambda) = \frac{\lambda^x e^{-\lambda}}{x!} \qquad (14\text{-}27)$$

and if the amount of rainfall, R, per event is exponentially distributed

$$P_R(k|u) = ue^{-uk} \qquad (14\text{-}28)$$

where $1/u$ is the mean rainfall per event, then the return period of k units of rain in a season, T, is

$$T_R(k|\lambda,u) = [1 - \exp(-\lambda e^{-uk})] \qquad (14\text{-}29)$$

By using a linear rainfall-runoff relationship

$$Q = C \ (R\text{-}A) \qquad (14\text{-}30)$$

where C is a coefficient depending upon the rainfall characteristics of a given watershed and A is a measure of initial abstraction , also depending on the watershed, then an expression for the probability density distribution of the flood return period, $T_Q$, can be given as

$$T_Q(y|\lambda,u) = [1 - \exp\{-\lambda + \lambda P_Q(y|u)\}]^{-1} \qquad (14\text{-}31)$$

where $P_Q(y|u)$ is the distribution function of runoff per event.

Uncertainty is included in the analysis (9) by considering the parameters $\lambda$, u and c as variables. Davis et al. assumed that $\lambda$ and u could be described by a 2-parameter gamma distribution while a beta distribution was used for c.

The results of this approach provide design flows relying only on rainfall data for watersheds with ungauged streams by taking into account the uncertainty of the site parameters. It was found (9) that a closed form solution was not possible and so data generation was used to derive the distribution of the flood return period.

To conclude this chapter on hydrologic risk, a final note on the accumulation of risk. On an individual basis a design return period of 1000 years is often considered safe. When it is considered, however, that there are now approximately 10,000 large dams in the world (7), 1000 of which can be thought of as in independent basins, then since 0.001 x 1000 = 1.0, the 1000-year event will be equalled or exceeded each year at at least one of the dam sites. Similarly, Alexander (1) has shown that in Japan, where there are about 1700 dams, the average design return period for spillway design floods is of the order of 200 years. It would be expected, therefore, that on the average 8 or 9 dams will incur design floods annually.

## References

1.  Alexander, G. N., 1969, Application of Probability to Spill-way Design Flood Estimation, Proc. IASH Symposium on Floods and their Computations, Leningrad, 1967, IASH-UNESCO-WMO Studies and Reports in Hydrology No. 3, pp. 536-543.

2.  ASCE, 1973, Re-evaluating Spillway Adequacy of Existing Dams, Report of the Task Committee on the Re-evaluation of the Adequacy of Spillways of Existing Dams of the Committee on Hydrometeorology of the Hydraulics Division, Proc. ASCE, Vol. 99, No. HY2, pp. 337-372.

3.  American Water Works Association, 1966, Spillway Design Practice, AWWA Manual No. M13, New York.

4.  Banerji, S. and D. K. Gupta, 1969, On a General Theory of Duration Curve and its Application to Evaluate the Plotting Position of Maximum Probable Precipitation or Discharge, Proc. IASH Symposium on Floods and their Computations, Leningrad, 1967, IASH-UNESCO-WMO Studies and Reports on Hydrology No. 3, pp. 183-193.

5.  Benson, M. A., 1960, Characteristics of Frequency Curves
    Based on a Theoretical 1000-Year Record, USGS Water
    Supply Paper No. 1543-A, pp. 51-73.

6.  Benson, M. A., 1967, Average Probability of Extreme Events,
    Water Resources Research, Vol. 3, No. 1, p. 225.

7.  Biswas, A. K., 1971, Some Thoughts on Estimating Spillway
    Design Flood, Bull. IASH, Vol. XVI, No. 4, pp. 63-72.

8.  Biswas, A. K. and S. Chatterjee, 1971, Dam Disasters: An
    Assessment, J. Engineering Institute of Canada, Vol.
    54, No. 3, pp. 3-8.

9.  Davis, D. R., L. Duckstein, C. C. Kisiel and M. M. Fogel,
    1973, A Decision - Theoretic Approach to Uncertainty
    in the Return Period of Maximum Flow Volumes Using
    Rainfall Data, Proc. UNESCO-WMO-IASH Symposium on the
    Design of Water Resources Projects with Inadequate
    Data, Madrid, Vol. 1, pp. 63-74.

10. Gill, M. A., 1972, Analysis of Probability and Risk Equa-
    tions, Proc. ASCE, Vol. 98, No. HY5, pp. 969-971.

11. Gumbel, E. J., 1955, The Calculated Risk in Flood Control,
    Appl. Sci. Res., Sec. A, Vol. 5, pp. 273-280.

12. McCaig, I. W. and O. M. Erickson, 1959, Spillway Capacity
    and Flood Flows, Proc. Symposium No. 1, Spillway
    Design Floods, NRC, Ottawa, pp. 262-287.

13. Moss, M. E., 1969, Maximization of Net Benefit from a Stream-
    gauge, Proc. Fiftieth Annual Meeting of the American
    Geophysical Union, Washington, D. C.

14. Prasad, T., 1971, Discussion of "Risks in Hydrologic Design
    of Engineering Projects", Proc. ASCE, Vol. 97, No. HY1,
    pp. 201-202.

15. Shane, R., 1966, A Statistical Analysis of Base-Flow Flood
    Discharge, PhD Thesis, Cornell University.

16. Thomas, R. B., 1971, Errors in Streamflow Estimates from
    Continuous Stage Records, Proc. Symposium on Statis-
    tical Hydrology, Tucson, Arizona.

17. Yen, B. C., 1971, Risks in Hydrologic Design of Engineering
    Projects, Proc. ASCE, Vol. 96, No. HY4, pp. 959-966.

18. Yen, B. C. and A. H. A. Ang, 1971, Risk Analysis in Design
    of Hydraulic Projects, Proc. Symp. on Stochastic
    Hydraulics, Univ. Pittsburgh, pp. 694-709.

19. Yevjevich, V., 1968, Misconceptions in Hydrology and their Consequences, Water Resources Research, Vol. 4, No. 2, pp. 225-232.

# CHAPTER 15
## SUMMARY AND CONCLUSIONS

Introduction

The magnitude and frequency of occurrence of extreme hydrologic events is of every day importance in most parts of the
world.  Since man has, for reasons of communication, water supply,
agriculture, etc., built most of his communities on the flood
plains of large rivers his life-style is extremely susceptible
to flood damage.  Today's pressure of population increases the
density of development along the river banks.  The flood of 1948
on the Fraser River in British Columbia caused $20 million of
damage.  It has been estimated that if the same magnitude of
flood occurred today the damage would be over $200 million.

At the opposite end of the water spectrum, the production
of sufficient food to feed the world's rapidly increasing population necessitates the increasing use of irrigation.  Mankind
thus becomes ever more susceptible to disaster through drought.

Proper use of existing hydrologic techniques could, through
flood plain zoning and efficient design techniques, eliminate
much of the present loss of life and damage caused by floods
and droughts.

In the 1920's and 1930's the introduction of simple statistical analysis gave an impetus to the science of hydrology.
Very soon, however, a general distrust of probability methods
began to grow because too many users knew too little about
statistical analysis and they apparently expected the methods
to overcome the lack of data (9).

A growing use of deterministic methods, replacing the fall
in popularity of probability techniques, led to the development
in the late 1930's of the unit hydrograph principle.  Advances
in meteorology enabled the conditions producing storm rainfall
to be analysed, with the result that maximum rain producing
storms could be synthesized.  This technique produced a large
number of new, rather vague, technical terms such as Probable
Maximum Precipitation, Maximum Possible Precipitation, Standard
Project Storm, Maximum Probable Flood, etc., based on the premise
that some definite limit existed for all the variables respons-
ible for flood events and that, subsequently, some limit must
apply to the flood runoff itself.  The drawback to this method
is that no probability level can be assigned to the "probable"
events, because of their deterministic origins.  Similarly no
confidence limits can be applied to these events and the non-
specialist is left with the impression that these estimates are
100% accurate with no risk involved.

The philosophical error in the Maximum Probable argument has been described by Yevjevich (27) and others. It is not reasonable to say that a precipitation of 300 mm in one hour can occur but a precipitation of 301 mm cannot. The probability approach states that the variable has a finite probability of reaching any value between zero and infinity. As the precipitation increases in magnitude so the associated probability of occurrence decreases and for very large events approaches zero. However, given enough time, even the improbable becomes certain. The essential stochasticity of precipitation has been recognized by Yevjevich (28) as being mainly due to the random nature of atmospheric variables such as opacity and transmitted radiation. The stochastic precipitation events are then somewhat attenuated by the water and energy storages of the oceans and continents to produce runoff events of mixed deterministic/stochastic nature.

Today, the necessity of producing economically designed projects has produced a need for hydrologic risk analysis and a corresponding upsurge in the use of probabilistic methods in hydrology. Compared to the 1920's and 1930's, however, more data are available and the theory of sampling errors and risk analysis is better understood. Problems still exist in the statistical techniques, however, and a certain amount of subjectivity is still involved, particularly in the choice of a frequency distribution to fit to the observed data.

## Data Abstraction, Graphs, and Plotting Positions

Data for frequency analyses may be abstracted from the recorded data using either annual series or partial duration series. Annual series consist of one event per year; partial duration series consist of all events above a base magnitude, regardless of time of occurrence. The partial-duration series method would initially seem advantageous in that more data, and hence information, are incorporated. However, these additional data increase the definition of event magnitudes only in the central part of the frequency curve which is the area of least interest. Use of the partial-duration series always involves the arbitrary establishment of a base flow and sometimes requires subjective decisions regarding the independence of adjacent events. For these reasons and because the annual series is simpler to abstract and analyse it is to be preferred in frequency studies.

Riggs (20) and Benson (2) have detailed many other reasons, but, to summarize, it is usually stated that mathematical fitting of a standard probability distribution is preferable to plotting a graph and fitting a curve by eye because this eliminates the subjectivity of individual judgement.

In fact, curve fitting by eye and by probability function are both empirical since the true distribution of the recorded events is not known. In addition the very lack of subjectivity of the mathematical procedure is sometimes a disadvantage; the inclusion or exclusion of one or two events may result in large changes in the resulting frequency relationship. The mathematical procedure incorporates all data whereas the individual drawing in a curve by eye may elect to ignore some events in order to get a better-fitting curve. A method is not better simply because it leads to uniform answers, if those answers are uniformly unsound (1).

An intermediary or semi-graphical method exists in which graphs are used to fit curves of standard probability distributions to the data. The procedure recommended is to use a mathematical method of fitting a standard probability distribution and to arrange for the output of the method always to include a plot of the data points and the fitted curve. In this way the procedure can be standardized and automated for machine computation while retaining the option of looking at a graph and reviewing the fitted line on the basis of engineering experience.

Any type of plot of extreme events requires the consideration of plotting positions. When all analysis computations were done by hand, graphical techniques of analysis were extremely attractive because of their brevity and simplicity and the choice of plotting positions was therefore of great importance. Today, when all calculations are performed by computer, graphs are used only as a pictorial form of output presentation and the problem of plotting position has faded somewhat in importance. With this in mind it would be difficult to justify use of any plotting position other than the mean frequency

$$p = m/(n+1) \qquad\qquad (15\text{-}1)$$

where m is the order of the event in the sorted series of n observed events. For the largest event in the series m = 1 and for the smallest event in the series, m = n. This method gives conservative results in that the return period conforms closely to the period of record. Benson (3) has demonstrated that this is the best plotting position to use for economic studies of hydrologic design.

## Frequency Distributions

*Selection of a Distribution* - The primary objectives of frequency analysis are to determine the return periods of recorded events of known magnitude and then to estimate the magnitude of events for design return periods beyond the recorded range. The intermediary between these two objectives is the theoretical

probability distribution. The sample data are used as an estimate of an unknown population to calculate the parameters of the
selected probability distribution. The fitted distribution is
then used to estimate event magnitudes corresponding to return
periods greater than or less than those of the recorded events.

There is no general agreement amongst hydrologists as to
which of the many theoretical distributions available are most
suitable to describe natural events, no agreement has been
reached on the best techniques of fitting a distribution (4) and
no agreement on design standards has been reached. As examples
of this divergence of choice, Spence (24) compared the fit of
the normal, 2-parameter lognormal, type I extremal and log-type
I extremal distributions to annual maximum flows on the Canadian
Prairies and found that the lognormal was the best fitting; Cruff
and Rantz (10) compared six probability distributions in Cali-
ornia and found that the Pearson type III was the most desirable.
In other studies, Santos (21) has found the lognormal distribution better than the Pearson type III, Gumbel (11) has explained
that "It seems that the rivers know the [extreme value] theory.
It only remains to convince the engineer...of the validity of
this analysis", and Benson (2) has found in a study of 100 long
term flood records that no one type of frequency distribution
gives consistently better results. In the United States, Reich
(19) conducted a survey of engineers and hydrologists and found
that of the extremal type I, log extremal type I and log-Pearson
type III was preferred. In Italy, Cicioni et al. (8) tested
the 2-parameter lognormal, 3-parameter lognormal, 2-parameter
gamma, Pearson type III and extremal type I distributions on 108
data sets and found the 2-parameter lognormal to be the most
suitable.

Acceptance of a certain distribution for analysis of flood
peaks must be based on the goals and conditions that are to be
fulfilled and satisfied by the distribution (29). Goodness of
fit is a necessary but not a sufficient condition for acceptance.
Goodness of fit tests often used include Chi-Square and Kolmogorov-Smirnov (26). If goodness of fit were the only criterion,
then high order polynomials would often provide a much better
fit than any of the standard distributions, and yet this method
is not used because there is no hydrologic justification. The
most important criteria in the selection of a model are that
there be a sound theory describing the phenomenon and that the
model should abstract the maximum information from the data
using proper estimation techniques.

This was realized by the recent U. S. Water Resources
Council Work Group (4) who wrote that "no single method of test-
ing the computed results against the original data was acceptable to all those on the Work Group, and the statistical consul-
tants could not offer a mathematically rigorous method." The
Work Group concluded that a frequency distribution could not be

chosen solely on statistical grounds but recommended that the
log-Pearson type III distribution be used because as a 3-param-
eter distribution it offers considerable flexibility, for a zero
skewness it reduced to the lognormal distribution, and finally
because it is in common use by U. S. Government agencies.

The problem of choosing a distribution is not restricted
to hydrology by any means.  In the field of biosciences, Katti
and Sly (12) summarized their findings as:

(a)  No single theoretical distribution has been found to
describe any large scale data.

(b)  For a number of data there could be two or more theoretical distributions that fit equally well and there is no way
to choose between them based on fit alone.

(c)  Two or more physical models could lead to the same
final statistical distribution and hence the estimation of the
parameters of the distribution may not have unique meaning.

(d)  "...Different methods of estimation lead to widely
differing estimates when the methods are consistent...there are
a number of empirical frequencies to which the same theoretical
frequency function has been fitted by different consistent
methods...."

Although statistical methods cannot by themselves determine
the correct frequency distribution, they can, in some cases,
provide reasons why distributions may not be suitable.  As an
example some distributions such as the type III extremal, Pearson type III and log-Pearson type III require the estimation
of the coefficient of skewness from the sample data.  It is
well known that the variability of sample estimates of the
coefficient of skew is large (10) and this may be sufficient
reason to prefer some other distribution.  On the other hand use
of 2-parameter distributions implies a fixed value of the coef-
fecient of skew which may not be valid either.

As a second example of this process of elimination the
following objections have been raised (2), to the use of the
type I extremal distribution (Gumbel) for flood flows:

(a)  It is assumed that the treatment derived for daily
discharges can also be applied to instantaneous flows.

(b)  The daily discharges are not independent events.

(c)  The 365 daily discharges in a year do not constitute
a large number as required by the theory of extreme values.

(d)  An assumption underlying the extreme value theory is
that all the events are part of the same statistical population.
Yet, in many cases, the annual maximum event may be due to a
variety of causes such as normal rainfall, snowmelt or hurricane.

There are different physical factors controlling each of these types of events. The assumption of one population therefore may not be valid.

Whether or not these objections to the use of the type I extremal distribution are sufficiently serious to deter use, is a matter of opinion.

Other statistical restrictions on the use of certain distributions which may help in choosing one distribution from many include:

(a)  The truncated normal distribution (with the probability of events being less than or equal to zero being replaced by a probability mass, P, at x = 0) may be used only if the sample coefficient of skew is very close to zero and the sample data are such that P is very small.

(b)  The 2-parameter and 3-parameter lognormal distributions may be satisfactory only if the coefficients of skew of the reduced data (ln x or ln x-a) are very close to zero.

(c)  The type I extremal distribution will be suitable only when the coefficient of skew of the recorded events is very close to 1.13.

(d)  The Pearson type III distribution is only unbounded at the upper end for positive coefficient of skew.

(e)  The log-Pearson type III distribution is only unbounded at the upper end for positive coefficients of skew of the logarithms and when the parameter $1/\alpha$ is greater than zero and $\beta$ is greater than one.

Of lower importance than the theoretical background is the ease of computation. Matalas and Wallis (15) found several computational difficulties in obtaining maximum likelihood solutions for the Pearson type III and log-Pearson type III distributions and recommended that the use of other distributions warrants consideration. Similarly Pentland and Cuthbert (18) found that use of the log-Pearson type III distribution for the Fraser River Basin, in British Columbia, led to large discontinuities and unnatural flood frequencies. They substituted the lognormal distribution in place of the log-Pearson type III.

*Estimation of Parameters* - There are four main methods of estimating distribution parameters: moments, maximum likelihood, least squares and graphical. It is often accepted (15) that the method of maximum likelihood is the most efficient method and should be used wherever possible. By computing relative efficiencies, Matalas (14) has shown that for low flow analysis the method of moments uses only one-half of the sample information extracted by maximum likelihood for a shape parameter of about 10.

*Frequency Factors* - For any distribution the T-year event magnitude can be computed (7) from a general equation of the form:

$$x(K) = \mu + K\sigma \qquad (15-2)$$

where $\mu$ and $\sigma$ are sample estimates of the population mean and standard deviation and K is a frequency factor specific to the chosen distribution. For each probability distribution the frequency factor K, can be derived from the same size, sample parameters, etc.

*Confidence Limits* - Once a suitable probability distribution has been chosen and the magnitude of the event, $x(K)$, at the required return period, T, has been computed from the general frequency equation then upper and lower confidence limits should be established for this event magnitude.

From the frequency curve, two methods are available to compute confidence limits. The analytical method uses numerical techniques to integrate the probability density function of the sample quantile. For practical distributions, however, this is very difficult and an empirical method is often used instead. The empirical method computes the standard error of $x(K)$, $s(K)$, and then assumes that the T-year event is normally distributed with mean $x(K)$ and standard deviation $s(K)$ so that the confidence interval is

$$x(K) \pm t\, s(K) \qquad (15-3)$$

where t is the standard normal deviate at the required confidence level.

The validity of the empirical method rests on the assumption of normality of the distribution of T-year events. This has been tested by data generation and the results have been described (13).

## Regional Analysis

Regional analysis techniques provide a means of combining records from many gauges. This provides the two advantages of reducing standard errors of estimates at gauge sites and enabling estimates to be prepared for ungauged sites.

The description of regional analysis techniques given in Chapter 13 divided the techniques into five main methods: index-flood, multiple regression, square grid, modified single station probability distributions, and regional record maxima. Each of the first four methods contains several variations, however, and these variations tend to overlap between methods so that the division becomes somewhat artificial. The index-flood method, as originally proposed and put into practice suffers from many defects. The most serious defects are that the method is totally empirical, the distribution of peak events is not known and all relationships are derived by graphical curve-fitting.

The square grid method has a logical ordering of basic data which, when combined with the convenience of automatic data processing, provides a data file of regularly spaced physiographic, meteorologic and hydrologic data easily and rapidly accessible for many types of study. The data file can be regularly updated and may be enlarged by the addition of more parameters as they become available. Using this data file as a base, an extremely versatile series of analyses becomes possible. Originally the square grid method was used to define the areal distribution of mean annual runoff for preliminary province-wide water resources studies (22), but additional steps now available include the generation of synthetic monthly flows at ungauged sites (17), and, with the addition of meteorologic forecasts, the use of a parametric model to provide monthly forecasts of streamflow (23), as well as the provision of flood frequency analyses at ungauged sites and modeling of sediment discharges.

Using the square grid method for frequency analyses combines the use of a standard frequency distribution with a more efficient data base. In recommending this square grid approach for regional frequency analysis the probability distribution to be used should be selected on the basis of the comments and recommendations previously made. In some areas the attribution of all the error to the precipitation data may limit the applicability of the square-grid method of estimating flows for ungauged basins.

Hydrologic series frequently consist of observations that are dependent upon one another, they are serially correlated. Thus each observation contains some information which has already been contained in previous observations. A serially or auto-correlated time series therefore contains less information than would an equal number of pure random observations. If a time series of a given length is correlated with a shorter time series, then the two-station correlation can be used to increase the information content of the shorter series. If either or both of the time series are autocorrelated then the increase in information will be less than if the time series were pure random. Similarly if a number of gauging stations are used to estimate regional parameters then the information content of those parameters will be a maximum when all the station records are independent and will decrease as the dependence between records increases.

Bayesian analysis (25) offers the possibility of combining information from regional analysis with single station frequency analyses.

253

## Risk

Most hydraulic structures are designed on the basis of deterministic Probable Maximum Floods for which the risk is totally unknown. For those structures designed on the basis of frequency analysis, the hydrologic risk is generally accounted for by a simple choice of return period for the design flood. For any given return period however, the risk of project failure is proportional to the expected project life.

An alternate procedure is to initially assign an acceptable overall risk to the project. This overall risk should be computed on the basis of the consequences and costs of failure. This total risk can then be subdivided into structural, hydraulic and hydrologic risk and the allowable hydrologic risk can be determined. On the basis of this allowable hydrologic risk and *the expected project life the required return* period can be found and by assuming a probability distribution the corresponding T-year event magnitude can be calculated. In this manner, Probable Maximum Floods or other deterministic estimates can often be assigned a risk and used in further analysis.

To account for the uncertainties in estimation of distribution parameters it is usual to compute the upper 95% confidence limit for the T-year event and design for that discharge.

It is well to keep in mind, however, that no matter how sophisticated a risk analysis is undertaken, the unexpected tends to occur with remarkable regularity. As an example (5), the Rincon de Bonete hydroelectric project on the Rio Negro in Uraguay was designed for the 1000-year flood of 325,000 cfs (9 200 $m^3$/s) on the basis of 27 years of streamflow records during which the peak flow was 135,000 cfs (3 800 $m^3$/s). Fourteen years after construction (1959) as a result of prolonged heavy rainfall a flood peak of 605,000 cfs (17 100 $m^3$/s) was measured. Using the original frequency analysis the 1959 flood has a theoretical return period of 500,000 years. Fortunately, the dam held and about one-half of the flood peak was absorbed by the reservoir which rose 15 feet (almost 5 metres) above its designed maximum level.

## References

1. American Water Works Association, 1966, Spillway Design Practice, AWWA Manual No. M13, New York.

2. Benson, M. A., 1962, Evolution of Methods for Evaluating the Occurrence of Floods, USGS Water Supply Paper 1580-A.

3. Benson, M. A., 1962, Plotting Positions and the Economics of Engineering Planning, Proc. ASCE, Vol. 88, No. HY6, pp. 57-72.

4. Benson, M. A., 1968, Uniform Flood Frequency Estimating Methods for Federal Agencies, Wat. Res. Res., Vol. 4, No. 5, pp. 891-908.

5. Biswas, A. K., 1971, Some Thoughts on Estimating Spillway Design Floods, Bull. IAHS, Vol. XVI, No. 4, pp. 63-72.

6. Blench, T., 1959, Empirical Methods, Proc. Symposium No. 1, Spillway Design Floods, NRC, Ottawa, pp. 36-48.

7. Chow, V. T., 1954, The Log-Probability Law and Its Engineering Applications, Proc. ASCE, Vol. 80, pp. 1-25.

8. Cicioni, G., G. Giuliano and F. M. Spaziani, 1973, Best Fitting of Probability Functions to a Set of Data for Flood Studies, Proc. Second International Symposium in Hydrology, Floods and Droughts, Water Resources Publications, Fort Collins, Colorado, pp. 304-314.

9. Clark, R. H., 1959, Opening Address, Symposium No. 1, Spillway Design Floods, NRC, Ottawa, pp. 1-3.

10. Cruff, R. W. and S. E. Rantz, 1965, A Comparison of Methods Used in Flood Frequency Studies for Coastal Basins in California, USGS Water Supply Paper 1580-E.

11. Gumbel, E. J., 1966, Extreme Value Analysis of Hydrologic Data, Proc. Hydrology Symposium No. 5, NRC, Ottawa, pp. 147-169.

12. Katti, S. K. and L. E. Sly, 1965, Analysis of Contagious Data Through Behaviouristic Models, in: Classical and Contagious Discrete Distributions, ed. G. P. Patil, Statistical Publishing Society, Calcutta, pp. 303-319.

13. Kite, G. W., 1975, Confidence Limits for Design Events, Wat. Res. Res., Vol. 11, No. 1, pp. 48-53.

14. Matalas, N. C., 1963, Probability Distribution of Low Flows, USGS Professional Paper 434-A.

15. Matalas, N. C. and J. R. Wallis, 1973, Eureka! It fits a Pearson Type III Distribution, Wat. Res. Res., Vol. 9, No. 2, pp. 281-289.

16. Nash, J. E. and J. Amorocho, 1966, The Accuracy of the Prediction of Floods of High Return Period, Wat. Res. Res., Vol. 2, No. 2, pp. 191-198.

17. Pentland, R. L. and D. R. Cuthbert, 1971, Operational Hydrology for Ungauged Streams by the Grid Square Technique, Wat. Res. Res., Vol. 7, No. 2, pp. 283-291.

18. Pentland, R. L. and D. R. Cuthbert, 1973, Multisite Daily Flow Generator, Wat. Res. Res., Vol. 9, No. 2, pp. 470-473.

19. Reich, B. M., 1973, Log Pearson Type III and Gumbel Analysis of Floods, Proc. Second International Symposium in Hydrology, Floods and Droughts, Water Resources Publications, Fort Collins, Colorado, pp. 291-303.

20. Riggs, H. C., 1968, Frequency Curves, Chapter A2, Book 4, in Techniques of Water Resources Investigations of the USGS, U. S. Government Printing Office, Washington, D. C.

21. Santos, A., 1970, The Statistical Treatment of Floods Flows, Water Power, Vol. 22, No. 2, pp. 63-67.

22. Solomon, S. I., T. P. Denouvilliez, C. Cadou and E. J. Chart, 1968, The Use of a Square Grid System for Computer Estimation of Precipitation, Temperature and Runoff in a Sparsely Gauged Area, Wat. Res. Res., Vol. 4, No. 5, pp. 919-930.

23. Solomon, S. I. and A. S. Qureshi, 1972, Application of a Parametric Model for Estimating Snow Accumulation and Flow Forecasting, Proc. IHD/UNESCO/WMO Symposia on the Role of Snow and Ice in Hydrology, Banff.

24. Spence, E. S., 1973, Theoretical Frequency Distributions for the Analysis of Plains Streamflow, Can. J. Earth Sciences, Vol. 10, pp. 130-139.

25. Wood, E. F. and I. Rodriguez-Iturbe, 1975, Bayesian Inference and Decision Making for Extreme Hydrologic Events, Wat. Res. Res., Vol. 11, No. 4, pp. 533-542.

27. Yevjevich, V., 1964, Fluctuations of Wet and Dry Years, Part II, Analysis by Serial Correlation, Hydrology Paper No. 4, Colorado State University, Fort Collins, Colorado.

27. Yevjevich, V., 1968, Misconceptions in Hydrology and Their Consequences, Wat. Res. Res., Vol. 4, No. 2, pp. 225-232.

28. Yevjevich, V., 1971, Stochasticity in Geophysical and Hydrological Time Series, Nordic Hydrology, Vol. II, pp. 217-242.

29. Zelenhasic, E., 1970, Theoretical Probability Distributions For Flood Flows, Hydrology Paper No. 42, Colorado State University, Fort Collins, Colorado.